Atlantis Studies in Computing

Volume 3

Series Editors

For further volumes:
www.atlantis-press.com

Aims and Scope of the Series

The series aims at publishing books in the areas of computer science, computer and network technology, IT management, information technology and informatics from the technological, managerial, theoretical/fundamental, social or historical perspective.

We welcome books in the following categories:

Technical monographs: these will be reviewed as to timeliness, usefulness, relevance, completeness and clarity of presentation.
Textbooks.
Books of a more speculative nature: these will be reviewed as to relevance and clarity of presentation.

For more information on this series and our other book series, please visit our website at:

www.atlantis-press.com/publications/books
Atlantis Press
29, avenue Laumière
75019 Paris, France

Fabio Mogavero

Logics in Computer Science

A Study on Extensions of Temporal and Strategic Logics

ATLANTIS PRESS

Fabio Mogavero
Università degli Studi di Napoli Federico II
Napoli
Italy

ISSN 2212-8557
ISSN 2212-8565 (electronic)
ISBN 978-94-6239-056-0
ISBN 978-94-91216-95-4 (eBook)
DOI 10.2991/978-94-91216-95-4

Softcover re-print of the Hardcover 1st edition 2013

Printed on acid-free paper

To Antonella,
my gentle love

To my parents and grandparents

"Thought is only a flash between two long nights, but this flash is everything."

Henri Poincare

Foreword

The Italian chapter of the European Association for Theoretical Computer Science (EATCS) was founded in 1988, and aims at facilitating the exchange of ideas and results among Italian theoretical computer scientists, and at stimulating cooperation between the theoretical and the applied communities in Italy.

One of the major activities of this chapter is to promote research in theoretical computer science, stimulating scientific excellence by supporting and encouraging the very best and creative young Italian theoretical computer scientists. This is done also by sponsoring a prize for the best Ph.D. thesis. This award has been usually presented every two years and an interdisciplinary committee selects the best two Ph.D. theses, among those defended in the previous two-year period, one on the themes of Algorithms, Automata, Complexity and Game Theory, and the other on the themes of Logics, Semantics and Programming Theory.

In 2012 we started a cooperation with Atlantis Press so that the selected Ph.D. theses would be published as volumes in the Atlantis Studies in Computing. The present volume contains one of the two theses selected for publication in 2012:

Logics in Computer Science by Fabio Mogavero (supervisor: Prof. Aniello Murano, University of Napoli Federico II, Napoli, Italy)

and

On Searching and Extracting Strings from Compressed Textual Data by Rossano Venturini (supervisor: Prof. Paolo Ferragina, University of Pisa, Italy).

The scientific committee which selected these theses was composed of Professors Marcella Anselmo (University of Salerno), Pierluigi Crescenzi (University of Florence) and Gianluigi Zavattaro (University of Bologna).

They gave the following reasons to justify the assignment of the award to the thesis by Fabio Mogavero

"The thesis by Fabio Mogavero addresses problems related to property specification and verification by means of temporal logics and model checking.

In particular, this work significantly contributes to two specific fields: temporal logics with graded modalities, as far as closed systems are concerned, and logics for strategies, as far as concurrent and multi-agent systems are concerned.

Graded modalities are particularly useful to express properties like: "there are at least n different computations allowing the system to reach a predefined state".

*Concerning this subject, the main contribution of the thesis is G*CTL*: an extension of the Computation Tree Logic* (CTL) *with such operators. The study of succinctness and complexity of satisfiability for G*CTL *is particularly valuable: satisfiability turns out to be an* EXPTIME-COMPLETE *problem, as it is the case for other logics with graded modalities in the literature, while G*CTL *is proved to be exponentially more succinct with respect to those logics.*

The logics for strategies have been proposed in the literature as tools to deal with interactions in open systems. About this subject, the main contribution of this thesis is SL, *a logic particularly useful to reason about multi-agent concurrent systems.* SL *strictly includes other similar logics such as the strategy logic in the literature, like the strategy logic by Chatterjee-Henzinger-Piterman. The model checking problem is proved to be 2*EXPTIME-COMPLETE *for* SL, *a result that significantly improves the complexity boundaries already known for the logic by Chatterjee-Henzinger-Piterman.*

The above results have been presented at conferences of excellent international level like the Symposium on Logic in Computer Science (LICS) *and the* Conference on Foundations of Software Technology and Theoretical Computer Science (FSTTCS)."

This year the Italian chapter of EATCS decided to dedicate the two prizes to the memory of Prof. Stefano Varricchio (the one on themes of Algorithms, Automata, Complexity and Game Theory), and Prof. Nadia Busi (the one on themes on Logics, Semantics and Programming Theory), both excellent researchers who prematurely passed away.

I hope that this initiative will further contribute to strengthen the sense of belonging to the same community of all the young researchers who have accepted the challenges posed by any branch of theoretical computer science.

Napoli, January 2013 Tiziana Calamoneri

Preface

For a supervisor, it is always a pleasure and an honor to write a foreword for the thesis of his Ph.D. student. For an excellent student like Fabio, this becomes enormously great.

Rather than talking about the remarkable results obtained by Fabio, which are absolutely well introduced and described in this thesis, I would like to talk about passion, tenacity, perseverance, and above all the difficulty that Fabio had to overcome in obtaining them. Also, I would love to report how everything began and some anecdotes, which I am sure, one day, by reading them here again, will give a remembrance of the good time we spent all together during the Ph.D. period. Moreover, I would like to mention some of the people who had a significant role along Fabio's scientific training.

I still remember the first time I met Fabio. It was the end of 2006. Fabio told me of his passion for the formal methods, in particular modal and temporal logics, the several books and articles he had read, and his will to do a Master thesis on these topics. I must admit that I was surprised and a little bit puzzled to see an engineering student interested in such topics and asking a computer science researcher to be his thesis advisor. At the same time I was happy, because during the years spent at the computer science course, I had never seen a student of mine with the same true interests.

At that time, I had just finished a tough work on mu-calculus enriched with graded and backwards modalities. I spoke to Fabio and mentioned to him the possibility of extending the same work to branching-time temporal logics. For a week, Fabio came back to me with a series of comments on my work and a first clear formalization of Graded Computation Tree Logic with backwards modalities. It was at that moment that I got fully convinced that Fabio could easily attend a Ph.D. program and achieve high level results. The first aim was to finish his Master thesis in time for the Ph.D. call, that is, within October 2007, but without letting the Master thesis to suffer in quality. We achieved the time target by a whisker! Fabio discussed his Master thesis at the end of October and, most important, he got a special encomium from the board that led me to be very proud of all the work done. Then, Fabio participated in the Ph.D. examination and got the best score. Thus, we immediately started the doctoral program. We decided together local courses and summer schools to attend. Mainly, we immediately went back to work on the topics of research undertaken during the Master's thesis and thanks to important new results, we wrote two articles, one on Graded Computation Tree Logic (GCTL, for short) and the other on Minimal Model Quantifiers. Both have required a lot of effort that has often led us to work without interruption for weeks (Sundays included). The results have been very satisfactory and after about a year both works have been published in conference proceedings. In particular the results on GCTL were presented at the outstanding conference LICS 2009 by Fabio, in the United States of America.

Meanwhile, I have to mention that, at the same Ph.D. competition that Fabio took, another engineering student participated and won: Alessandro Bianco. Immediately after the competition, Alessandro joined our group, working under my supervision as well. It was Fabio who encouraged and persuaded Alessandro to participate in the Ph.D. competition and with pleasure I can say that his presence in the laboratory has been a great enrichment for the entire group. In this, Fabio has proved to be an excellent talent scout. Fabio has been a sincere friend toward Alessandro and always spurred him on the research activity. It was a great pleasure to get in the lab and see them discuss deeply on research topics of any sort. In three years, we achieved many important results together with Alessandro but, unfortunately, many of them did not find space in this thesis. I think that period would not have been the same without having Alessandro with us.

I promised at the beginning some anecdotes. Well, here they are. Mostly, they are related with Fabio's traveling. Let me say that every trip Fabio has taken during his Ph.D., be it a school, a conference, or a visiting researcher period, there has always been something disastrous: a flood, a train trip canceled, an airplane that would have changed the landing airport at the very last moment, to mention just a few. For example, the first school Fabio attended was in Germany. While he was moving from the school location to the hotel, a deluge fell on Fabio, without giving him any opportunity of covering. The second school was in Italy, at Bertinoro, the return home was an odyssey because of a train crash. But, my favorite stories are related to the four months spent by Fabio as a visiting scholar in Houston to visit Moshe Vardi at Rice University, in the Autumn of 2008. Well, it was me who pushed Fabio to visit Moshe. Overall, I visited Moshe for more than a year during my Ph.D. and I enjoyed very much my stay there. On the contrary, Fabio's visiting stay has been full of adventure, starting with the outgoing journey: due to a violent storm in Houston, the airplane had to finally land in Dallas. After about a month of permanence in Houston, a violent hurricane came and swept away half of the city. Most of the population remained without electricity for almost a month with all the difficulties that could arise, and so it happened for Fabio. You can imagine my apprehension being in Italy! Well, at the end Fabio returned to Italy 17 pounds lighter, but with a formidable research idea: a new Strategy Logic (SL, for short)! At least the research goal was achieved.

At Rice, Moshe first introduced Fabio to the logics for strategic reasoning through ATL*, then he suggested him to investigate this logic along with some sort of extensions, including the past modalities and the concept of relentfulness. Then, it was decided to move to Strategy Logic introduced by Chatterjee, Henzinger, and Piterman. The aim was to borrow their idea of having strategies quantified as first order objects and to define with this a new logic, much more powerful and versatile. The intuition has been excellent, as it is witnessed by the results reported in this thesis and the related articles published so far. However, most of the work was left remaining when Fabio came back to Italy and it required a year of hard work to come up with some results regarding the satisfiability and model checking problems. Unfortunately, they were not so "low" in complexity as we expected and showed this logic to be not so useful in practical verification issues. We realized that we needed to talk with Moshe again to get directions on how to proceed. It was at the beginning of 2010 and, fortunately, Moshe was moving to Israel, where he was going to spend a few months for his sabbatical. So, we decided to reach him there. Also, I had just got a project prize with Orna Kupferman, in which Fabio was involved as well, and the occasion was perfect to visit both. The visit was absolutely productive. We spent a fabulous month in Jerusalem, Moshe gave his talented touch to the work and his invaluable suggestions for future directions about the research. Also, we had several meetings with Orna and Moshe that opened our mind on several new research subjects. Of course we did more than just work in Israel. How can you do this with such a unique blend and having a landlord like Orna? We will never forget the warm hospitality of Orna and her family, as well as the magic trips with them at the old City.

The visit in Israel also gave the occasion to visit another friend of ours, Benjamin Aminof, and to work with him on another interesting subject: Synthesis of Hierarchical Structures. Benjamin is a very good friend and with him we felt at home all the time. We spent very pleasant days at his house working hard on the research subjects, but also having fun walking around in Jerusalem. The results we got on the synthesis were collected in a work, ready by the time we came back from Israel, but presented at the conference FACS 2011 only a year and a half later because we got absurd comments of rejection from a few conferences we submitted the paper to before. Among others, we had one remark that "the paper was too good for that specific conference," but my favorite is "I did not get the review by my sub-reviewer." Unforgettable!

After our coming back to Italy, we explored all ideas, suggestions, and advice we got in Israel. Overall, one month of visiting gave us more than six months of full work. We achieved all the goals. In particular, we wrote a very nice paper on SL that was accepted and presented by Fabio at FSTTCS in India, toward the end of 2010. This time just coincided with the end of the Ph.D. program as well. Fabio got interesting feedback from that conference that led to improved future works on the subject.

When Fabio came back from FSTTCS, he had just the time to get ready and defend his thesis. The discussion was absolutely perfect. Fabio showed full maturity and deep knowledge on all topics presented at the board. Deservedly, he obtained the maximal grade of "eccellente," which is rarely used by a Ph.D. board in Italy.

At the time of writing this preface, two years had already passed from Fabio's Ph.D. thesis defense. During this period, Fabio has kept working under my supervision, collaborating with Moshe Vardi, and started several new collaborations. We have also written several journal papers extending most of the conferences' results. All these papers have been accepted by high level journals and, among others, I would like to report the ACM Journal of Transaction in Computer Science and the Journal of Theoretical Computer Science. At the moment, Fabio has a post-doctoral research position and helps me to follow Master and Ph.D. students with success and autonomy.

All this says that during all the years that I have known Fabio, he has consistently demonstrated the highest standard of maturity, responsibility, and dedication to his work. His presence and contribution in all the topics in which he has been involved has been invaluable and I am proud of the feedback and comments I received about his work. Fabio has fully demonstrated himself to be truly an outstanding and unique researcher. In my career I have never met another student like him. He is a very talented scientist, with a wonderful ability to capture the intuition about any work and make the subject come alive. Last but not least, Fabio is a very friendly person, loved and appreciated by all his collaborators. A person with whom anyone can collaborate. My wishes for Fabio is all the best for his career, which I am sure will be gorgeous.

Napoli, January 2013 Aniello Murano

Contents

Part I Logics for Computations

Abstract

In this book, we collect four different research works on the introduction and examination of new temporal logic formalisms that can be used as specification languages for the automated verification of the reliability of hardware and software designs with respect to a desired behavior [Mogavero (2011)].

The work is organized into two parts. In the first part, we reason about two logics for computations, GCTL* and MCTL*, which are useful to describe a correct execution of monolithic closed systems. In the second part, instead, we focus on two logics for strategies, SL and mATL*, which are useful to formalize several interesting properties about interactive plays in multi-entity systems modeled as multi-agent games.

In the "Logics for Computations" part, we first study the embedding of the concept of graded quantifications into the temporal-logic framework. In first order logic, existential and universal quantifiers express the concepts of the existence of at least one individual object satisfying a formula, or that all individual objects satisfy a formula. In other logics, these quantifiers have been generalized to express that, for a given non-negative integer *n*, at least *n* or all but *n* individuals satisfy a particular formula. Here, we consider GCTL, a temporal logic with *graded path quantifiers*, which allows to describe properties like "there exist at least *n* different classes of computational fluxes in which a system reaches a predetermined state," where the classes over paths are computed by means of a predetermined equivalence relation. More precisely, we uniformly extend the classic concept of graded quantifiers from states to paths, through the use of a concept of path equivalence with respect to a given path formula. About this logic, in particular, we study the expressiveness and succinctness relationships with respect to GμCALCULUS and the complexity of the satisfiability problem, which results to be EXPTIME-COMPLETE. This research is based on the articles [Bianco et al. (2009)] "*Graded Computation Tree Logic*" and [Bianco et al. (2010)] "*Graded Computation Tree Logic with Binary Coding*" published in the proceedings of the "*IEEE Symposium on Logic in Computer Science, 2009*" and "*EACSL Annual Conference on Computer Science Logic, 2010*," respectively. A full and extended version of these two articles is presented in Bianco et al. (2012). Preliminary results can be also found in Mogavero (2007).

Furthermore, we consider special quantifiers over substructures, which allow to select, using parametric criteria, small critical parts of a system to be successively verified. In the literature, there are some attempts to define a logic that allows to modify the underlying structure under exam and then to verify on it some assigned property. However, as far as we know, none of them is able to select minimal submodels of a given property describing the criteria on which they execute the verification process. Here, we base our work on the search of a new operator that merges the concept of quantifiers on structures with that derived by a generalization of the concept of pruning. The results of this work, is a class of three different extensions of CTL* with *minimal model quantifiers*, which we name MCTL*. Regarding these logics, we study several reductions among them, as well as the satisfiability problem that we prove to be highly undecidability, i.e., $\sum_1^1$-HARD, for two out of the three cases. This research is partially based on the paper [Mogavero and Murano (2009)] "*Branching-Time Temporal Logics with Minimal Model Quantifiers*" published in the proceedings of the "*International Conference on Developments in Language Theory, 2009.*"

In the "Logics for Strategies" part, we first study the problem of defining a new specification language through which it is possible to express several important properties of multi-entity systems that are neither expressible using classical monolithic temporal logics, such as CTL*, nor using two-agent-team temporal logics, such as ATL*. In the literature, we can find some proposal of logics that try to achieve this goal, but unfortunately, none of them succeeds completely on all these aspects. Among them, one of the most important attempts is CHP-SL, a logics in which one can use variables over strategies. However, this logic has a deep weakness, since it does not allow to describe games with more than two players and even two-player concurrent games. Here, we introduce SL, a logic with a syntax similar in some aspects to the first order logic, in which the strategies of the agent building the game are treated as first order objects on which we can quantify. This logic generalizes CHP-SL, by allowing the specification of the correct behavior of multi-agent concurrent games. In SL, for example, we are able to express very complex but useful Nash equilibria that are not expressible with CHP-SL. We enlighten that Nash equilibrium is one of the most important concepts in game theory. For the introduced logic, we solve two problems left open in the work on CHP-SL. Precisely, we show that the related model-checking problem, under a suitable restricted semantics, is 2EXPTIME-COMPLETE, thus not harder of the same problem for several subsumed logics, while we prove that its satisfiability problem is highly undecidable, i.e., $\sum_1^1$-HARD. This research is partially based on the paper [Mogavero et al. (2010a)] "*Reasoning About Strategies*" published in the proceedings of the "*IARCS Annual Conference on Foundations of Software Technology and Theoretical Computer Science, 2010.*" More recent results can also be found in Mogavero et al. (2011, 2012a,b).

Finally, we consider the concept of relentful strategic reasoning, i.e., a formalism that expresses the ability of a strategy to be used not only to achieve a first given goal, but also to change its final goal independent of the history of the play. In the context of planning, memoryful quantification, i.e., quantification over computations that does not lose information about the past along the time, is one of the principal ways to express the fact that a system is able to achieve a desired result, while leaving the possibility to shift to a different goal if some event happens. However, this kind of quantification was not considered before in the context of multi-agent planning. Here, we introduce mATL*, a merging of the classic alternating temporal logic ATL* with memoryful quantification, with the aim of covering the explained idea. About this logics, we prove that, although it is equivalent to ATL*, it is exponentially more succinct. Nevertheless, we prove that both the model-checking and the satisfiability problems remain 2EXPTIME-COMPLETE, as for ATL*. This research is based on the paper [Mogavero et al. (2010b)] "*Relentful Strategic Reasoning in Alternating-Time Temporal Logic*" published in the proceedings of the "*International Conference on Logic for Programming Artificial Intelligence and Reasoning, 2010.*"

Part I
Logics for Computations

General Preliminaries I

In this section, we introduce some more preliminary definitions and further notation used in the first part of the book.

Kripke structures A *Kripke structure* (Ks, for short) (Kripke 1963) is a tuple $\mathcal{K} \triangleq \langle \mathrm{AP}, \mathrm{W}, \mathrm{R}, \mathrm{L}, \mathrm{w}_0 \rangle$, where AP is a finite non-empty set of *atomic propositions*, W is an enumerable non-empty set of *worlds*, $w_0 \in \mathrm{W}$ is a designated *initial world*, $R \subseteq \mathrm{W} \times \mathrm{W}$ is a *transition relation*, and $\mathsf{L} : \mathrm{W} \to 2^{\mathrm{AP}}$ is a *labeling function* that maps each world to the set of atomic propositions true in that world. A Ks is said *total* iff it has a *total* transition relation R, i.e., for all $w \in \mathrm{W}$, there is $w' \in \mathrm{W}$ such that $(w, w') \in R$. By $\|\mathcal{K}\| \triangleq |R| \leq |\mathrm{W}|^2$ we denote the *size* of $\mathcal{K}$, which also corresponds to the size of the transition relation. A *finite* Ks is a structure of finite size.

Kripke trees A *Kripke tree* (KT, for short) is a Ks $\mathcal{T} \triangleq \langle \mathrm{AP}, \mathrm{W}, \mathrm{R}, \mathrm{L}, \varepsilon \rangle$, where (i) $\mathrm{W} \subseteq \Delta^*$ is a Δ-tree for a given set Δ of directions and (ii) for all $t \in \mathrm{W}$ and $d \in \Delta$, it holds that $t \cdot d \in \mathrm{W}$ iff $(t, t \cdot d) \in R$.

Tracks and paths A *track* in $\mathcal{K}$ is a finite sequence of worlds $\rho \in \mathrm{W}^*$ such that, for all $i \in [0, |\rho| - 1[$, it holds that $((\rho)_i, (\rho)_{i+1}) \in R$. Furthermore, a *path* in $\mathcal{K}$ is a finite or infinite sequence of worlds $\pi \in \mathrm{W}^\infty$ such that, for all $i \in [0, |\pi| - 1[$, it holds that $((\pi)_i, (\pi)_{i+1}) \in R$ and if $|\pi| < \infty$ then there is no world $w \in \mathrm{W}$ such that $(\mathsf{lst}(\pi), w) \in R$, i.e., it is *maximal*. Intuitively, tracks and paths of a Ks $\mathcal{K}$ are legal sequences of reachable worlds in $\mathcal{K}$ that can be seen as a partial or complete description of the possible *computations* of the system modeled by $\mathcal{K}$. A track ρ is said *non-trivial* iff $|\rho| > 0$, i.e., $\rho \neq \varepsilon$. We use $\mathrm{Trk}(\mathcal{K}) \subseteq \mathrm{W}^+$ and $\mathrm{Pth}(\mathcal{K}) \subseteq \mathrm{W}^\infty$ to indicate, respectively, the sets of all non-trivial tracks and paths of the Ks $\mathcal{K}$. Moreover, by $\mathrm{Trk}(\mathcal{K}, w) \subseteq \mathrm{Trk}(\mathcal{K})$ and $\mathrm{Pth}(\mathcal{K}, w) \subseteq \mathrm{Pth}(\mathcal{K})$ we denote the subsets of tracks and paths starting at the world w.

Bisimulation Let $\mathcal{K}_1 = \langle \mathrm{AP}, \mathrm{W}_1, \mathrm{R}_1, \mathsf{L}_1, w_{0_1} \rangle$ and $\mathcal{K}_2 = \langle \mathrm{AP}, \mathrm{W}_2, \mathrm{R}_2, \mathsf{L}_2, w_{0_2} \rangle$ be two Kss. Then, $\mathcal{K}_1$ and $\mathcal{K}_2$ are *bisimilar* iff there is a relation $\sim \subseteq \mathrm{W}_1 \times \mathrm{W}_2$ between worlds, called *bisimulation relation*, such that $w_{0_1} \sim w_{0_2}$ and if $w_1 \sim w_2$ then (i) $\mathsf{L}_1(w_1) = \mathsf{L}_2(w_2)$, (ii) for all $v_1 \in \mathrm{W}_1$ such that $(w_1, v_1) \in R_1$, there is $v_2 \in \mathrm{W}_2$ such that $(w_2, v_2) \in R_2$ and $v_1 \sim v_2$, and (iii) for all $v_2 \in \mathrm{W}_2$ such that $(w_2, v_2) \in R_2$, there is $v_1 \in \mathrm{W}_1$ such that $(w_1, v_1) \in R_1$ and $v_1 \sim v_2$.

Unwinding Let $\mathcal{K} = \langle \mathrm{AP}, \mathrm{W}, \mathrm{R}, \mathsf{L}, w_0 \rangle$ be a Ks. Then, the *unwinding* of $\mathcal{K}$ is the KT $\mathcal{K}_U \triangleq \langle \mathrm{AP}, \mathrm{W}', \mathrm{R}', \mathsf{L}', \varepsilon \rangle$, where (i) W is the set of directions, (ii) the states in $\mathrm{W}' \triangleq \{\rho \in \mathrm{W}^* : w_0 \cdot \rho \in \mathrm{Trk}(\mathcal{K})\}$ are the suffixes of the tracks starting in w_0, (iii) $(\rho, \rho \cdot w) \in R'$ iff $(\mathsf{lst}(w_0 \cdot \rho), w) \in R$, and (iv) there is a surjective function $\mathsf{unw} : \mathrm{W}' \to \mathrm{W}$, called *unwinding function*, such that (iv.i) $\mathsf{unw}(\rho) \triangleq \mathsf{lst}(w_0 \cdot \rho)$ and (iv.ii) $\mathsf{L}'(\rho) \triangleq \mathsf{L}(\mathsf{unw}(\rho))$, for all $\rho \in \mathrm{W}'$ and $w \in \mathrm{W}$. It is easy to note that a Ks is always bisimilar to its unwinding, since the unwinding function is a particular relation of bisimulation.

Chapter 1
Graded Computation Tree Logic

Abstract In modal logics, *graded (world) modalities* have been deeply investigated as a useful framework for generalizing standard existential and universal modalities in such a way that they can express statements about a given number of immediately accessible worlds. These modalities have been recently investigated with respect to the μCALCULUS, which have provided succinctness, without affecting the satisfiability of the extended logic, i.e., it remains solvable in EXPTIME. A natural question that arises is how logics that allow reasoning about paths could be affected by considering *graded path modalities*. In this work, we investigate this question in the case of the branching-time temporal logic CTL (GCTL, for short). We prove that, although GCTL is more expressive than CTL, the satisfiability problem for GCTL remains solvable in EXPTIME, even in the case that the graded numbers are coded in binary. This result is obtained by exploiting an automata-theoretic approach, which involves a model of alternating automata with satellites. The satisfiability result turns out to be even more interesting as we show that GCTL is at least exponentially more succinct than graded μCALCULUS.

1.1 Introduction

Temporal logics are a special kind of *modal logics* that provide a formal framework for qualitatively describing and reasoning about how the truth values of assertions change over time. First pointed out by Pnueli (1977), these logics turn out to be particularly suitable for reasoning about correctness of concurrent programs (Pnueli 1981).

Depending on the view of the underlying nature of time, two types of temporal logics are mainly considered (Lamport 1980). In *linear-time temporal logics*, such as LTL (Pnueli 1977), time is treated as if each moment in time has a unique possible future. Conversely, in *branching-time temporal logics*, such as CTL (Clarke and Emerson 1981) and CTL* (Emerson and Halpern 1986), each moment in time may split into various possible futures and *existential* and *universal quantifiers* are used

F. Mogavero, *Logics in Computer Science*, Atlantis Studies in Computing 3, DOI: 10.2991/978-94-91216-95-4_1,

to express properties along one or all the possible futures. In modal logics, such as ALC (Schmidt-Schauß and Smolka 1991) and μCALCULUS (Kozen 1983), these kinds of quantifiers have been generalized by means of *graded (worlds) modalities* (Fine 1972; Tobies 2001), which allow to express properties such as "there exist at least n accessible worlds satisfying a certain formula" or "all but n accessible worlds satisfy a certain formula". For example, in a multitasking scheduling specification, we can express properties such as every time a computation is invoked, immediately next there are at least two empty records in the task allocation table available for the allocation of two tasks that take care of the computation, without expressing exactly which records they are. This generalization has been proved to be very powerful as it allows to express system specifications in a very succinct way. In some cases, the extension makes the logic much more complex. An example is the guarded fragment of the first order logic, which becomes undecidable when extended with a very weak form of counting quantifiers (Grädel 1999). In some other cases, one can extend a logic with very strong forms of counting quantifiers without increasing the computational complexity of the obtained logic. For example, this is the case for μALCQ (see Baader et al. 2003 for a recent handbook) and GμCALCULUS (Kupferman et al. 2002; Bonatti et al. 2008), for which the decidability problem is EXPTIME-COMPLETE.

Despite its high expressive power, the μCALCULUS is considered in some sense a low-level logic, making it "unfriendly" for users, whereas simpler logics, such as CTL, can naturally express complex properties of computation trees. Therefore, an interesting and natural question that arises is how the extension of CTL with graded modalities can affect its expressiveness and decidability. There is a technical challenge involved in such an extension, which makes this task non-trivial. In the μCALCULUS, and other modal logics studied in the graded context so far, the existential and universal quantifiers range over the set of successors, thus it is easy to count the domain and its elements. In CTL, on the other hand, the underlying objects are both states and paths. Thus, the concept of graded must relapse on both of them. We solve this problem by introducing *graded path modalities* that extend to classes of paths the generalization induced to successor worlds by classical graded modalities, i.e., they allow to express properties such as "there are at least n classes of paths satisfying a formula" and "all but at most less than n classes of paths satisfy a formula". We call the logic CTL extended with graded path modalities GCTL, for short.

A point that requires few considerations here is how we count paths along the model. We address this question by embedding in our framework a generic equivalence relation on the set of paths, but satisfying specific consistency properties. Therefore, the decisional algorithm we propose is very general and can be applied to different definitions of GCTL, along with different ways to identify the classes of paths. Along this line, one can observe that a state in a model can have only one direct successor, but possibly different paths going through it. This must be taken into account while satisfying a given graded path property. To deal this difficulty, we introduce a combinatorial tool which applies to a wide class of interesting equivalences. The tool is the partitioning of a natural number, i.e., we consider all possible decompositions of a number into its summands (e.g., $3 = 3+0 = 2+1 = 1+1+1$).

This is used to distribute a set of different paths emerging from a state onto all its direct successors. Note that, while CTL linearly translates to μCALCULUS, the above complication makes the translation of GCTL to GμCALCULUS not easy at all. Indeed, we show such a translation with a double-exponential blow-up, by taking into account the above path partitioning.

As a special equivalence class over paths, we also consider that one induced by the minimality and conservativeness requirements along the paths (Mogavero 2007; Bianco et al. 2009, 2010). The minimality property allows to decide GCTL formulas on a restricted but significant space domain, i.e., the set of paths of interest, in a very natural way. In more detail, it is enough to consider only the part of a system behavior that is effectively responsible for the satisfiability of a given formula, whenever each of its extensions satisfies the formula as well. So, we only take into account a set of non-comparable paths satisfying the same property using in practice a particular equivalence relation on the set of all paths. Moreover, if we drop the minimality, it may happen that to discuss the existence of a path in a structure does not make sense anymore. This is the case, for example, when the existence of a non-minimal path satisfying a formula may induce also the existence of an infinite number of paths satisfying it.

The ability of GCTL to reason about numbers of paths turns out to be suitable in several contexts. For example, it can be useful to query XML documents (Arenas et al. 2007; Libkin and Sirangelo 2008). These documents, indeed, can be viewed as labeled unranked trees (Barceló and Libkin 2005) and GCTL allows reasoning about a number of links among tags of descendant nodes, in a very succinct way, without naming any of the intermediate ones. We also note that our framework of graded path quantifiers has some similarity with the concept of *cyclomatic complexity*, as it was defined by McCabe in a seminal work in software engineering (McCabe 1976). McCabe studied a way to measure the complexity of a program, identifying it in the number of independent instruction flows. From an intuitive point of view, since graded path quantifiers allow to specify how many classes of computational paths satisfying a given property reside in a program, GCTL subsumes the cyclomatic complexity, where the independence concept can be embedded into an apposite equivalence class. As another and more practical example of an application of GCTL, consider again the above multitasking scheduling, where we may want to check that every time a non-atomic (i.e., non one-step) computation is required, then there are at least n distinct (i.e., non completely equivalent) computational flows that can be executed. This property can be easily expressed in GCTL. There are also several other practical examples that show the usefulness of GCTL and we refer to (Ferrante et al. 2008, 2009) for a list of them.

The introduced framework of graded path modalities turns out to be very efficient in terms of expressiveness and complexity. Indeed, we prove that GCTL is more expressive than CTL, it retains the tree and the finite model properties, and its satisfiability problem is solvable in EXPTIME, therefore not harder than that for CTL (Emerson and Halpern 1985). This, along with the fact that GCTL is at least exponentially more succinct than GμCALCULUS, makes GCTL even more appealing. The upper bound for the satisfiability complexity result is obtained by exploiting an automata-theoretic approach (Kupferman et al. 2000). To develop a decision procedure for a logic with the tree model property, one first develops an appropriate notion of tree automata and studies their emptiness problem. Then, the satisfiability problem for the logic is reduced to the emptiness problem of the automata.

In Bianco et al. (2009), we have first addressed the specific case of GCTL where numbers are coded in unary. In particular, it has first shown that unary GCTL indeed has the tree model property, by showing that any formula φ is satisfiable on a Kripke structure iff it has a tree model whose branching degree is polynomial in the size of φ. Then, a corresponding tree automaton model named *partitioning alternating Büchi tree automata* (PABT) has been introduced and shown that, for each unary GCTL formula φ, it is always possible to build in linear time a PABT accepting all tree models of φ. Then, by using a nontrivial extension of the Miyano and Hayashi technique (Miyano and Hayashi 1984) it has been shown an exponential translation of a PABT into a non-deterministic Büchi tree automata (NBT). Since the emptiness problem for NBT is solvable in polynomial time (in the size of the transition function that is polynomial in the number of states and exponential in the width of the tree in input) (Vardi and Wolper 1986b), we obtain that the satisfiability problem for unary GCTL is solvable in EXPTIME.

A detailed analysis on the above technique shows two points where it fails to give a single exponential-time algorithm when applied to binary GCTL. First, the tree model property shows for binary GCTL the necessity to consider also tree models with a branching degree exponential in the highest degree of the formula. Second, the number of states of the NBT derived from the PABT is double-exponential in the coding of the highest degree g of the formula. These two points reflect directly in the transition relation of the NBT, which turns to be double exponential in the coding of the degree g. To take care of the first point, we develop a sharp binary encoding of each tree model. In practice, for a given model $\mathscr{T}$ of φ we build a binary encoding $\mathscr{T}_D$ of $\mathscr{T}$, called *delayed generation tree*, such that, for each node x in $\mathscr{T}$ having $m+1$ children $x \cdot 0, \ldots, x \cdot m$, there is a corresponding node y of x in $\mathscr{T}_D$ and nodes $y \cdot 0^i$ having $x \cdot i$ as right child and $y \cdot 0^{(i+1)}$ as left child, for $0 \leq i \leq m$. To address the second point, we exploit a careful construction of the alternating automaton accepting all models of the formula, in a way that the graded numbers do not give any exponential blow-up in the translation of the automaton in an NBT.

We now describe the main idea behind the automata construction. Basically, we use alternating tree automata enriched with *satellites* (ATAS) as an extension of that introduced in Kupferman and Vardi (2006). In particular, we use the Büchi acceptance condition (ABTS). The satellite is a nondeterministic tree automaton and is used to ensure that the tree model satisfies some structural properties along its paths and it is kept apart from the main automaton. This separation, as it has been proved in Kupferman and Vardi (2006), allows to solve the emptiness problem for Büchi automata in a time exponential in the number of states of the main automaton and polynomial in the number of states of the satellite. Then, we obtain the desired complexity by forcing the satellite to take care of the graded modalities and by noting that the main automaton is polynomial in the size of the formula.

The achieved result is even more appealing as we also show here that binary GCTL is much more succinct than GμCALCULUS. In particular, the best known translation from GCTL to GμCALCULUS is double-exponential in the degree of the formula (Bianco et al. 2010).

Related works Graded modalities along with CTL have been also studied in Ferrante et al. (2008, 2009), but under a different semantics. There, the authors consider overlapping paths (as we do) as well as disjoint paths, but they neither consider the general framework of equivalence classes over paths nor the particular concepts of minimality and conservativeness, which we deeply analyze in our work. In Ferrante et al. (2008) the model-checking problem for non-minimal and non-conservative unary GCTL has been investigated. In particular, by opportunely extending the classical algorithm for CTL (Clarke and Emerson 1981), they show that, in the case of overlapping paths, the model-checking problem is PTIME-COMPLETE (thus not harder than CTL), while in the case of disjoint paths, it is in PSPACE and both NPTIME-HARD and CONPTIME-HARD. The work continues in Ferrante et al. 2009), by showing a symbolic model-checking algorithm for the binary coding and, limited to the unary case, a satisfiability procedure. Regarding the comparison between GCTL and graded CTL with overlapping paths studied in Ferrante et al. (2008), it can be shown that they are equivalent by using an exponential reduction in both ways, whereas we do not know whether any of the two blow-ups can be avoid. However, it is important to note that our general technique can be also adapted to obtain an EXPTIME satisfiability procedure for the binary graded CTL under the semantics proposed in Ferrante et al. (2008). Indeed, it is needed only to slightly modify the transition function of the main automaton (w.r.t. until and release formulas), without changing the structure of the whole satellite. Moreover, it can be used to prove that, in the case of unary GCTL, the complexity of the satisfiability problem is only polynomial in the degree. Finally, our method can be also applied to the satisfiability of the GμCALCULUS while the technique developed in Kupferman et al. (2002) cannot be used for GCTL.

Outline In Sect. 1.2, we recall the basic notions regarding Kripke structures and trees, bisimulation, unwinding, and numeric partitions. Then, we have Sect. 1.3, in which we introduce GCTL* and define its syntax and semantics, followed by Sects. 1.4 and 1.5, in which there are studied the main properties of path equivalence relations and the particular case of the prefix path equivalence based on the concepts of minimality and conservativeness. In Sect. 1.6, we describe the ATAS automaton model. Finally, in Sect. 1.7 we construct the binary tree encoding of a Kripke structure and in Sect. 1.8 we describe the procedure used to solve the related satisfiability problem. Note that in the accompanying Appendix we recall the classical mathematical notation and some basic definitions that are used along the whole work.

1.2 Preliminaries

Numeric partitions Let $n \in [1, \omega[$. We define $\mathrm{P}(n)$ as the set of all *partition solutions* $p \in \mathbb{N}^n$ of the *linear Diophantine equation* $1 \cdot (p)_1 + 2 \cdot (p)_2 + \cdots + n \cdot (p)_n = n$ and $\mathrm{C}(n)$ as the set of all the *cumulative solutions* $c \in \mathbb{N}^{n+1}$

obtained by summing increasing sets of elements from p. Formally, $\mathrm{P}(n) \triangleq \{p \in \mathbb{N}^n : \sum_{i=1}^{n} i \cdot (p)_i = n\}$ and $\mathrm{C}(n) \triangleq \{c \in \mathbb{N}^{n+1} : \exists p \in \mathrm{P}(n). \forall i \in [1, n+1]. (c)_i = \sum_{j=i}^{n} (p)_j\}$. It is easy to verify that all cumulative solutions satisfy the simple equation $(c)_1 + (c)_2 + \cdots + (c)_n = n$. Moreover, $(c)_i \geq (c)_{i+1}$, for all $i \in [1, n]$, and $(c)_{n+1} = 0$. Hence, there is just one cumulative solution $c \in \mathrm{C}(n)$, with $(c)_i = 1$, for all $i \in [1, n]$, which also corresponds to the unique solution $p \in \mathrm{P}(n)$, with $(p)_n = 1$. We use to define the cumulative solutions to be tuples of $n + 1$ and not only of n elements only to simplify the notation when we use this concept. As an example of these sets, consider the case $n = 4$. Then, we have that $\mathrm{P}(n) = \{(4, 0, 0, 0), (2, 1, 0, 0), (0, 2, 0, 0), (1, 0, 1, 0), (0, 0, 0, 1)\}$ and $\mathrm{C}(n) = \{(4, 0, 0, 0, 0), (3, 1, 0, 0, 0), (2, 2, 0, 0, 0), (2, 1, 1, 0, 0), (1, 1, 1, 1, 0)\}$. Note that $|\mathrm{C}(n)| = |\mathrm{P}(n)|$ and, since for each solution p of the above Diophantine equation there is exactly one *partition* of n, we have that $|\mathrm{C}(n)| = \mathsf{p}(n)$, where $\mathsf{p}(n)$ is function returning the number of partitions of n. By Apostol (1976) (see also Sloane and Plouffe 1995), it holds that $\mathsf{p}(n) \to \frac{k_1}{n} \cdot 2^{k_2 \cdot \sqrt{n}}$, where $k_1 = 4 \cdot \sqrt{3}$ and $k_2 = \sqrt{2/3} \cdot \pi \cdot \log e$, for $n \to \infty$. Hence, $|\mathrm{C}(n)| = \Theta(\frac{1}{n} \cdot 2^{k_2 \cdot \sqrt{n}})$.

1.3 Graded Computation Tree Logics

In this section, we introduce a class of extensions of the classical branching-time temporal logics CTL (Clarke and Emerson 1981) with graded path quantifiers. We show, in the next sections, that these extensions allow to gain expressiveness without paying any extra cost on deciding their satisfiability. To formally define the extended logics, we use the CTL* (Emerson and Halpern 1986) state and path formulas framework.

1.3.1 Syntax

The *graded full computation tree logic* (GCTL*, for short) extends CTL* by using two special path quantifiers, the existential $\mathsf{E}^{\geq g}$ and the universal $\mathsf{A}^{<g}$, where the finite or infinite number $g \in \widehat{\mathbb{N}}$ denotes the corresponding *degree*. As in CTL*, these quantifiers can prefix a linear-time formula composed of an arbitrary Boolean combination and nesting of temporal operators X *"next"*, U *"until"*, and R *"release"* together with their weak version $\tilde{\mathsf{X}}$, $\tilde{\mathsf{U}}$, and $\tilde{\mathsf{R}}$. The quantifiers $\mathsf{E}^{\geq g}$ and $\mathsf{A}^{<g}$ can be informally read as *"there are at least g paths"* and *"all but less than g paths"*, respectively. The formal syntax of GCTL* follows.

Definition 1.1 (GCTL* **Syntax**) GCTL* state *(φ) and* path *(ψ)* formulas *are built inductively from the sets of atomic propositions* AP *in the following way, where* $p \in \mathrm{AP}$ *and* $g \in \widehat{\mathbb{N}}$*:*

(1) $\varphi ::= p \mid \neg\varphi \mid \varphi \wedge \varphi \mid \varphi \vee \varphi \mid \mathsf{E}^{\geq g}\psi \mid \mathsf{A}^{<g}\psi$;
(2) $\psi ::= \varphi \mid \neg\psi \mid \psi \wedge \psi \mid \psi \vee \psi \mid \mathsf{X}\psi \mid \psi\,\mathsf{U}\psi \mid \psi\,\mathsf{R}\psi \mid \tilde{\mathsf{X}}\psi \mid \psi\,\tilde{\mathsf{U}}\psi \mid \psi\,\tilde{\mathsf{R}}\psi$.

The class of GCTL* *formulas is the set of state formulas generated by the above grammar. In addition, the simpler class of* GCTL *formulas is obtained by forcing each temporal operator occurring into a formula to be coupled with a path quantifier, as in the classical case of* CTL.

We now introduce some auxiliary syntactical notation. For a formula φ, we define the *degree* $\mathsf{deg}(\varphi)$ of φ as the maximum natural number g occurring among the degrees of all its path quantifiers. Formally, *(i)* $\mathsf{deg}(p) \triangleq 0$, for $p \in \mathrm{AP}$, *(ii)* $\mathsf{deg}((\mathsf{Op}\ \psi)) \triangleq \mathsf{deg}(\psi)$, for all $\mathsf{Op} \in \{\neg, \mathsf{X}, \tilde{\mathsf{X}}\}$, *(iii)* $\mathsf{deg}((\psi_1 \mathsf{Op}\ \psi_2)) \triangleq \max\{\mathsf{deg}(\psi)_1, \mathsf{deg}(\psi)_2\}$, for all $\mathsf{Op} \in \{\wedge, \vee, \mathsf{U}, \mathsf{R}, \tilde{\mathsf{U}}, \tilde{\mathsf{R}}\}$, *(iv)* $\mathsf{deg}((\mathsf{Qn}\ \psi)) \triangleq \max\{g, \mathsf{deg}(\psi)\}$, for all $\mathsf{Qn} \in \{\mathsf{E}^{\geq g}, \mathsf{A}^{<g}\}$ with $g \in \mathbb{N}$, and *(v)* $\mathsf{deg}((\mathsf{Qn}\ \psi)) \triangleq \mathsf{deg}(\psi)$, for all $\mathsf{Qn} \in \{\mathsf{E}^{\geq\omega}, \mathsf{A}^{<\omega}\}$. We assume that the degree is coded in binary. The *length* of φ, denoted by $\mathsf{lng}(\varphi)$, is defined as for CTL* and does not consider the degrees at all. Formally, *(i)* $\mathsf{lng}(p) \triangleq 1$, for $p \in \mathrm{AP}$, *(ii)* $\mathsf{lng}(\mathsf{Op}\ \psi) \triangleq 1+\mathsf{lng}(\psi)$, for all $\mathsf{Op} \in \{\neg, \mathsf{X}, \tilde{\mathsf{X}}\}$, *(iii)* $\mathsf{lng}(\psi_1 \mathsf{Op}\ \psi_2) \triangleq 1+\mathsf{lng}(\psi_1)+\mathsf{lng}(\psi_2)$, for all $\mathsf{Op} \in \{\wedge, \vee, \mathsf{U}, \mathsf{R}, \tilde{\mathsf{U}}, \tilde{\mathsf{R}}\}$, and *(iv)* $\mathsf{lng}(\mathsf{Qn}\ \psi) \triangleq 1+\mathsf{lng}(\psi)$, for all $\mathsf{Qn} \in \{\mathsf{E}^{\geq g}, \mathsf{A}^{<g}\}$. Accordingly, the *size* of φ, denoted by $\mathsf{siz}(\varphi)$, is defined in the same way of the length, by considering $\mathsf{siz}(\mathsf{E}^{\geq g}\psi)$ and $\mathsf{siz}(\mathsf{A}^{<g}\psi)$ to be equal to $1+\lceil\log(g)\rceil+\mathsf{siz}(\psi)$, for $g \in [1,\omega[$, and to $1+\mathsf{siz}(\psi)$, otherwise. Clearly, it holds that $\lceil\log(\mathsf{deg}(\varphi))\rceil \leq \mathsf{siz}(\varphi)$ and $\mathsf{lng}(\varphi) \leq \mathsf{siz}(\varphi)$. We also use $\mathsf{cl}(\psi)$ to denote the classical Fischer-Ladner *closure* (Fischer and Ladner 1979) of ψ defined recursively as for CTL* in the following way: $\mathsf{cl}(\varphi) \triangleq \{\varphi\} \cup \mathsf{cl}'(\varphi)$, for all state formulas φ and $\mathsf{cl}(\psi) \triangleq \mathsf{cl}'(\psi)$, for all path formulas ψ, where *(i)* $\mathsf{cl}'(p) \triangleq \emptyset$, for $p \in \mathrm{AP}$, *(ii)* $\mathsf{cl}'(\mathsf{Op}\ \psi) \triangleq \mathsf{cl}(\psi)$, for all $\mathsf{Op} \in \{\neg, \mathsf{X}, \tilde{\mathsf{X}}\}$, *(iii)* $\mathsf{cl}'(\psi_1 \mathsf{Op}\ \psi_2) \triangleq \mathsf{cl}(\psi_1) \cup \mathsf{cl}(\psi_2)$, for all $\mathsf{Op} \in \{\wedge, \vee, \mathsf{U}, \mathsf{R}, \tilde{\mathsf{U}}, \tilde{\mathsf{R}}\}$, and *(iv)* $\mathsf{cl}'(\mathsf{Qn}\ \psi) \triangleq \mathsf{cl}(\psi)$, for all $\mathsf{Qn} \in \{\mathsf{E}^{\geq g}, \mathsf{A}^{<g}\}$. Intuitively, $\mathsf{cl}(\varphi)$ is the set of all the state formulas that are subformulas of φ. Finally, by $\mathsf{rcl}(\psi)$ we denote the *reduced closure* of ψ, i.e., the set of the maximal states formulas contained in ψ. Formally, (i) $\mathsf{rcl}(\varphi) \triangleq \{\varphi\}$, for all state formulas φ, *(ii)* $\mathsf{rcl}(\mathsf{Op}\ \psi) \triangleq \mathsf{rcl}(\psi)$ when $\mathsf{Op}\ \psi$ is a path formula, for all $\mathsf{Op} \in \{\neg, \mathsf{X}, \tilde{\mathsf{X}}\}$, and *(iii)* $\mathsf{rcl}(\psi_1 \mathsf{Op}\ \psi_2) \triangleq \mathsf{rcl}(\psi_1) \cup \mathsf{rcl}(\psi_2)$ when $\psi_1 \mathsf{Op}\ \psi_2$ is a path formula, for all $\mathsf{Op} \in \{\wedge, \vee, \mathsf{U}, \mathsf{R}, \tilde{\mathsf{U}}, \tilde{\mathsf{R}}\}$. It is immediate to see that $\mathsf{rcl}(\psi) \subseteq \mathsf{cl}(\psi)$ and $|\mathsf{cl}(\psi)| = \mathsf{O}(\mathsf{lng}(\psi))$.

1.3.2 Semantics

We now define the semantics of GCTL* w.r.t. a KS $\mathcal{K} = \langle \mathrm{AP}, \mathrm{W}, R, \mathsf{L}, w_0 \rangle$. For a world $w \in \mathrm{W}$, we write $\mathcal{K}, w \models \varphi$ to indicate that a state formula φ holds on $\mathcal{K}$ at w. Moreover, for a path $\pi \in \mathrm{Pth}(\mathcal{K})$, we write $\mathcal{K}, \pi \models \psi$ to indicate that a path formula ψ holds on π. The semantics of GCTL* state formulas simply extends that of CTL* and is reported in the following. In particular, for the definition of graded quantifiers, we deeply make use of a generic equivalence relation $\equiv^{\psi}_{\mathcal{K}}$ on the set of paths $\mathrm{Pth}(\mathcal{K})$ that may depend on both the KS $\mathcal{K}$ and the path formula ψ.

This equivalence is used to reasonably count the number of ways a structure has to satisfy a path formula starting from a given node, w.r.t. an a priori fixed criterion. The semantics of the GCTL* path formulas is defined as usual for LTL on both finite and infinite paths and, for sake of simplicity, is omitted here. We recall that the weak temporal operators are used to deal with finite paths on which their strong version may result unsatisfiable (see Eisner et al. 2003) for a definition of the LTL semantics with strong and weak temporal operators).

Definition 1.2 (GCTL* **Semantics**) *Given a* KS $\mathcal{K} = \langle \mathrm{AP}, \mathrm{W}, R, \mathsf{L}, w_0 \rangle$, *for all* GCTL* *state formulas* φ *and worlds* $w \in \mathrm{W}$, *the relation* $\mathcal{K}, w \models \varphi$ *is inductively defined as follows.*

(1) $\mathcal{K}, w \models p$ *iff* $p \in \mathsf{L}(w)$, *with* $p \in \mathrm{AP}$.
(2) For all state formulas φ, φ_1, *and* φ_2, *it holds that:*

  (a) $\mathcal{K}, w \models \neg\varphi$ *iff not* $\mathcal{K}, w \models \varphi$, *that is* $\mathcal{K}, w \not\models \varphi$;
  (b) $\mathcal{K}, w \models \varphi_1 \wedge \varphi_2$ *iff* $\mathcal{K}, w \models \varphi_1$ *and* $\mathcal{K}, w \models \varphi_2$;
  (c) $\mathcal{K}, w \models \varphi_1 \vee \varphi_2$ *iff* $\mathcal{K}, w \models \varphi_1$ *or* $\mathcal{K}, w \models \varphi_2$.

(3) For a number $g \in \widehat{\mathbb{N}}$ *and a path formula* ψ, *it holds that:*

  (a) $\mathcal{K}, w \models \mathsf{E}^{\geq g}\psi$ *iff* $|(\mathrm{Pth}(\mathcal{K}, w, \psi)/\equiv_{\mathcal{K}}^{\psi})| \geq g$;
  (b) $\mathcal{K}, w \models \mathsf{A}^{< g}\psi$ *iff* $|(\mathrm{Pth}(\mathcal{K}, w, \neg\psi)/\equiv_{\mathcal{K}}^{\neg\psi})| < g$;

  where $\mathrm{Pth}(\mathcal{K}, w, \psi) \triangleq \{\pi \in \mathrm{Pth}(\mathcal{K}, w) : \mathcal{K}, \pi \models \psi\}$ *is the set of paths of* $\mathcal{K}$ *starting in* w *that satisfy the path formula* ψ *and* $(\mathrm{Pth}(\mathcal{K}, w, \psi)/\equiv_{\mathcal{K}}^{\psi})$ *denotes the* quotient *set of* $\mathrm{Pth}(\mathcal{K}, w, \psi)$ *w.r.t. the equivalence relation* $\equiv_{\mathcal{K}}^{\psi}$, *i.e., the set of all the related equivalence classes.*

For all GCTL* *path formulas* ψ *and paths* $\pi \in \mathrm{Pth}(\mathcal{K})$, *the relation* $\mathcal{K}, \pi \models \psi$ *is defined as follows.*

(4) $\mathcal{K}, \pi \models \psi$ *iff* $\varpi_{\mathcal{K},\psi}(\pi) \models \psi$, *where* ψ *is considered as an* LTL *formula over its restricted closure* $\mathsf{rcl}(\psi)$ *and* $\varpi_{\mathcal{K},\psi}(\pi) \in (2^{\mathsf{rcl}(\psi)})^{|\pi|}$ *is the trace such that* $\varphi \in (\varpi_{\mathcal{K},\psi}(\pi))_k$ *iff* $\mathcal{K}, (\pi)_k \models \varphi$, *for all* $\varphi \in \mathsf{rcl}(\psi)$ *and* $k \in [0, |\pi|[$.

Intuitively, by using the graded existential quantifier $\mathsf{E}^{\geq g}\psi$, we can count how many different equivalence classes w.r.t. $\equiv_{\mathcal{K}}^{\psi}$ there are over the set $\mathrm{Pth}(\mathcal{K}, w, \psi)$ of paths satisfying ψ. The universal quantifier $\mathsf{A}^{< g}\psi$ is simply the dual of $\mathsf{E}^{\geq g}\psi$ and it allows to count how many classes w.r.t. $\equiv_{\mathcal{K}}^{\neg\psi}$ there are over the set $\mathrm{Pth}(\mathcal{K}, w, \neg\psi)$ of paths not satisfying ψ. It is important to note that, since $(\mathrm{Pth}(\mathcal{K}, w, \psi)/\equiv_{\mathcal{K}}^{\psi}) \neq \emptyset$ and $(\mathrm{Pth}(\mathcal{K}, w, \neg\psi)/\equiv_{\mathcal{K}}^{\neg\psi}) \neq \emptyset)$ are equivalent to $\mathrm{Pth}(\mathcal{K}, w, \psi) \neq \emptyset$ and $\mathrm{Pth}(\mathcal{K}, w, \neg\psi) \neq \emptyset$, respectively, it holds that all GCTL* formulas with degree 1 are CTL* formulas too, and vice versa.

Observe that, in the definition of the semantics, we introduced a transformation $\varpi_{\mathcal{K},\psi}(\cdot)$, for each path formula ψ, that maps each path π of the KS $\mathcal{K}$ to a trace $\varpi_{\mathcal{K},\psi}(\pi) \in (2^{\mathsf{rcl}(\psi)})^{|\pi|}$ given by the sequence of sets of state formulas in $\mathsf{rcl}(\psi)$

satisfied at the worlds of π. Hence, we interpret the path formula ψ on AP evaluated on π as an LTL formula on $\mathsf{rcl}(\psi)$ evaluated on $\varpi_{\mathcal{K},\psi}(\pi)$.

Let $\mathcal{K}$ be a KS and φ be a GCTL* formula. Then, $\mathcal{K}$ is a *model* for φ, in symbols $\mathcal{K} \models \varphi$, iff $\mathcal{K}, w_0 \models \varphi$, where we recall that w_0 is the initial state of $\mathcal{K}$. In this case, we also say that $\mathcal{K}$ is a model for φ on w_0. A formula φ is said *satisfiable* iff there exists a model for it. Moreover, it is an *invariant* for the two KSs $\mathcal{K}_1$ and $\mathcal{K}_2$ iff either $\mathcal{K}_1 \models \varphi$ and $\mathcal{K}_2 \models \varphi$ or $\mathcal{K}_1 \not\models \varphi$ and $\mathcal{K}_2 \not\models \varphi$. For all state formulas φ_1 and φ_2, we say that φ_1 *implies* φ_2, in symbols $\varphi_1 \Rightarrow \varphi_2$, iff, for all KS $\mathcal{K}$, it holds that if $\mathcal{K} \models \varphi_1$ then $\mathcal{K} \models \varphi_2$.Consequently, we say that φ_1 is *equivalent* to φ_2, in symbols $\varphi_1 \equiv \varphi_2$, iff $\varphi_1 \Rightarrow \varphi_2$ and $\varphi_2 \Rightarrow \varphi_1$. In the following, when we say that two GCTL* paths formulas ψ_1 and ψ_2 are equivalent, in symbols $\psi_1 \equiv \psi_2$, we mean that they are equivalent if considered as LTL formulas over the union $\mathsf{rcl}(\psi_1) \cup \mathsf{rcl}(\psi_2)$ of their restricted closures.

For technical reasons, we also define the relation of satisfiability of path formulas on tracks, by simply setting $\mathcal{K}, \rho \models \psi$ iff $\varpi_{\mathcal{K},\psi}(\rho) \models \psi$, for all $\rho \in \mathrm{Trk}(\mathcal{K})$. We now show the basic properties of the satisfiability relation $\models$ on paths and tracks directly inherited by the LTL semantics.

Proposition 1.1 (Path Satisfiability Properties) *Let φ be a state formula, ψ, ψ_1, and ψ_2 be path formulas, and $\pi \in (\mathrm{Pth}(\mathcal{K}, w) \cup \mathrm{Trk}(\mathcal{K}, w))$ be a path/track starting at the world w of the* KS *$\mathcal{K}$. Then, the following properties hold:* (i) *if $\psi_1 \equiv \psi_2$ then $\mathcal{K}, \pi \models \psi_1$ iff $\mathcal{K}, \pi \models \psi_2$;* (ii) *$\mathcal{K}, w \models \varphi$ iff $\mathcal{K}, \pi \models \varphi$;* (iii) *$\mathcal{K}, \pi \models \psi_1 \wedge \psi_2$ iff $\mathcal{K}, \pi \models \psi_1$ and $\mathcal{K}, \pi \models \psi_2$;* (iv) *$\mathcal{K}, \pi \models \psi_1 \vee \psi_2$ iff $\mathcal{K}, \pi \models \psi_1$ or $\mathcal{K}, \pi \models \psi_2$;* (v) *$\mathcal{K}, \pi \models \mathsf{X}\psi$ iff $\pi_{\geq 1} \neq \varepsilon$ and $\mathcal{K}, \pi_{\geq 1} \models \psi$;* (vi) *$\mathcal{K}, \pi \models \tilde{\mathsf{X}}\psi$ iff $\pi_{\geq 1} = \varepsilon$ or $\mathcal{K}, \pi_{\geq 1} \models \psi$;* (vii) *$\mathcal{K}, \pi \models \psi_1 \mathsf{U} \psi_2$ iff $\mathcal{K}, \pi \models \psi_2 \vee \psi_1 \wedge \mathsf{X}\psi_1 \mathsf{U} \psi_2$;* (viii) *$\mathcal{K}, \pi \models \psi_1 \mathsf{R} \psi_2$ iff $\mathcal{K}, \pi \models \psi_2 \wedge (\psi_1 \vee \mathsf{X}\psi_1 \mathsf{R} \psi_2)$;* (ix) *$\mathcal{K}, \pi \models \psi_1 \tilde{\mathsf{U}} \psi_2$ iff $\mathcal{K}, \pi \models \psi_2 \vee \psi_1 \wedge \tilde{\mathsf{X}}\psi_1 \tilde{\mathsf{U}} \psi_2$;* (x) *$\mathcal{K}, \pi \models \psi_1 \tilde{\mathsf{R}} \psi_2$ iff $\mathcal{K}, \pi \models \psi_2 \wedge (\psi_1 \vee \tilde{\mathsf{X}}\psi_1 \tilde{\mathsf{R}} \psi_2)$.*

Proof First note that in this proof, we make use of a slightly more general map of $\varpi_{\mathcal{K},\psi}(\cdot)$ that associates each path in $\mathcal{K}$ with the sequence of state formulas, belonging to a given set Z, satisfied at the worlds of π. Formally, by $\varpi_{\mathcal{K},\mathrm{Z}}(\pi)$ we denote the trace in $(2^{\mathrm{Z}})^{|\pi|}$ such that, for all $\varphi \in \mathrm{Z}$ and $k \in [0, |\pi|[$, it holds that $\varphi \in (\varpi_{\mathcal{K},\mathrm{Z}}(\pi))_k$ iff $\mathcal{K}, (\pi)_k \models \varphi$. Observe that, for every GCTL* path formula ψ and set Z of state formulas containing $\mathsf{rcl}(\psi)$, when ψ is interpreted as an LTL formula on $\mathsf{rcl}(\psi)$, it is satisfied on a trace $\varpi_{\mathcal{K},\psi}(\pi)$ iff it is satisfied on all traces $\varpi_{\mathcal{K},\mathrm{Z}}(\pi)$ as well. We can now start with the proofs of all items.

(1) Let $\mathrm{Z} = \mathsf{rcl}(\psi_1) \cup \mathsf{rcl}(\psi_2)$. For $i \in \{1, 2\}$, if $\mathcal{K}, \pi \models \psi_i$, then $\varpi_{\mathcal{K},\psi_i}(\pi) \models \psi_i$. Now, since $\mathsf{rcl}(\psi_i) \subseteq \mathrm{Z}$, we have that $\varpi_{\mathcal{K},\mathrm{Z}}(\pi) \models \psi_i$. By the equivalence $\psi_1 \equiv \psi_2$, we obtain then that $\varpi_{\mathcal{K},\mathrm{Z}}(\pi) \models \psi_{3-i}$. So, since $\mathsf{rcl}(\psi_{3-i}) \subseteq \mathrm{Z}$, we have that $\varpi_{\mathcal{K},\psi_{3-i}}(\pi) \models \psi_{3-i}$ and consequently $\mathcal{K}, \pi \models \psi_{3-i}$.
(2) Since φ is a state formula, by definition of the transformation map $\varpi_{\mathcal{K},\varphi}(\cdot)$, we have that $\mathcal{K}, w \models \varphi$ iff $\varphi \in (\varpi_{\mathcal{K},\varphi}(\pi))_0$ and so $\varpi_{\mathcal{K},\varphi}(\pi) \models \varphi$, from which we derive $\mathcal{K}, \pi \models \varphi$ and vice versa.

(3) Let $\psi = \psi_1 \wedge \psi_2$. Then, it holds that $\mathcal{K}, \pi \models \psi$ iff $\varpi_{\mathcal{K},\psi}(\pi) \models \psi$, which is equivalent to $\varpi_{\mathcal{K},\psi}(\pi) \models \psi_i$, for $i \in \{1, 2\}$. At this point, since $\mathsf{rcl}(\psi_i) \subseteq \mathsf{rcl}(\psi)$, we have that $\mathcal{K}, \pi \models \psi$ is equivalent to $\varpi_{\mathcal{K},\psi_i}(\pi) \models \psi_i$, for $i \in \{1, 2\}$. Hence, $\mathcal{K}, \pi \models \psi$ iff $\mathcal{K}, \pi \models \psi_1$ and $\mathcal{K}, \pi \models \psi_2$.

(4) Mutatis mutandis, the proof is the same of the previous item.

(5) Note that $\mathsf{rcl}(\mathsf{X}\psi) = \mathsf{rcl}(\psi)$. Then, it holds that $\mathcal{K}, \pi \models \mathsf{X}\psi$ iff $\varpi_{\mathcal{K},\psi}(\pi) \models \mathsf{X}\psi$, which is equivalent to $(\varpi_{\mathcal{K},\psi}(\pi))_{\geq 1} \neq \varepsilon$, i.e., $\pi_{\geq 1} \neq \varepsilon$, and $(\varpi_{\mathcal{K},\psi}(\pi))_{\geq 1} \models \psi$, i.e., $\varpi_{\mathcal{K},\psi}(\pi_{\geq 1}) \models \psi$. Hence, $\mathcal{K}, \pi \models \mathsf{X}\psi$ iff $\pi_{\geq 1} \neq \varepsilon$ and $\mathcal{K}, \pi_{\geq 1} \models \psi$.

(6) Mutatis mutandis, the proof is the same of the previous item.

vii-x. These items can be directly derived by Item i and the classical LTL one step unfolding equivalences $\psi_1 \mathsf{U} \psi_2 \equiv \psi_2 \vee \psi_1 \wedge \mathsf{X}\psi_1 \mathsf{U}\psi_2$, $\psi_1 \mathsf{R} \psi_2 \equiv \psi_2 \wedge (\psi_1 \vee \mathsf{X}\psi_1 \mathsf{R}\psi_2)$, $\psi_1 \tilde{\mathsf{U}} \psi_2 \equiv \psi_2 \vee \psi_1 \wedge \tilde{\mathsf{X}}\psi_1 \tilde{\mathsf{U}}\psi_2$, and $\psi_1 \tilde{\mathsf{R}} \psi_2 \equiv \psi_2 \wedge (\psi_1 \vee \tilde{\mathsf{X}}\psi_1 \tilde{\mathsf{R}}\psi_2)$. □

In the rest of the work, we only consider formulas in *positive normal form* (*pnf*, for short), i.e., the negation is applied only to atomic propositions. In fact, it is to this aim that we have considered in the syntax of GCTL* both the Boolean connectives $\wedge$ and $\vee$, the path quantifiers $\mathsf{A}^{<g}$ and $\mathsf{E}^{\geq g}$, and temporal operators X, U, and R together with their weak version $\tilde{\mathsf{X}}$, $\tilde{\mathsf{U}}$, and $\tilde{\mathsf{R}}$. Indeed, all formulas can be linearly translated in *pnf* by using De Morgan's laws and the following equivalences, which directly follow from the semantics of the logic: $\neg \mathsf{E}^{\geq g}\psi \equiv \mathsf{A}^{<g}\neg\psi$; $\neg \mathsf{X}\psi \equiv \tilde{\mathsf{X}}\neg\psi$; $\neg(\psi_1 \mathsf{U} \psi_2) \equiv (\neg\psi_1)\tilde{\mathsf{R}}(\neg\psi_2)$; $\neg(\psi_1 \mathsf{R} \psi_2) \equiv (\neg\psi_1)\tilde{\mathsf{U}}(\neg\psi_2)$. Under this assumption, we consider $\neg\varphi$ as the *pnf* formula equivalent to the negation of φ. Finally, as abbreviations we use the Boolean values $\mathfrak{t}$ ("*true*") and $\mathfrak{f}$ ("*false*") and the path quantifiers $\mathsf{E}^{>g}\psi \triangleq \mathsf{E}^{\geq g+1}\psi$ ("there exist more than g paths"), $\mathsf{A}^{\leq g}\psi \triangleq \mathsf{A}^{<g+1}\psi$ ("all but at most g paths"), $\mathsf{E}^{=g}\psi \triangleq \mathsf{E}^{\geq g}\psi \wedge \neg \mathsf{E}^{>g}\psi$ ("there exist just g paths"), and $\mathsf{A}^{=g}\psi \triangleq \mathsf{A}^{\leq g}\psi \wedge \neg \mathsf{A}^{<g}\psi$ ("all but exactly g paths"), with $g \in [0, \omega[$.

We now report some basic equivalences that are directly derived from the definition of the logic and Proposition 1.1, and are independent from the particular path equivalence relation $\equiv$ considered.

Proposition 1.2 (Basic Equivalences) *Let φ and ψ be a state and a path formula, respectively, and $g \in \widehat{\mathbb{N}}$. Then, the following equivalences hold:* (i) $\mathsf{E}^{\geq 0}\psi \equiv \mathfrak{t}$; (ii) $\mathsf{E}^{\geq 1}\varphi \equiv \varphi$; (iii) $\mathsf{E}^{\geq 1}(\varphi \wedge \psi) \equiv \varphi \wedge \mathsf{E}^{\geq 1}\psi$; (iv) $\mathsf{E}^{\geq 1}(\varphi \vee \psi) \equiv \varphi \vee \mathsf{E}^{\geq 1}\psi$; (v) $\mathsf{E}^{\geq 1}\mathsf{X}\psi \equiv \mathsf{E}^{\geq 1}\mathsf{X}\mathsf{E}^{\geq 1}\psi$; (vi) $\mathsf{E}^{\geq 1}\tilde{\mathsf{X}}\psi \equiv \mathsf{E}^{\geq 1}\tilde{\mathsf{X}}\mathfrak{f} \vee \mathsf{E}^{\geq 1}\mathsf{X}\psi$; (vii) $\mathsf{E}^{>g}\psi \Rightarrow \mathsf{E}^{\geq g}\psi$; (viii) $\mathsf{A}^{<0}\psi \equiv \mathfrak{f}$; (ix) $\mathsf{A}^{<1}\varphi \equiv \varphi$; (x) $\mathsf{A}^{<1}(\varphi \wedge \psi) \equiv \varphi \wedge \mathsf{A}^{<1}\psi$; (xi) $\mathsf{A}^{<1}(\varphi \vee \psi) \equiv \varphi \vee \mathsf{A}^{<1}\psi$; (xii) $\mathsf{A}^{<1}\mathsf{X}\psi \equiv \mathsf{A}^{<1}\mathsf{X}\mathfrak{t} \wedge \mathsf{A}^{<1}\tilde{\mathsf{X}}\psi$; (xiii) $\mathsf{A}^{<1}\tilde{\mathsf{X}}\psi \equiv \mathsf{A}^{<1}\tilde{\mathsf{X}}\mathsf{A}^{<1}\psi$; (xiv) $\mathsf{A}^{<g}\psi \Rightarrow \mathsf{A}^{\leq g}\psi$.

Finally, we list the classical CTL fixpoint equivalences embedded in the GCTL framework, for the four binary temporal operators U, R, $\tilde{\mathsf{U}}$, and $\tilde{\mathsf{R}}$.

Proposition 1.3 (CTL Fixpoint Equivalences) *Let φ_1 and φ_2 be two state formulas. Then, the following hold:*

(1) $\mathsf{E}^{\geq 1}\varphi_1 \mathsf{U} \varphi_2 \equiv \varphi_2 \vee \varphi_1 \wedge \mathsf{E}^{\geq 1}\mathsf{X}\mathsf{E}^{\geq 1}\varphi_1 \mathsf{U}\varphi_2$;

(2) $\mathsf{E}^{\geq 1}\varphi_1 \mathsf{R} \varphi_2 \equiv \varphi_2 \wedge (\varphi_1 \vee \mathsf{E}^{\geq 1}\mathsf{X}\mathsf{E}^{\geq 1}\varphi_1 \mathsf{R}\varphi_2)$;

(3) $\mathsf{E}^{\geq 1}\varphi_1\tilde{\mathsf{U}}\varphi_2 \equiv \varphi_2 \vee \varphi_1 \wedge (\mathsf{E}^{\geq 1}\tilde{\mathsf{X}}\mathsf{f} \vee \mathsf{E}^{\geq 1}\mathsf{X}\mathsf{E}^{\geq 1}\varphi_1\tilde{\mathsf{U}}\varphi_2)$;
(4) $\mathsf{E}^{\geq 1}\varphi_1\tilde{\mathsf{R}}\varphi_2 \equiv \varphi_2 \wedge (\varphi_1 \vee \mathsf{E}^{\geq 1}\tilde{\mathsf{X}}\mathsf{f} \vee \mathsf{E}^{\geq 1}\mathsf{X}\mathsf{E}^{\geq 1}\varphi_1\tilde{\mathsf{R}}\varphi_2)$;
(5) $\mathsf{A}^{<1}\varphi_1\mathsf{U}\varphi_2 \equiv \varphi_2 \vee \varphi_1 \wedge (\mathsf{A}^{<1}\mathsf{X}\mathsf{t} \wedge \mathsf{A}^{<1}\tilde{\mathsf{X}}\mathsf{A}^{<1}\varphi_1\mathsf{U}\varphi_2)$;
(6) $\mathsf{A}^{<1}\varphi_1\mathsf{R}\varphi_2 \equiv \varphi_2 \wedge (\varphi_1 \vee \mathsf{A}^{<1}\mathsf{X}\mathsf{t} \wedge \mathsf{A}^{<1}\tilde{\mathsf{X}}\mathsf{A}^{<1}\varphi_1\mathsf{R}\varphi_2)$;
(7) $\mathsf{A}^{<1}\varphi_1\tilde{\mathsf{U}}\varphi_2 \equiv \varphi_2 \vee \varphi_1 \wedge \mathsf{A}^{<1}\tilde{\mathsf{X}}\mathsf{A}^{<1}\varphi_1\tilde{\mathsf{U}}\varphi_2$;
(8) $\mathsf{A}^{<1}\varphi_1\tilde{\mathsf{R}}\varphi_2 \equiv \varphi_2 \wedge (\varphi_1 \vee \mathsf{A}^{<1}\tilde{\mathsf{X}}\mathsf{A}^{<1}\varphi_1\tilde{\mathsf{R}}\varphi_2)$.

1.4 Path Equivalence Properties

In the definition of GCTL* semantics, we make use of an arbitrary equivalence relation on paths. It is useful to investigate what properties can make such an equivalence a reasonable one for our purposes. In this section, we present a detailed exposition of its principal properties. Note that, in order to be not too repetitive, when we talk about "number of paths", we always mean the number of equivalence classes of paths w.r.t. a path formula, which is clear from the context. Moreover, every equivalence concerning the universal quantifier, if not otherwise specified, is obtained through the dualization ($\mathsf{A}^{<g}\psi \equiv \neg\mathsf{E}^{\geq g}\neg\psi$) of the related existential one.

To help the reader in following the exposition of the several properties we are going to introduce, we now give an outline of this section. First, in Sect. 1.4.1, we introduce two basic properties of equivalences on paths, namely *syntax independence* and *state focus*. In Sect. 1.4.2, we further give the properties of *next consistency*, along with its weak form, and that of *source dependence*. From this we derive two fundamental expansion constructions that allow to generalize, to the case of graded quantifiers, the classical CTL* expansion equivalences $\mathsf{EX}\psi \equiv \mathsf{EXE}\psi$ and $\mathsf{A}\tilde{\mathsf{X}}\psi \equiv \mathsf{A}\tilde{\mathsf{X}}\mathsf{A}\psi$. In Sect. 1.4.3, we introduce two consistency properties on the Boolean connectives $\wedge$ and $\vee$ and a satisfiability constraint. Finally, in Sect. 1.4.4, we define the concept of *adequacy* of an equivalence on paths and use it to show a set of fixpoint equivalences for GCTL, which are used, in Sect. 1.5, to prove the existence of a translation of a fragment of GCTL into GμCALCULUS.

1.4.1 Elementary Requirements

Suppose we have two equivalent path formulas ψ_1 and ψ_2. Then, we would like to have them to be exchangeable in a GCTL* path quantification, obtaining in this way that two state formulas $\mathsf{Qn}\ \psi_1$ and $\mathsf{Qn}\ \psi_2$ are equivalent, for all $\mathsf{Qn} \in \{\mathsf{E}^{\geq n}, \mathsf{A}^{<n}\}$ and $n \in \widehat{\mathbb{N}}$. Hence, what we need to require is that, whenever two paths are equivalent w.r.t. ψ_1, they are equivalent w.r.t. ψ_2, too. Before introducing the formal definition of this concept, we want to enlighten on the equivalence relation between path formulas we consider here is not just the classical LTL equivalence $\equiv$, but rather a generic equivalence $\cong^{w}_{\mathcal{K}}$ that may depend on both a KS and one of its worlds. More motivations for this choice are given in the next subsection.

Definition 1.3 (Syntax Independence) *An equivalence relation* $\equiv_{\mathcal{K}}^{\cdot}$ *on paths is said* syntax independent *iff, for all pairs of equivalent path formulas* ψ_1 *and* ψ_2 *w.r.t* $\cong_{\mathcal{K}}^{w}$, *it holds that* $\pi_1 \equiv_{\mathcal{K}}^{\psi_1} \pi_2$ *iff* $\pi_1 \equiv_{\mathcal{K}}^{\psi_2} \pi_2$, *for all* $\pi_1, \pi_2 \in \mathrm{Pth}(\mathcal{K}, w)$.

Theorem 1.1 (Equivalent Quantifications) *Let* $\equiv_{\cdot}^{\cdot}$ *be a syntax-independent equivalence relation. Moreover, let* ψ_1 *and* ψ_2 *be two equivalent path formulas and* $g \in \widehat{\mathbb{N}}$. *Then, the following holds:* (i) $\mathsf{E}^{\geq g}\psi_1 \equiv \mathsf{E}^{\geq g}\psi_2$ *and* (ii) $\mathsf{A}^{<g}\psi_1 \equiv \mathsf{A}^{<g}\psi_2$.

Proof Let $\mathcal{K}$ be a KS and w_0 its initial world. Since $\psi_1 \equiv \psi_2$, by Item i of Proposition 1.1, it is immediate to see that $\mathrm{Pth}(\mathcal{K}, w_0, \psi_1) = \mathrm{Pth}(\mathcal{K}, w_0, \psi_2)$ and consequently $(\mathrm{Pth}(\mathcal{K}, w_0, \psi_1)/{\equiv_{\mathcal{K}}^{\psi_1}}) = (\mathrm{Pth}(\mathcal{K}, w_0, \psi_2)/{\equiv_{\mathcal{K}}^{\psi_1}})$. Now, by the syntax-independence property, we have that $\pi_1 \equiv_{\mathcal{K}}^{\psi_1} \pi_2$ iff $\pi_1 \equiv_{\mathcal{K}}^{\psi_2} \pi_2$, for all $\pi_1, \pi_2 \in \mathrm{Pth}(\mathcal{K})$. Thus, we have that $(\mathrm{Pth}(\mathcal{K}, w_0, \psi_2)/{\equiv_{\mathcal{K}}^{\psi_1}}) = (\mathrm{Pth}(\mathcal{K}, w_0, \psi_2)/{\equiv_{\mathcal{K}}^{\psi_2}})$. Hence the thesis. □

The following corollary is directly derived by using the classical LTL equivalences for the four binary temporal operators.

Corollary 1.1 (One Step Unfolding) *Let* $\equiv_{\cdot}^{\cdot}$ *be a syntax-independent equivalence relation. Moreover, let* ψ_1 *and* ψ_2 *be two path formulas and* $g \in \widehat{\mathbb{N}}$. *Then, the following equivalences hold:* (i) $\mathsf{E}^{\geq g}\psi_1 \mathsf{U} \psi_2 \equiv \mathsf{E}^{\geq g}(\psi_2 \vee \psi_1 \wedge \mathsf{X}\psi_1 \mathsf{U} \psi_2)$; (ii) $\mathsf{E}^{\geq g}\psi_1 \mathsf{R} \psi_2 \equiv \mathsf{E}^{\geq g}(\psi_2 \wedge (\psi_1 \vee \mathsf{X}\psi_1 \mathsf{R} \psi_2))$; (iii) $\mathsf{E}^{\geq g}\psi_1 \tilde{\mathsf{U}} \psi_2 \equiv \mathsf{E}^{\geq g}(\psi_2 \vee \psi_1 \wedge \tilde{\mathsf{X}}\psi_1 \tilde{\mathsf{U}} \psi_2)$; (iv) $\mathsf{E}^{\geq g}\psi_1 \tilde{\mathsf{R}} \psi_2 \equiv \mathsf{E}^{\geq g}(\psi_2 \wedge (\psi_1 \vee \tilde{\mathsf{X}}\psi_1 \tilde{\mathsf{R}} \psi_2))$; (v) $\mathsf{A}^{<g}\psi_1 \mathsf{U} \psi_2 \equiv \mathsf{A}^{<g}(\psi_2 \vee \psi_1 \wedge \mathsf{X}\psi_1 \mathsf{U} \psi_2)$; (vi) $\mathsf{A}^{<g}\psi_1 \mathsf{R} \psi_2 \equiv \mathsf{A}^{<g}(\psi_2 \wedge (\psi_1 \vee \mathsf{X}\psi_1 \mathsf{R} \psi_2))$; (vii) $\mathsf{A}^{<g}\psi_1 \tilde{\mathsf{U}} \psi_2 \equiv \mathsf{A}^{<g}(\psi_2 \vee \psi_1 \wedge \tilde{\mathsf{X}}\psi_1 \tilde{\mathsf{U}} \psi_2)$; (viii) $\mathsf{A}^{<g}\psi_1 \tilde{\mathsf{R}} \psi_2 \equiv \mathsf{A}^{<g}(\psi_2 \wedge (\psi_1 \vee \tilde{\mathsf{X}}\psi_1 \tilde{\mathsf{R}} \psi_2))$.

Consider now a state formula φ on which we have to verify the equivalence between paths. Then, we may want to have that, when a world satisfies φ, all paths starting from that world are counted just once. This is because, after all, we have only one way to practically satisfy the formula.

Definition 1.4 (State Focus) *An equivalence relation* $\equiv_{\mathcal{K}}^{\cdot}$ *is said* state focused *iff, given a state formula* φ, *if* $\mathcal{K}, w \models \varphi$ *then* $\pi_1 \equiv_{\mathcal{K}}^{\varphi} \pi_2$, *for all* $\pi_1, \pi_2 \in \mathrm{Pth}(\mathcal{K}, w)$.

Theorem 1.2 (State Quantification) *Let* $\equiv_{\cdot}^{\cdot}$ *be a state-focused equivalence relation. Moreover, let* φ *be a state formula and* $g \in [2, \omega]$. *Then, the following holds:* (i) $\mathsf{E}^{\geq g}\varphi \equiv \mathfrak{f}$ *and* (ii) $\mathsf{A}^{<g}\varphi \equiv \mathfrak{t}$.

Proof Suppose by contradiction that $\mathsf{E}^{\geq g}\varphi \not\equiv \mathfrak{f}$, i.e., that there is a KS $\mathcal{K}$ such that $\mathcal{K}, w_0 \models \mathsf{E}^{\geq g}\varphi$, where w_0 is the initial world of $\mathcal{K}$. This means that $|(\mathrm{Pth}(\mathcal{K}, w_0, \varphi)/{\equiv_{\mathcal{K}}^{\varphi}})| \geq g$, so $\mathrm{Pth}(\mathcal{K}, w_0, \varphi) \neq \emptyset$ and then, by Item ii of Proposition 1.1, it holds that $\mathcal{K}, w_0 \models \varphi$. Now, by the state-focus property, we have that $\pi_1 \equiv_{\mathcal{K}}^{\varphi} \pi_2$, for all paths $\pi_1, \pi_2 \in \mathrm{Pth}(\mathcal{K}, w_0)$. Hence, $|(\mathrm{Pth}(\mathcal{K}, w_0, \varphi)/{\equiv_{\mathcal{K}}^{\varphi}})| = 1 < g$, but this contradict the hypothesis. □

1.4.2 Temporal Requirements

Consider a path formula ψ. We would like that the number of paths satisfying $X\psi$ at a world w is equal to the sum of the number of paths that satisfy ψ on all successor worlds w' of w. This requires that two paths π_1 and π_2 are distinct w.r.t. $X\psi$ iff the paths $(\pi_1)_{\geq 1}$ and $(\pi_2)_{\geq 1}$ are also distinct w.r.t. ψ.

Definition 1.5 (Next Consistency) *An equivalence relation* $\equiv^{\cdot}_{\mathcal{K}}$ *on paths is said* next consistent *iff it holds that* $\pi_1 \equiv^{X\psi}_{\mathcal{K}} \pi_2$ *iff* $(\pi_1)_{\geq 1} \equiv^{\psi}_{\mathcal{K}} (\pi_2)_{\geq 1}$, *for all* $\pi_1, \pi_2 \in \mathrm{Pth}(\mathcal{K}, w)$.

By the state focus and next-consistency properties, it is immediate to derive the following first accessory lemma.

Lemma 1.1 (Next Equivalence I) *Let* $\equiv^{\cdot}_{\mathcal{K}}$ *be a state-focused and next-consistent equivalence relation. Moreover, let* $\pi_1, \pi_2 \in \mathrm{Pth}(\mathcal{K}, w)$ *be two paths starting in a common world* w *and* φ *be a state formula. Then,* $(\pi_1)_1 = (\pi_2)_1 = w'$ *and* $\mathcal{K}, w' \models \varphi$ *imply* $\pi_1 \equiv^{X\varphi}_{\mathcal{K}} \pi_2$.

Proof By the state-focus property, it holds that $(\pi_1)_{\geq 1} \equiv^{\varphi}_{\mathcal{K}} (\pi_2)_{\geq 1}$. Now, by the next-consistency property, we obtain that $\pi_1 \equiv^{X\varphi}_{\mathcal{K}} \pi_2$. □

For a $\tilde{X}\psi$ formula, the only difference w.r.t. $X\psi$ is that the formula can be satisfied on a path because there are no successor worlds. In such a situation there is only one path satisfying the formula. In the other cases $\tilde{X}\psi$ behaves just like $X\psi$, hence, we would like the first to satisfy a similar property w.r.t. the latter. However, when ψ is a tautology, we have that $\tilde{X}\psi$, differently from $X\psi$, is equivalent to t, i.e., the formula is always satisfied. For this reason all choices are indifferent and may be regarded as equivalent.

However, there may be other reasons to consider a given path formula ψ as tautological: one may consider that at w in $\mathcal{K}$, there are some "physical boundaries" on what paths starting from w can achieve. Then, it makes sense to take into account such limitations when evaluating whether a path formulas is tautological or not. For example, there may be a three-valued property, which is encoded by means of two binary variables a and b. However, the two binary variables actually encode a four-valued property. Then, it makes sense to assume, as hypothesis in the system, that one of the value, say $\neg a \wedge \neg b$, is never assumed by a KS. In such a case, the formula $\neg(\neg a \wedge \neg b)$ can be considered equivalent to true. As a further example there may be an world w modeling an end state with a property *end* and a self loop. In such a case, the formula $G\mathit{end}$ can be considered equivalent to true too.

We highlight the tautological nature of a path formula by means of a generalization of the LTL equivalence relation that depends also on the context the formulas are evaluated in, i.e., on a particular KS $\mathcal{K}$ and on one of its world w. Two path formulas should be equivalent at $\mathcal{K}, w$ only if they cannot distinguish paths in $\mathrm{Pth}(\mathcal{K}, w)$.

We now formally define the general notion of equivalence among path formulas.

Definition 1.6 (Equivalence Structure) *An* equivalence structure $\cong^{\cdot}_{\cdot}$ *is a parametric equivalence relation among path formulas depending on a* KS $\mathcal{K}$ *and one of its worlds* w *such that, for all path formulas* ψ_1 *and* ψ_2, (i) $\psi_1 \equiv \psi_2$ *implies that* $\psi_1 \cong^{w}_{\mathcal{K}} \psi_2$ *and* (ii) $\psi_1 \cong^{w}_{\mathcal{K}} \psi_2$ *implies that* $\mathcal{K}, \pi \models \psi_1$ *iff* $\mathcal{K}, \pi \models \psi_2$*, for all* $\pi \in \mathrm{Pth}(\mathcal{K}, w)$.

Observe that the LTL equivalence relation $\equiv$ is a particular equivalence structure $\cong^{w}_{\mathcal{K}}$ that does not depend on the KS $\mathcal{K}$ and world w.

At this point, we are ready to define a generic tautology.

Definition 1.7 (Tautology Structure) *Given a* KS $\mathcal{K}$ *and one of its worlds* w*, a path formula* ψ *is a* $\cong^{w}_{\mathcal{K}}$*-tautology iff* $\psi \cong^{w}_{\mathcal{K}}$ t.

Using the above concept we can state the consistency property required by the weak next operator.

Definition 1.8 (Weak Next Consistency) *An equivalence relation* $\equiv^{\cdot}_{\mathcal{K}}$ *on paths is said* weak next consistent *w.r.t. an equivalence structure* $\cong^{\cdot}_{\mathcal{K}}$ *iff it holds that* $\pi_1 \equiv^{\tilde{\mathsf{X}}\psi}_{\mathcal{K}} \pi_2$ *iff* $\tilde{\mathsf{X}}\psi$ *is an* $\cong^{w}_{\mathcal{K}}$*-tautology or* $(\pi_1)_{\geq 1} \equiv^{\psi}_{\mathcal{K}} (\pi_2)_{\geq 1}$*, for all* $\pi_1, \pi_2 \in \mathrm{Pth}(\mathcal{K}, w)$.

By the next and weak next-consistency properties, we can derive the simplification theorem for the quantifications of the weak next temporal operator.

Theorem 1.3 (Weak Next Simplification) *Let* $\equiv^{\cdot}_{\cdot}$ *be a next-consistent and weak next-consistent equivalence relation w.r.t.* $\cong^{\cdot}_{\cdot}$*. Moreover, let* $\mathcal{K}$ *be a* KS, ψ *be a path formula and* $g \in [2, \omega]$*. Then, the following holds:* (i) $\mathcal{K} \models \mathsf{E}^{\geq g}\tilde{\mathsf{X}}\psi$ *iff* $\tilde{\mathsf{X}}\psi$ *is not an* $\cong^{w_0}_{\mathcal{K}}$*-tautology and* $\mathcal{K} \models \mathsf{E}^{\geq g}\mathsf{X}\psi$ *and* (ii) $\mathcal{K} \models \mathsf{A}^{<g}\mathsf{X}\psi$ *iff* $\neg\mathsf{X}\psi$ *is an* $\cong^{w_0}_{\mathcal{K}}$*-tautology or* $\mathcal{K} \models \mathsf{A}^{<g}\tilde{\mathsf{X}}\psi$*, where* w_0 *is the initial world of* $\mathcal{K}$.

Proof By hypotheses, it holds that $\pi_1 \equiv^{\tilde{\mathsf{X}}\psi}_{\mathcal{K}} \pi_2$ iff $\tilde{\mathsf{X}}\psi$ is an $\cong^{w_0}_{\mathcal{K}}$-tautology or $\pi_1 \equiv^{\mathsf{X}\psi}_{\mathcal{K}} \pi_2$, for all $\pi_1, \pi_2 \in \mathrm{Pth}(\mathcal{K}, w_0)$, where w_0 is the initial world of $\mathcal{K}$.

[Only if]. If $\mathcal{K}, w_0 \models \mathsf{E}^{\geq g}\tilde{\mathsf{X}}\psi$ then $|(\mathrm{Pth}(\mathcal{K}, w_0, \tilde{\mathsf{X}}\psi)/\equiv^{\tilde{\mathsf{X}}\psi}_{\mathcal{K}})| \geq g$. Since there are at least two different classes w.r.t. $\equiv^{\tilde{\mathsf{X}}\psi}_{\mathcal{K}}$ and so, at least two non-equivalent paths starting in w_0, it holds that $\tilde{\mathsf{X}}\psi$ cannot be an $\cong^{w_0}_{\mathcal{K}}$-tautology. Consequently, we have that $\pi_1 \equiv^{\tilde{\mathsf{X}}\psi}_{\mathcal{K}} \pi_2$ iff $\pi_1 \equiv^{\mathsf{X}\psi}_{\mathcal{K}} \pi_2$, for all $\pi_1, \pi_2 \in \mathrm{Pth}(\mathcal{K}, w_0)$. Moreover, since w_0 has necessarily a successor, by Items v and vi of Proposition 1.1, it holds that $\mathrm{Pth}(\mathcal{K}, w_0, \tilde{\mathsf{X}}\psi) = \mathrm{Pth}(\mathcal{K}, w_0, \mathsf{X}\psi)$. Thus, we obtain that $(\mathrm{Pth}(\mathcal{K}, w_0, \tilde{\mathsf{X}}\psi)/\equiv^{\tilde{\mathsf{X}}\psi}_{\mathcal{K}}) = (\mathrm{Pth}(\mathcal{K}, w_0, \mathsf{X}\psi)/\equiv^{\mathsf{X}\psi}_{\mathcal{K}})$. Hence, the thesis holds.

[If]. If $\mathcal{K}, w_0 \models \mathsf{E}^{\geq g}\mathsf{X}\psi$ then $|(\mathrm{Pth}(\mathcal{K}, w_0, \mathsf{X}\psi)/\equiv^{\mathsf{X}\psi}_{\mathcal{K}})| \geq g$. Since there are at least two different classes w.r.t. $\equiv^{\mathsf{X}\psi}_{\mathcal{K}}$, w_0 has necessarily a successor and so, by Items v and vi of Proposition 1.1, it holds that $\mathrm{Pth}(\mathcal{K}, w_0, \mathsf{X}\psi) = \mathrm{Pth}(\mathcal{K}, w_0, \tilde{\mathsf{X}}\psi)$. Moreover, $\tilde{\mathsf{X}}\psi$ is not an $\cong^{w_0}_{\mathcal{K}}$-tautology. Consequently, we have that $\pi_1 \equiv^{\mathsf{X}\psi}_{\mathcal{K}} \pi_2$

iff $\pi_1 \equiv_{\mathcal{K}}^{\tilde{X}\psi} \pi_2$, for all $\pi_1, \pi_2 \in \mathrm{Pth}(\mathcal{K}, w_0)$. Thus, we obtain that $(\mathrm{Pth}(\mathcal{K}, w_0, X\psi)/\equiv_{\mathcal{K}}^{X\psi}) = (\mathrm{Pth}(\mathcal{K}, w_0, \tilde{X}\psi)/\equiv_{\mathcal{K}}^{\tilde{X}\psi})$. Hence, the thesis holds. □

In general, there are no GCTL* formulas expressing the fact that $\tilde{X}\psi$ and $\neg X\psi$ are or are not an $\cong_{\mathcal{K}}^{w_0}$-tautology. However, in the case that a particular $\cong_{\mathcal{K}}^{w_0}$-tautology of the previous formulas can be expressed with the two apposite formulas $\varphi_{\tilde{X}\psi}$ and $\varphi_{\neg X\psi}$, we can easily state $E^{\geq g}\tilde{X}\psi \equiv (E^{\geq g}X\psi) \wedge \neg\varphi_{\tilde{X}\psi}$ and $A^{<g}X\psi \equiv (A^{<g}\tilde{X}\psi) \vee \varphi_{\neg X\psi}$, for $g \in [2, \omega]$. Moreover, we recall that Items vi and xii of Proposition 1.2 assert that $E^{\geq g}\tilde{X}\psi \equiv E^{\geq 1}\tilde{X}f \vee E^{\geq 1}X\psi$ and $A^{<g}X\psi \equiv A^{<1}Xt \wedge A^{<1}\tilde{X}\psi$, for $g = 1$. Then, we introduce the two macros $E\tilde{X}(g, \psi, \varphi)$ and $AX(g, \psi, \varphi)$, defined below, to represent in short the expansion formula for $E\tilde{X}$ and AX.

- $E\tilde{X}(g, \psi, \varphi) \triangleq \begin{cases} E^{\geq 1}\tilde{X}f \vee E^{\geq 1}X\psi, & \text{if } g = 1; \\ (E^{\geq g}X\psi) \wedge \varphi, & \text{otherwise.} \end{cases}$
- $AX(g, \psi, \varphi) \triangleq \begin{cases} A^{<1}Xt \wedge A^{<1}\tilde{X}\psi, & \text{if } g = 1; \\ (A^{\geq g}\tilde{X}\psi) \vee \varphi, & \text{otherwise.} \end{cases}$

It is immediate to see that $\mathrm{lng}(E\tilde{X}(g, \psi, \varphi)) = \mathrm{lng}(AX(g, \psi, \varphi)) = \Theta(\mathrm{lng}(\varphi) + \mathrm{lng}(\psi))$.

The above properties for the next and the weak next operators allow us to say that the number of paths that satisfy $X\psi$ or $\tilde{X}\psi$ at world w is equal to the number of paths that satisfy ψ on some successor world w' of w. Since two paths π_1 and π_2 passing through two distinct successors may represent two different ways to satisfy $X\psi$, we would like to consider them as distinct w.r.t. $X\psi$. So, we should have that the two paths $(\pi_1)_{\geq 1}$ and $(\pi_2)_{\geq 1}$ are not-equivalent just because they start from different nodes. Consequently, we may want to ensure that paths starting at different successors are never counted just as one.

Definition 1.9 (Source Dependence) *An equivalence relation $\equiv_{\mathcal{K}}^{\cdot}$ on paths is said* source-dependent *iff $\pi_1 \equiv_{\mathcal{K}}^{\psi} \pi_2$ implies $(\pi_1)_0 = (\pi_2)_0$, for all $\pi_1, \pi_2 \in \mathrm{Pth}(\mathcal{K})$.*

At this point, by the next-consistency and source-dependence properties it is immediate to derive the following second accessory lemma.

Lemma 1.2 (Next Equivalence II) *Let $\equiv_{\mathcal{K}}^{\cdot}$ be a next-consistent and source-dependent equivalence relation. Moreover, let $\pi_1, \pi_2 \in \mathrm{Pth}(\mathcal{K}, w)$ be two paths starting in a common world w. Then, $\pi_1 \equiv_{\mathcal{K}}^{X\psi} \pi_2$ implies $(\pi_1)_1 = (\pi_2)_1$.*

Proof By the next-consistency property, it holds that $(\pi_1)_{\geq 1} \equiv_{\mathcal{K}}^{\psi} (\pi_2)_{\geq 1}$. Now, by the source-dependence property, we obtain that $(\pi_1)_1 = (\pi_2)_1$. □

Before continuing with the discussion of the remaining properties, we have to make an important remark on our choice to define the semantics of GCTL* on both finite and infinite paths and, consequently, to have both the strong and weak versions of the temporal operators (see also [Eisner et al. (2003)], for further non-technical motivations for logics over the so-called truncated paths). Suppose, for a moment, to define the GCTL* semantics only on infinite paths, i.e., to consider only total KS. Under this assumption, it is immediate to see that strong and weak temporal operators are equivalent, i.e., $\mathsf{X}\psi \equiv \tilde{\mathsf{X}}\psi$, $\psi_1 \mathsf{U} \psi_2 \equiv \psi_1 \tilde{\mathsf{U}} \psi_2$, and $\psi_1 \mathsf{R} \psi_2 \equiv \psi_1 \tilde{\mathsf{R}} \psi_2$. In particular, it holds that $\mathsf{X}\mathsf{t} \equiv \mathsf{t}$ and so, for the syntax-independence and state-focus (specifically, here we need only that all paths are equivalent w.r.t. t) properties, we obtain that $\pi_1 \equiv^{\mathsf{Xt}}_{\mathcal{K}} \pi_2$, for all $\pi_1, \pi_2 \in \mathrm{Pth}(\mathcal{K})$. Hence, if we want to preserve the syntax independence, we are not able to simply count the number of successors of a given world, by using the formula $\mathsf{E}^{\geq g}\mathsf{X}\mathsf{t}$, without asserting any stronger property. However, all the classical graded logics, such as the GμCALCULUS, allow such a counting. Moreover, consider two paths $\pi_1, \pi_2 \in \mathrm{Pth}(\mathcal{K}, w)$ such that $(\pi_1)_1 \neq (\pi_2)_1$. By the previous lemma, we have that $\pi_1 \not\equiv^{\mathsf{Xt}}_{\mathcal{K}} \pi_2$, reaching in this way a contradiction. Hence, it is evident that it is impossible to cast together the three properties of syntax independence, next consistency, and source dependence in the framework of logics on infinite paths only. If we want to restrict ourselves to such a framework, we have to drop at least one property between the last two, changing completely the semantics of the logic and indirectly the interesting relationship with the GμCALCULUS shown in the next section. We can now return to the main track of thought of this section. In particular, we can enunciate a fundamental result on the loss of the bisimulation invariance, since the operation of counting is not bisimilar invariant at all, and, consequently, on the more expressiveness of the graded w.r.t. the related ungraded logics.

Theorem 1.4 (Bisimilarity Variance) *Let $\equiv^{\cdot}_{\cdot}$ be a next-consistent and source-dependent equivalence relation. Then* GCTL *and* GCTL* *are not invariant under bisimilarity. Moreover, they are more expressive than* CTL *and* CTL**, respectively.*

Proof We show that GCTL distinguishes between bisimilar models. Consider the two KTs $\mathcal{T}_1$ and $\mathcal{T}_2$ such as $\mathcal{T}_1$ contains only the root and one successor, while $\mathcal{T}_2$ contains also another successor of the root (see Fig. 1.1). Formally, $\mathcal{T}_1 = \langle \mathrm{AP}, \mathrm{W}_1, R_1, \mathsf{L}_1, \varepsilon \rangle$, with $\mathrm{AP} = \emptyset$, $\mathrm{W}_1 = \{\varepsilon, 0\}$, and $R_1 = \{(\varepsilon, 0)\}$, and $\mathcal{T}_2 = \langle \mathrm{AP}, \mathrm{W}_2, R_2, \mathsf{L}_2, \varepsilon \rangle$, with $\mathrm{W}_2 = \mathrm{W}_1 \cup \{1\}$, and $R_2 = R_1 \cup \{(\varepsilon, 1)\}$. By the definition of bisimilarity, it is immediate to see that $\mathcal{T}_1$ and $\mathcal{T}_2$ are bisimilar. Now, consider the formula $\varphi = \mathsf{E}^{\geq 2}\mathsf{X}\mathsf{t}$. It is evident that $\mathrm{Pth}(\mathcal{T}_1, \varepsilon, \mathsf{Xt}) = \{\pi_1\}$ with $\pi_1 = \varepsilon \cdot 0$, so $|(\mathrm{Pth}(\mathcal{T}_1, \varepsilon, \mathsf{Xt})/{\equiv^{\mathsf{Xt}}_{\mathcal{T}_1}})| = 1$ and then $\mathcal{T}_1 \not\models \varphi$.On the contrary, $\mathrm{Pth}(\mathcal{T}_2, \varepsilon, \mathsf{Xt}) = \{\pi_1, \pi_2\}$ with $\pi_2 = \varepsilon \cdot 1$. Since $(\pi_1)_1 \neq (\pi_2)_2$, by Lemma 1.2, we have that $\pi_1 \not\equiv^{\mathsf{Xt}}_{\mathcal{T}_2} \pi_2$, so $|(\mathrm{Pth}(\mathcal{T}_2, \varepsilon, \mathsf{Xt})/{\equiv^{\mathsf{Xt}}_{\mathcal{T}_2}})| = 2$ and then $\mathcal{T}_2 \models \varphi$. Hence, φ is not an invariant for the two KTs $\mathcal{T}_1$ and $\mathcal{T}_2$ and so, it can distinguish between bisimilar models. Now, it is known that both CTL and CTL* are invariant under bisimulation, so, they cannot distinguish between $\mathcal{T}_1$ and $\mathcal{T}_2$. Moreover, CTL and CTL* are sublogics of GCTL and GCTL*,

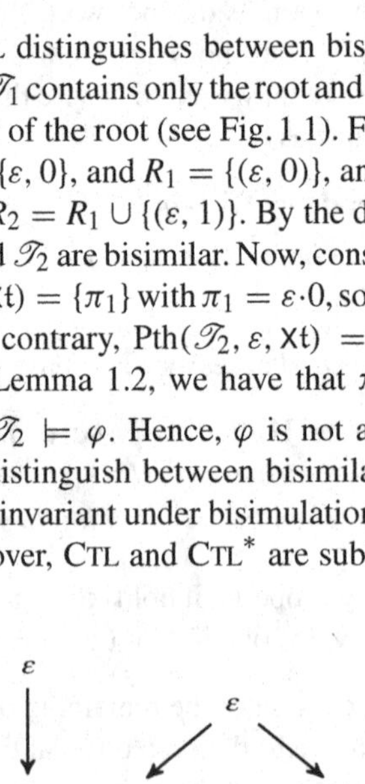

Fig. 1.1 The KTs $\mathcal{T}_1$ and $\mathcal{T}_2$

respectively. Thus, we have that the latter can characterize more models than those characterizable by the former logic. Consequently, the theses hold. □

As third and last accessory lemma, we derive an important and completely general combinatorial property on the dimension of groupings of equivalence classes depending on their size.

Lemma 1.3 (Classes Counting) *Let $\equiv$ be an equivalence relation on a finite set* S. *Moreover, let* $\mathrm{M}_n = \{\mathrm{D} \in (\mathrm{S}/\equiv) : |\mathrm{D}| = n\}$ *be the set of equivalence classes w.r.t.* $\equiv$ *having size n, for each* $n \in [1, |\mathrm{S}|]$. *Then, there is a partition solution* $p \in \mathrm{P}(|\mathrm{S}|)$ *such that* $|\mathrm{M}_n| = (p)_n$, *for each* $n \in [1, |\mathrm{S}|]$.

Proof First note that, by definition, $\mathrm{M}_{n_1} \cap \mathrm{M}_{n_2} = \emptyset$, for all $n_1, n_2 \in [1, |\mathrm{S}|]$ with $n_1 \neq n_2$. Moreover, for all $\mathrm{D}_1, \mathrm{D}_2 \in \mathrm{M}_n$ with $\mathrm{D}_1 \neq \mathrm{D}_2$, it holds that $\mathrm{D}_1 \cap \mathrm{D}_2 = \emptyset$, since they are different equivalence classes. Furthermore, it is evident that $\mathrm{S} = \bigcup_{n=1}^{|\mathrm{S}|} \bigcup_{\mathrm{D} \in \mathrm{M}_n} \mathrm{D}$. So, we have that $|\mathrm{S}| = |\bigcup_{n=1}^{|\mathrm{S}|} \bigcup_{\mathrm{D} \in \mathrm{M}_n} \mathrm{D}| = \sum_{n=1}^{|\mathrm{S}|} \sum_{\mathrm{D} \in \mathrm{M}_n} |\mathrm{D}| = \sum_{n=1}^{|\mathrm{S}|} \sum_{\mathrm{D} \in \mathrm{M}_n} n = \sum_{n=1}^{|\mathrm{S}|} n \cdot |\mathrm{M}_n|$. Hence, by the definition of partition solution, the thesis holds. □

Finally, we can enunciate two theorems that generalize to the case of graded quantifiers the classical CTL* expansion equivalences $\mathsf{EX}\psi \equiv \mathsf{EXE}\psi$ and $\mathsf{A}\tilde{\mathsf{X}}\psi \equiv \mathsf{A}\tilde{\mathsf{X}}\mathsf{A}\psi$. The first property is of crucial importance for the characterization of GCTL, without quantifiers with infinite degrees (i.e., without $\mathsf{E}^{\geq\omega}\psi$ and $\mathsf{A}^{<\omega}\psi$), as a fragment of the GμCALCULUS, as showed in the next section.

Theorem 1.5 (Next Expansion I) *Let* $\equiv_{:}$ *be a state-focused, next-consistent, and source-dependent equivalence relation. Moreover, let* ψ *be a path formula and* $g \in [1, \omega[$. *Then, the following equivalences hold:* (i) $\mathsf{E}^{\geq g}\mathsf{X}\psi \equiv \bigvee_{c \in \mathrm{C}(g)} \bigwedge_{i=1}^{g} \mathsf{E}^{\geq (c)_i}\mathsf{XE}^{\geq i}\psi$ *and* (ii) $\mathsf{A}^{<g}\tilde{\mathsf{X}}\psi \equiv \bigvee_{c \in \mathrm{C}(g-1)} \bigwedge_{i=1}^{g} \mathsf{A}^{\leq (c)_i}\tilde{\mathsf{X}}\mathsf{A}^{<i}\psi$.

Proof *[Only if]*. If $\mathcal{K}, w_0 \models \mathsf{E}^{\geq g}\mathsf{X}\psi$ then $|(\mathrm{Pth}(\mathcal{K}, w_0, \mathsf{X}\psi)/\equiv_{\mathcal{K}}^{\mathsf{X}\psi})| \geq g$, where $\mathcal{K} = \langle \mathrm{AP}, \mathrm{W}, R, \mathsf{L}, w_0 \rangle$. Thus, there is a set $\mathrm{S} \subseteq \mathrm{Pth}(\mathcal{K}, w_0, \mathsf{X}\psi)$ of g non-equivalent paths w.r.t. $\equiv_{\mathcal{K}}^{\mathsf{X}\psi}$. Each path in S is a representative of a different class, so $|\mathrm{S}| = |(\mathrm{S}/\equiv_{\mathcal{K}}^{\mathsf{X}\psi})| = g$.

Let now $\overset{succ}{\equiv}$ be the equivalence relation on $\mathrm{Pth}(\mathcal{K})$ such that $\pi_1 \overset{succ}{\equiv} \pi_2$ iff $(\pi_1)_1 = (\pi_2)_1$. Moreover, let $\mathrm{M}_n \triangleq \{\mathrm{D} \in (\mathrm{S}/\overset{succ}{\equiv}) : |\mathrm{D}| = n\}$ be the set of equivalence classes w.r.t. $\overset{succ}{\equiv}$ having size $n \in [1, g]$. By Lemma 1.3, there is a partition solution $p \in \mathrm{P}(g)$ such that $|\mathrm{M}_n| = (p)_n$, for all $n \in [1, g]$. At this point, we can write $\mathrm{M}_n = \{\mathrm{D}_{n,1}, \ldots, \mathrm{D}_{n,(p)_n}\}$. Furthermore, we can associate to each class $\mathrm{D}_{n,j}$ a different successor $w_{n,j}$ of the initial world w_0 such that $w_{n,j} = (\pi)_1$, for all $\pi \in \mathrm{D}_{n,j}$.

Since $\mathrm{D}_{n,j} \subseteq \mathrm{S}$, we have that $\mathcal{K}, \pi \models \mathsf{X}\psi$ and so, by Item v of Proposition 1.1, $\mathcal{K}, \pi_{\geq 1} \models \psi$, for all $\pi \in \mathrm{D}_{n,j}$. Hence, let $\mathrm{D}'_{n,j} \triangleq \{\pi_{\geq 1} : \pi \in \mathrm{D}_{n,j}\}$, we obtain that $\mathrm{D}'_{n,j} \subseteq \mathrm{Pth}(\mathcal{K}, w_{n,j}, \psi)$. Note that $|\mathrm{D}'_{n,j}| = |\mathrm{D}_{n,j}| = n$. Moreover, by the next-consistency property, since $\pi_1 \not\equiv^{X\psi}_{\mathscr{K}} \pi_2$, for all $\pi_1, \pi_2 \in D_{n,j}$ with $\pi_1 \neq \pi_2$, we obtain that $(\pi_1)_{\geq 1} \not\equiv^{\psi}_{\mathscr{K}} (\pi_2)_{\geq 1}$ and so $|(D'_{n,j}/\equiv^{\psi}_{\mathscr{K}})| = |D'_{n,j}| = n$. Thus, we have that $|(\mathrm{Pth}(\mathscr{K}, w_{n,j}, \psi)/\equiv^{\psi}_{\mathscr{K}})| \geq n$. Hence, $\mathscr{K}, w_{n,j} \models \mathsf{E}^{\geq i}\psi$, for all $i \in [1, n]$. By Items ii and v of Proposition 1.1, the last statement implies that $\mathscr{K}, \pi \models \mathsf{XE}^{\geq i}\psi$, for all $\pi \in D_{n,j}$ with $n \in [i, g]$ and $j \in [1, (p)_n]$.

By Lemma 1.1, it holds that $\pi_1 \equiv^{\mathsf{XE}^{\geq i}\psi}_{\mathscr{K}} \pi_2$, for all $\pi_1, \pi_2 \in D_{n,j}$, and thus $|(D_{n,j}/\equiv^{\mathsf{XE}^{\geq i}\psi}_{\mathscr{K}})| = 1$. On the contrary, by Lemma 1.2, for all $\pi_1 \in D_{n_1,j_1}$ and $\pi_2 \in D_{n_2,j_2}$ with $n_1 \neq n_2$ or $j_1 \neq j_2$, since $(\pi_1)_1 = w_{n_1,j_1} \neq w_{n_2,j_2} = (\pi_2)_1$, it holds that $\pi_1 \not\equiv^{\mathsf{XE}^{\geq i}\psi}_{\mathscr{K}} \pi_2$ and thus $((D_{n_1,j_1} \cup D_{n_2,j_2})/\equiv^{\mathsf{XE}^{\geq i}\psi}_{\mathscr{K}}) = (D_{n_1,j_1}/\equiv^{\mathsf{XE}^{\geq i}\psi}_{\mathscr{K}}) \cup (D_{n_2,j_2}/\equiv^{\mathsf{XE}^{\geq i}\psi}_{\mathscr{K}})$.

Now, we can estimate the number of equivalence classes w.r.t. $\equiv^{\mathsf{XE}^{\geq i}\psi}_{\mathscr{K}}$ of the set of paths $\mathrm{Pth}(\mathscr{K}, w_0, \mathsf{XE}^{\geq i}\psi)$. Since, as previously proved, $\bigcup_{n=i}^{g}\bigcup_{j=1}^{(p)_n} D_{n,j} \subseteq \mathrm{Pth}(\mathscr{K}, w_0, \mathsf{XE}^{\geq i}\psi)$, we have that $|(\mathrm{Pth}(\mathscr{K}, w_0, \mathsf{XE}^{\geq i}\psi)/\equiv^{\mathsf{XE}^{\geq i}\psi}_{\mathscr{K}})| \geq |((\bigcup_{n=i}^{g}\bigcup_{j=1}^{(p)_n} D_{n,j})/\equiv^{\mathsf{XE}^{\geq i}\psi}_{\mathscr{K}})| = |\bigcup_{n=i}^{g}\bigcup_{j=1}^{(p)_n}(D_{n,j}/\equiv^{\mathsf{XE}^{\geq i}\psi}_{\mathscr{K}})| = \sum_{n=i}^{g}\sum_{j=1}^{(p)_n}|(D_{n,j}/\equiv^{\mathsf{XE}^{\geq i}\psi}_{\mathscr{K}})| = \sum_{n=i}^{g}\sum_{j=1}^{(p)_n} 1 = \sum_{n=i}^{g}(p)_n$. Let now $c \in \mathbb{N}^n$ be the vector such that $(c)_i = \sum_{n=i}^{g}(p)_n$. At this point, it is immediate to see that $\mathscr{K}, w_0 \models \mathsf{E}^{\geq (c)_i}\mathsf{XE}^{\geq i}\psi$. Since the previous reasoning can be done for every $i \in [1, g]$, we also have $\mathscr{K}, w_0 \models \bigwedge_{i=1}^{g}\mathsf{E}^{\geq (c)_i}\mathsf{XE}^{\geq i}\psi$. Now, by definition of cumulative-partition solution, we have that $c \in \mathrm{C}(g)$. So, $\mathscr{K}, w_0 \models \bigvee_{c \in \mathrm{C}(g)}\bigwedge_{i=1}^{g}\mathsf{E}^{\geq (c)_i}\mathsf{XE}^{\geq i}\psi$.

[If]. If $\mathscr{K}, w_0 \models \bigvee_{c \in \mathrm{C}(g)}\bigwedge_{i=1}^{g}\mathsf{E}^{\geq (c)_i}\mathsf{XE}^{\geq i}\psi$ then there is a cumulative-partition solution $c \in \mathrm{C}(g)$ such that, for all $i \in [1, g]$, it holds that $\mathscr{K}, w_0 \models \mathsf{E}^{\geq (c)_i}\mathsf{XE}^{\geq i}\psi$ and so $|(\mathrm{Pth}(\mathscr{K}, w_0, \mathsf{XE}^{\geq i}\psi)/\equiv^{\mathsf{XE}^{\geq i}\psi}_{\mathscr{K}})| \geq (c)_i$, where $\mathscr{K} = \langle \mathrm{AP}, \mathrm{W}, R, \mathsf{L}, w_0\rangle$. Let now $p \in \mathbb{N}^n$ be a vector such that $(p)_g = (c)_g$ and $(p)_i = (c)_i - (c)_{i+1}$, for all $i \in [1, g[$. By definition of cumulative-partition solution, it is immediate to see that p is a partition solution, i.e., $p \in \mathrm{P}(g)$.

First note that the set $\mathrm{V}_i \triangleq \{w \in \mathrm{W} : (w_0, w) \in R \wedge \mathscr{K}, w \models \mathsf{E}^{\geq i}\psi\}$ of successors of the initial world w_0 satisfying $\mathsf{E}^{\geq i}\psi$ has cardinality greater than or equal to $(c)_i$. Indeed, let $\pi_1, \pi_2 \in \mathrm{Pth}(\mathscr{K}, w_0, \mathsf{XE}^{\geq i}\psi)$ be two paths such that $\pi_1 \not\equiv^{\mathsf{XE}^{\geq i}\psi}_{\mathscr{K}} \pi_2$. Then, by Lemma 1.1, we have that $(\pi_1)_1 \neq (\pi_2)_1$. So, since, as shown before, there exist at least $(c)_i$ non-equivalent paths w.r.t. $\equiv^{\mathsf{XE}^{\geq i}\psi}_{\mathscr{K}}$, we obtain that there are at least $(c)_i$ different successors of w_0.

Now, for each $i \in [1, g[$, let $\mathrm{U}_i \subseteq \mathrm{V}_i$ be a set of $(p)_i$ worlds such that $\mathrm{U}_i \cap \mathrm{U}_j = \emptyset$, for all $j \in]i, g]$. By finite induction, it is immediate to see that we can effectively construct such sets, since $|\mathrm{V}_i \setminus \bigcup_{j=i+1}^{g}\mathrm{U}_j| \geq (c)_i - \sum_{j=i+1}^{g}|\mathrm{U}_j| = (c)_i - \sum_{j=i+1}^{g}(p)_j = (c)_i - (c)_{i+1} = (p)_i$. At this point, we can write $\mathrm{U}_i = \{w_{i,1}, \ldots, w_{i,(p)_i}\}$. Furthermore, since $\mathscr{K}, w_{i,j} \models \mathsf{E}^{\geq i}\psi$, we can associate to each world $w_{i,j}$ a set $D'_{i,j} \subseteq \mathrm{Pth}(\mathscr{K}, w_{i,j}, \psi)$ of i non-equivalent paths w.r.t. $\equiv^{\psi}_{\mathscr{K}}$. Now, let $D_{i,j} \triangleq \{\pi \in \mathrm{Pth}(\mathscr{K}, w_0) : \pi_{\geq 1} \in D'_{i,j}\}$. By Item v of Proposition 1.1,

$D_{i,j} \subseteq \mathrm{Pth}(\mathcal{K}, w_0, X\psi)$. Note that $|D_{i,j}| = |D'_{i,j}| = i$. By the next-consistency property, since $(\pi_1)_{\geq 1} \not\equiv^{\psi}_{\mathcal{K}} (\pi_2)_{\geq 1}$, for all $\pi_1, \pi_2 \in D_{n,j}$ with $\pi_1 \neq \pi_2$, we obtain that $\pi_1 \not\equiv^{X\psi}_{\mathcal{K}} \pi_2$ and so $|(D_{i,j}/\equiv^{X\psi}_{\mathcal{K}})| = |D_{i,j}| = i$. Moreover, by Lemma 1.2, for all $\pi_1 \in D_{i_1,j_1}$ and $\pi_2 \in D_{i_2,j_2}$ with $i_1 \neq i_2$ or $j_1 \neq j_2$, since $(\pi_1)_1 = w_{i_1,j_1} \neq w_{i_2,j_2} = (\pi_2)_1$, it holds that $\pi_1 \not\equiv^{X\psi}_{\mathcal{K}} \pi_2$ and thus $((D_{i_1,j_1} \cup D_{i_2,j_2}) /\equiv^{X\psi}_{\mathcal{K}}) = (D_{i_1,j_1}/\equiv^{X\psi}_{\mathcal{K}}) \cup ((D_{i_2,j_2}/\equiv^{X\psi}_{\mathcal{K}}))$.

Now, we can estimate the number of equivalence classes w.r.t. $\equiv^{X\psi}_{\mathcal{K}}$ of the set of paths $\mathrm{Pth}(\mathcal{K}, w_0, X\psi)$. Since, as previously proved, $\bigcup_{i=1}^{g} \bigcup_{j=1}^{(p)_i} D_{i,j} \subseteq \mathrm{Pth}(\mathcal{K}, w_0, X\psi)$, we have that $|(\mathrm{Pth}(\mathcal{K}, w_0, X\psi)/\equiv^{X\psi}_{\mathcal{K}})| \geq |((\bigcup_{i=1}^{g} \bigcup_{j=1}^{(p)_i} D_{i,j}) /\equiv^{X\psi}_{\mathcal{K}})| = |\bigcup_{i=1}^{g} \bigcup_{j=1}^{(p)_i} (D_{i,j}/\equiv^{X\psi}_{\mathcal{K}})| = \sum_{i=1}^{g} \sum_{j=1}^{(p)_i} |(D_{i,j}/\equiv^{X\psi}_{\mathcal{K}})| = \sum_{i=1}^{g} \sum_{j=1}^{(p)_i} i = \sum_{i=1}^{g} i \cdot (p)_i = g$. The last equality is due to the fact that p is a partition solution. Hence, we have that $\mathcal{K}, w_0 \models E^{\geq g} X\psi$. □

Theorem 1.6 (Next Expansion II) *Let* $\equiv_{\cdot}^{\cdot}$ *be a state-focused, next-consistent, and source-dependent equivalence relation. Moreover, let* ψ *be a path formula. Then, the following equivalences hold:* (i) $E^{\geq\omega} X\psi \equiv E^{\geq\omega} X E^{\geq 1}\psi \vee E^{\geq 1} X E^{\geq\omega}\psi$ *and* (ii) $A^{<\omega}\tilde{X}\psi \equiv A^{<\omega}\tilde{X}A^{<1}\psi \wedge A^{<1}\tilde{X}A^{<\omega}\psi$.

Proof [Only if]. If $\mathcal{K}, w_0 \models E^{\geq\omega} X\psi$ then $|(\mathrm{Pth}(\mathcal{K}, w_0, X\psi)/\equiv^{X\psi}_{\mathcal{K}})| \geq \omega$, where $\mathcal{K} = \langle \mathrm{AP}, \mathrm{W}, R, \mathsf{L}, w_0 \rangle$. Thus, there is an infinite set $S \subseteq \mathrm{Pth}(\mathcal{K}, w_0, X\psi)$ of non-equivalent paths w.r.t. $\equiv^{X\psi}_{\mathcal{K}}$.

Let now $\overset{succ}{\equiv}$ be the equivalence relation on $\mathrm{Pth}(\mathcal{K})$ such that $\pi_1 \overset{succ}{\equiv} \pi_2$ iff $(\pi_1)_1 = (\pi_2)_1$. Moreover, let $M \triangleq (S/\overset{succ}{\equiv})$. To each class $D \in M$ we can associate a different successor w_D of the initial world w_0 such that $w_D = (\pi)_1$, for all $\pi \in D$.

Since $D \subseteq S$, we have that $\mathcal{K}, \pi \models X\psi$ and so, by Item v of Proposition 1.1, $\mathcal{K}, \pi_{\geq 1} \models \psi$, for all $\pi \in D$. Hence, let $D' \triangleq \{\pi_{\geq 1} : \pi \in D\}$, we obtain that $D' \subseteq \mathrm{Pth}(\mathcal{K}, w_{n,j}, \psi)$. Note that $|D'| = |D|$. Moreover, by the next-consistency property, since $\pi_1 \not\equiv^{X\psi}_{\mathcal{K}} \pi_2$, for all $\pi_1, \pi_2 \in D$ with $\pi_1 \neq \pi_2$, we obtain that $(\pi_1)_{\geq 1} \not\equiv^{\psi}_{\mathcal{K}} (\pi_2)_{\geq 1}$ and so $|(D'/\equiv^{\psi}_{\mathcal{K}})| = |D'|$. Consequently, it holds that $|(\mathrm{Pth}(\mathcal{K}, w_D, \psi)/\equiv^{\psi}_{\mathcal{K}})| \geq |D|$. Thus, $\mathcal{K}, w_D \models E^{\geq |D|}\psi$. The last statement implies that $\mathcal{K}, \pi \models X E^{\geq |D|}\psi$, for all $\pi \in D$.

At this point, we have two possibilities, each implying the truth of one of the two disjuncts in the formula $E^{\geq\omega} X E^{\geq 1}\psi \vee E^{\geq 1} X E^{\geq\omega}\psi$: either $|M| = \omega$ or $|M| < \omega$.

In the first case, each class $D \in M$ may be finite, so we can assert at most that $|D| \geq 1$, which implies $\mathcal{K}, \pi \models X E^{\geq 1}\psi$, for all $\pi \in D$. By Lemma 1.2, for all $\pi_1 \in D_1$ and $\pi_2 \in D_2$ with $D_1 \neq D_2$, since $(\pi_1)_1 = w_{D_1} \neq w_{D_2} = (\pi_2)_1$, it holds that $\pi_1 \not\equiv^{XE^{\geq 1}\psi}_{\mathcal{K}} \pi_2$ and thus $((D_1 \cup D_2)/\equiv^{XE^{\geq 1}\psi}_{\mathcal{K}}) = (D_1/\equiv^{XE^{\geq 1}\psi}_{\mathcal{K}}) \cup (D_2/\equiv^{XE^{\geq 1}\psi}_{\mathcal{K}})$. Now, since $\bigcup_{D \in M} D \subseteq \mathrm{Pth}(\mathcal{K}, w_0, XE^{\geq 1}\psi)$, we have that $|(\mathrm{Pth}(\mathcal{K}, w_0, XE^{\geq 1}\psi)/\equiv^{XE^{\geq 1}\psi}_{\mathcal{K}})| \geq |((\bigcup_{D \in M} D)/\equiv^{XE^{\geq 1}\psi}_{\mathcal{K}})| = |\bigcup_{D \in M}(D/\equiv^{XE^{\geq 1}\psi}_{\mathcal{K}})| = \sum_{D \in M} |(D/\equiv^{XE^{\geq 1}\psi}_{\mathcal{K}})| \geq \sum_{D \in M} 1 = |M| = \omega$. Hence, $\mathcal{K}, w_0 \models E^{\geq\omega} X E^{\geq 1}\psi$.

In the second case, since $S = \bigcup_{D \in M} D$ and so $|S| = \sum_{D \in M} |D|$, we have that there is a class $D \in M$ such that $|D| = \omega$. Thus, $\mathcal{K}, \pi \models \mathsf{XE}^{\geq \omega}\psi$, for all $\pi \in D$. This implies that $|\mathrm{Pth}(\mathcal{K}, w_0, \mathsf{XE}^{\geq \omega}\psi)| \geq 1$ and so $|(\mathrm{Pth}(\mathcal{K}, w_0, \mathsf{XE}^{\geq \omega}\psi)/\equiv_{\mathcal{K}}^{\mathsf{XE}^{\geq \omega}\psi})| \geq 1$, which means that $\mathcal{K}, w_0 \models \mathsf{E}^{\geq 1}\mathsf{XE}^{\geq \omega}\psi$.

[If]. On one hand, if $\mathcal{K}, w_0 \models \mathsf{E}^{\geq \omega}\mathsf{XE}^{\geq 1}\psi$ then $|(\mathrm{Pth}(\mathcal{K}, w_0, \mathsf{XE}^{\geq 1}\psi)/\equiv_{\mathcal{K}}^{\mathsf{XE}^{\geq 1}\psi}| \geq \omega$, where $\mathcal{K} = \langle \mathrm{AP}, \mathrm{W}, R, \mathsf{L}, w_0 \rangle$. Now, let $V \triangleq \{w \in W : (w_0, w) \in R \wedge \mathcal{K}, w \models \mathsf{E}^{\geq 1}\psi\}$ be the set of successors of the initial world w_0 satisfying $\mathsf{E}^{\geq 1}\psi$. It is immediate to see that $|V| = \omega$. Indeed, let $\pi_1, \pi_2 \in \mathrm{Pth}(\mathcal{K}, w_0, \mathsf{XE}^{\geq 1}\psi)$ be two paths such that $\pi_1 \not\equiv_{\mathcal{K}}^{\mathsf{XE}^{\geq 1}\psi} \pi_2$. Then, by Lemma 1.1, we have that $(\pi_1)_1 \neq (\pi_2)_1$. So, since there exist infinite non-equivalent paths w.r.t. $\equiv_{\mathcal{K}}^{\mathsf{XE}^{\geq 1}\psi}$, we obtain that there are infinite different successors of w_0. At this point, by Item v of Proposition 1.1, we can associate a path $\pi_w \in \mathrm{Pth}(\mathcal{K}, w_0, \mathsf{X}\psi)$ with $(\pi_w)_1 = w$ to each world $w \in V$. Let $D \triangleq \{\pi_w : w \in V\}$ be the set of all such paths. It is evident that $|D| = |V| = \omega$. Now, by Lemma 1.2, for all $\pi_{w_1}, \pi_{w_2} \in D$ with $w_1 \neq w_2$, it holds that $\pi_{w_1} \not\equiv_{\mathcal{K}}^{\mathsf{X}\psi} \pi_{w_2}$ and thus $|(D/\equiv_{\mathcal{K}}^{\mathsf{X}\psi})| = |D|$. Since $D \subseteq \mathrm{Pth}(\mathcal{K}, w_0, \mathsf{X}\psi)$, we have that $|(\mathrm{Pth}(\mathcal{K}, w_0, \mathsf{X}\psi)/\equiv_{\mathcal{K}}^{\mathsf{X}\psi})| \geq |(D/\equiv_{\mathcal{K}}^{\mathsf{X}\psi})| = |D| = \omega$. Hence, $\mathcal{K}, w_0 \models \mathsf{E}^{\geq \omega}\mathsf{X}\psi$.

On the other hand, if $\mathcal{K}, w_0 \models \mathsf{E}^{\geq 1}\mathsf{XE}^{\geq \omega}\psi$, by Items ii and v of Proposition 1.1, there is a successor $w \in W$ with $(w_0, w) \in R$ of the initial world w_0 satisfying $\mathsf{E}^{\geq \omega}\psi$. Hence, $|(\mathrm{Pth}(\mathcal{K}, w, \psi)/\equiv_{\mathcal{K}}^{\psi})| \geq \omega$. Moreover, let $D' \subseteq \mathrm{Pth}(\mathcal{K}, w, \psi)$ be a set of infinite of non-equivalent paths w.r.t. $\equiv_{\mathcal{K}}^{\psi}$ and $D \triangleq \{\pi \in \mathrm{Pth}(\mathcal{K}, w_0) : \pi_{\geq 1} \in D'\}$ be the set of their extensions with w_0. It is evident that $|D| = |D'| = \omega$. By the next-consistency property, since $(\pi_1)_{\geq 1} \not\equiv_{\mathcal{K}}^{\psi} (\pi_2)_{\geq 1}$, for all $\pi_1, \pi_2 \in D$ with $\pi_1 \neq \pi_2$, we obtain that $\pi_1 \not\equiv_{\mathcal{K}}^{\mathsf{X}\psi} \pi_2$ and so $|(D/\equiv_{\mathcal{K}}^{\mathsf{X}\psi})| = |D|$. Now, by Item v of Proposition 1.1, $D \subseteq \mathrm{Pth}(\mathcal{K}, w_0, \mathsf{X}\psi)$. Thus, we have that $|(\mathrm{Pth}(\mathcal{K}, w_0, \mathsf{X}\psi)/\equiv_{\mathcal{K}}^{\mathsf{X}\psi})| \geq |(D/\equiv_{\mathcal{K}}^{\mathsf{X}\psi})| = |D| = \omega$. Hence, $\mathcal{K}, w_0 \models \mathsf{E}^{\geq \omega}\mathsf{X}\psi$. □

In the following, we use the four expressions $\mathsf{EX}(g, \psi)$, $\mathsf{A\tilde{X}}(g, \psi)$, $\mathsf{EX}'(g, \psi)$, and $\mathsf{A\tilde{X}}'(g, \psi)$ defined below to represent in short the expansion formulas for the X and $\tilde{\mathsf{X}}$ temporal operators derived in the previous two theorems.

- $\mathsf{EX}(g, \psi) \triangleq \begin{cases} \bigvee_{c \in \mathrm{C}(g)} \bigwedge_{i=1}^{g} \mathsf{E}^{\geq (c)_i}\mathsf{XE}^{\geq i}\psi, & \text{if } g < \omega; \\ \mathsf{E}^{\geq \omega}\mathsf{XE}^{\geq 1}\psi \vee \mathsf{E}^{\geq 1}\mathsf{XE}^{\geq \omega}\psi, & \text{otherwise.} \end{cases}$
- $\mathsf{A\tilde{X}}(g, \psi) \triangleq \begin{cases} \bigvee_{c \in \mathrm{C}(g-1)} \bigwedge_{i=1}^{g} \mathsf{A}^{\leq (c)_i}\tilde{\mathsf{X}}\mathsf{A}^{< i}\psi, & \text{if } g < \omega; \\ \mathsf{A}^{< \omega}\mathsf{XA}^{< 1}\psi \wedge \mathsf{A}^{< 1}\mathsf{XA}^{< \omega}\psi, & \text{otherwise.} \end{cases}$
- $\mathsf{EX}'(g, \psi) \triangleq \begin{cases} \bigvee_{c \in \mathrm{C}(g)}^{(c)_g = 0} \bigwedge_{i=1}^{g-1} \mathsf{E}^{\geq (c)_i}\mathsf{XE}^{\geq i}\psi, & \text{if } g < \omega; \\ \mathsf{E}^{\geq \omega}\mathsf{XE}^{\geq 1}\psi, & \text{otherwise.} \end{cases}$
- $\mathsf{A\tilde{X}}'(g, \psi) \triangleq \begin{cases} \bigvee_{c \in \mathrm{C}(g-1)} \bigwedge_{i=1}^{g-1} \mathsf{A}^{\leq (c)_i}\tilde{\mathsf{X}}\mathsf{A}^{< i}\psi, & \text{if } g < \omega; \\ \mathsf{A}^{< \omega}\mathsf{XA}^{< 1}\psi, & \text{otherwise.} \end{cases}$

In this way, we obtain that $\mathsf{E}^{\geq g}\mathsf{X}\psi \equiv \mathsf{EX}(g, \psi) \equiv \mathsf{EX}'(g, \psi) \vee \mathsf{E}^{\geq 1}\mathsf{XE}^{\geq g}\psi$ and $\mathsf{A}^{<g}\tilde{\mathsf{X}}\psi \equiv \mathsf{A}\tilde{\mathsf{X}}(g, \psi) \equiv \mathsf{A}\tilde{\mathsf{X}}'(g, \psi) \wedge \mathsf{A}^{<1}\tilde{\mathsf{X}}\mathsf{A}^{<g}\psi$, for all $g \in \widehat{\mathbb{N}}$. For the existential case, the second equivalence for finite degree is due to the fact that, when $(c)_g = 1$, it holds that $\bigwedge_{i=1}^{g} \mathsf{E}^{\geq (c)_i}\mathsf{XE}^{\geq i}\psi = \bigwedge_{i=1}^{g} \mathsf{E}^{\geq 1}\mathsf{XE}^{\geq i}\psi \equiv \mathsf{E}^{\geq 1}\mathsf{XE}^{\geq g}\psi$. For the universal case, instead, the same equivalence is derived by the observation that, since $(c)_g = 0$, each disjunct necessarily contains the conjunct $\mathsf{A}^{\leq 0}\tilde{\mathsf{X}}\mathsf{A}^{<g}\psi$.

Now, it is interesting to note that, for finite degrees, the formula $\mathsf{EX}(g, \psi)$ allows to partition at least g paths through $c_1 \leq g$ successor worlds, for a given vector $c \in \mathrm{C}(g)$. Indeed, c_i is the number of successor worlds from which at least i paths satisfying ψ start. Therefore, c_1 is a sufficient bound on the number of successor worlds we have to consider to ensure the satisfiability of the formula. A similar dual reasoning can be done for the universal formula $\mathsf{A}\tilde{\mathsf{X}}(g, \psi)$.

Observe that $\mathsf{EX}(1, \psi)$ and $\mathsf{A}\tilde{\mathsf{X}}(1, \psi)$ are equal to the classical CTL* expansions $\mathsf{EXE}\psi$ and $\mathsf{A}\tilde{\mathsf{X}}\mathsf{A}\psi$, respectively.

By a simple calculation, it follows that $(g-1) \cdot (|\mathrm{C}(g)| - 1) \cdot (\mathsf{lng}(\psi) + 4) - 1 = \mathsf{lng}(\mathsf{EX}'(g, \psi)) < \mathsf{lng}(\mathsf{EX}(g, \psi)) = g \cdot |\mathrm{C}(g)| \cdot (\mathsf{lng}(\psi) + 4) - 1$ and $(g-1) \cdot |\mathrm{C}(g-1)| \cdot (\mathsf{lng}(\psi) + 4) - 1 = \mathsf{lng}(\mathsf{A}\tilde{\mathsf{X}}'(g, \psi)) < \mathsf{lng}(\mathsf{A}\tilde{\mathsf{X}}(g, \psi)) = g \cdot |\mathrm{C}(g-1)| \cdot (\mathsf{lng}(\psi) + 4) - 1$. So, both the lengths of $\mathsf{EX}(g, \psi)$ and $\mathsf{EX}'(g, \psi)$ are $\Theta((\mathsf{lng}(\psi) + 4) \cdot 2^{k \cdot \sqrt{g}})$, while those of $\mathsf{A}\tilde{\mathsf{X}}(g, \psi)$ and $\mathsf{A}\tilde{\mathsf{X}}'(g, \psi)$ are $\Theta((\mathsf{lng}(\psi) + 4) \cdot 2^{k \cdot \sqrt{g-1}})$, for a constant k. Furthermore, the degree of $\mathsf{EX}(g, \psi)$, $\mathsf{A}\tilde{\mathsf{X}}(g, \psi)$, $\mathsf{EX}'(g, \psi)$, and $\mathsf{A}\tilde{\mathsf{X}}'(g, \psi)$ is $\max\{g, \mathsf{deg}(\psi)\}$. As an example, consider the formula $\varphi = \mathsf{E}^{\geq g}\mathsf{XX}p$. It is evident that $\mathsf{lng}(\varphi) = 4$, $\mathsf{deg}(\varphi) = g$, and $\mathsf{siz}(\varphi) = 4 + \lceil \log(g) \rceil$. Moreover, $\mathsf{lng}(\mathsf{EX}(g, \mathsf{X}p)) = \Theta(2^{k \cdot \sqrt{g}}) = \Theta(2^{k \cdot \sqrt{2^{\mathsf{siz}(\varphi) - 4}}})$. Hence, the length of an expansion $\mathsf{EX}(g, \psi)$ can be, in general, double exponential in the size of the original formula, also in the case its length is constant. The same thing happens for the expansion $\mathsf{A}\tilde{\mathsf{X}}(g, \psi)$.

1.4.3 Boolean Requirements

At this point, we can reason about the properties that an equivalence has to satisfy w.r.t. the positive Boolean combination of formulas.

Suppose we have two path formulas ψ_1 and ψ_2. We would like to have that, from a given world, both the number of paths that satisfy ψ_1 and ψ_2 are not less than those satisfying their conjunction. Hence, we need that paths equivalent w.r.t. both ψ_1 and ψ_2 are equivalent w.r.t. $\psi_1 \wedge \psi_2$ too, otherwise, each equivalence class for ψ_1 and ψ_2 may provide more than one equivalence class for $\psi_1 \wedge \psi_2$ allowing the latter formula to have more paths. Moreover, we would like that, among the paths that satisfy ψ_1 (resp., ψ_2), the number of those satisfying ψ_2 (resp., ψ_1) is equal to those satisfying $\psi_1 \wedge \psi_2$. Hence, we need that paths equivalent w.r.t. $\psi_1 \wedge \psi_2$ are also equivalent w.r.t. both ψ_1 and ψ_2.

Definition 1.10 (Conjunction Consistency) *An equivalence relation* $\equiv^{\cdot}_{\mathscr{K}}$ *on paths is said* conjunction consistent *iff it holds that* $\pi_1 \equiv^{\psi_1 \wedge \psi_2}_{\mathscr{K}} \pi_2$ *iff* $\pi_1 \equiv^{\psi_1}_{\mathscr{K}} \pi_2$ *and* $\pi_1 \equiv^{\psi_2}_{\mathscr{K}} \pi_2$, *for all* $\pi_1, \pi_2 \in \mathrm{Pth}(\mathscr{K})$.

By the state-focus and conjunction-consistency properties, we can derive an equivalence on the existential quantification of a conjunction between a state and a path formula that allow to extract the first one from the scope of the quantifier. Similarly, we can extract a state formula from a universal quantification of a disjunction between this and a path formula. This property is simply an extension of what we have in the case of ungraded quantifications.

Theorem 1.7 (Local Conjunction Quantification) *Let* $\equiv^{\cdot}_{\cdot}$ *be a state-focused and conjunction-consistent equivalence relation. Moreover, let* φ *and* ψ *be a state and a path formula, respectively, and* $g \in [1, \omega]$. *Then, the following holds:* (i) $\mathsf{E}^{\geq g}(\varphi \wedge \psi) \equiv \varphi \wedge \mathsf{E}^{\geq g}\psi$ *and* (ii) $\mathsf{A}^{<g}(\varphi \vee \psi) \equiv \varphi \vee \mathsf{A}^{<g}\psi$.

Proof *[Only if].* If $\mathscr{K}, w_0 \models \mathsf{E}^{\geq g}\varphi \wedge \psi$ then $|(\mathrm{Pth}(\mathscr{K}, w_0, \varphi \wedge \psi)/\equiv^{\varphi \wedge \psi}_{\mathscr{K}})| \geq g$, where w_0 is the initial world of $\mathscr{K}$. The inequality implies $\mathrm{Pth}(\mathscr{K}, w_0, \varphi \wedge \psi) \neq \emptyset$, so, by Item iii of Proposition 1.1, there is a path $\pi \in \mathrm{Pth}(\mathscr{K}, w_0)$ such that $\mathscr{K}, \pi \models \varphi$ and, by Item ii of the same proposition, this means that $\mathscr{K}, w_0 \models \varphi$. Then, again by Item iii of Proposition 1.1, it is immediate to see that $\mathrm{Pth}(\mathscr{K}, w_0, \varphi \wedge \psi) = \mathrm{Pth}(\mathscr{K}, w_0, \psi)$. Moreover, by the state-focus property, we have that $\pi_1 \equiv^{\varphi}_{\mathscr{K}} \pi_2$, for all paths $\pi_1, \pi_2 \in \mathrm{Pth}(\mathscr{K}, w_0)$. Now, by the conjunction-consistency property, we obtain that $\pi_1 \equiv^{\varphi \wedge \psi}_{\mathscr{K}} \pi_2$ iff $\pi_1 \equiv^{\psi}_{\mathscr{K}} \pi_2$. At this point, $(\mathrm{Pth}(\mathscr{K}, w_0, \varphi \wedge \psi)/\equiv^{\varphi \wedge \psi}_{\mathscr{K}}) = (\mathrm{Pth}(\mathscr{K}, w_0, \psi)/\equiv^{\varphi \wedge \psi}_{\mathscr{K}}) = (\mathrm{Pth}(\mathscr{K}, w_0, \psi)/\equiv^{\psi}_{\mathscr{K}})$. Hence, $\mathscr{K}, w_0 \models \mathsf{E}^{\geq g}\psi$ and consequently $\mathscr{K}, w_0 \models \varphi \wedge \mathsf{E}^{\geq g}\psi$.

[If]. If $\mathscr{K}, w_0 \models \varphi \wedge \mathsf{E}^{\geq g}\psi$, we have that $\mathscr{K}, w_0 \models \varphi$ and $|(\mathrm{Pth}(\mathscr{K}, w_0, \psi) / \equiv^{\psi}_{\mathscr{K}})| \geq g$. Then, by Items ii and iii of Proposition 1.1, it is immediate to see that $\mathrm{Pth}(\mathscr{K}, w_0, \psi) = \mathrm{Pth}(\mathscr{K}, w_0, \varphi \wedge \psi)$. Moreover, by the state-focus property, we have that $\pi_1 \equiv^{\varphi}_{\mathscr{K}} \pi_2$, for all paths $\pi_1, \pi_2 \in \mathrm{Pth}(\mathscr{K}, w_0)$. Now, by the conjunction-consistency property, we obtain that $\pi_1 \equiv^{\varphi \wedge \psi}_{\mathscr{K}} \pi_2$ iff $\pi_1 \equiv^{\psi}_{\mathscr{K}} \pi_2$. At this point, $(\mathrm{Pth}(\mathscr{K}, w_0, \psi)/\equiv^{\psi}_{\mathscr{K}}) = (\mathrm{Pth}(\mathscr{K}, w_0, \varphi \wedge \psi)/\equiv^{\psi}_{\mathscr{K}}) = (\mathrm{Pth}(\mathscr{K}, w_0, \varphi \wedge \psi)/\equiv^{\varphi \wedge \psi}_{\mathscr{K}})$. Hence, $\mathscr{K}, w_0 \models \mathsf{E}^{\geq g}\varphi \wedge \psi$. □

It is interesting to note that, in order to prove the previous result, we do not need the full power of the conjunction consistency but a weaker property, which we denote *local conjunction consistency*, that only links the equivalence w.r.t. a conjunction of a state and a path formula to the equivalences w.r.t. the conjuncts. However, as we show later, we need the full power of the property when we have to reason about complex CTL* path formulas.

Consider again the two path formulas ψ_1 and ψ_2. We would like that, from a given world, the sum of the number of paths that satisfy ψ_1 together with that satisfying ψ_2 is not less than the number of paths that satisfy their disjunction. Suppose that there are only two paths that satisfy ψ_1 (resp., ψ_2) and are equivalent w.r.t. the same formula. Then, the two paths need to be equivalent w.r.t. $\psi_1 \vee \psi_2$, too. Hence, one way to ensure such a property is to ask that, whenever two paths are equivalent w.r.t. one formula, they are also equivalent w.r.t. its disjunctions. Moreover, we would like that both the number of paths that satisfy ψ_1 and ψ_2 are not greater than those satisfying $\psi_1 \vee \psi_2$. Hence, we need that paths satisfying ψ_1 (resp., ψ_2) and equivalent w.r.t. $\psi_1 \vee \psi_2$ are also equivalent w.r.t. ψ_1 (resp., ψ_2). So, we would like that two paths are equivalent w.r.t. a disjunction iff they are equivalent w.r.t. one of the two disjuncts.

Definition 1.11 (Disjunction Consistency) *An equivalence relation* $\equiv^{\cdot}_{\mathcal{K}}$ *on paths is said* disjunction consistent *iff it holds that* $\pi_1 \equiv^{\psi_1 \vee \psi_2}_{\mathcal{K}} \pi_2$ *iff* $\pi_1 \equiv^{\psi_1}_{\mathcal{K}} \pi_2$ *or* $\pi_1 \equiv^{\psi_2}_{\mathcal{K}} \pi_2$, *for all* $\pi_1, \pi_2 \in \text{Pth}(\mathcal{K})$.

In general, however, such a property contradicts the syntax-independence, state-focus, and the next- and weak next-consistency properties. Indeed, let $\psi_1 = \mathsf{X}p$ and $\psi_2 = \neg\,\mathsf{X}p$, for an atomic proposition $p \in \text{AP}$. Then, $\psi_1 \vee \psi_2$ is equivalent to $\mathfrak{t}$. Consider now two paths $\pi_1, \pi_2 \in \text{Pth}(\mathcal{K}, w)$ such that $\mathcal{K}, (\pi_1)_1 \models p$ and $\mathcal{K}, (\pi_2)_1 \not\models p$, and so $(\pi_1)_1 \neq (\pi_2)_1$. Since the two paths have, in their second position, different successors of the origin, they are distinct w.r.t. ψ_1 and ψ_2 but they are identical w.r.t. $\psi_1 \vee \psi_2$, because of the state-focus and syntax-independence properties. In this example, the contradiction rises from the fact that the disjunction turns out to be a weaker property (a tautology) than the two base formulas. Hence, the formula is always satisfied and, since all choices over the paths are indifferent, they may be regarded as equivalent. Now, one may think that this is a problem related only to tautologies that rise from the disjunction. Unfortunately, this is not the case. Indeed, the disjunction may contain an hidden tautology that reveals itself only at some later points on the paths. For example, let $\psi_1 = \mathsf{X}\mathsf{X}p$ and $\psi_2 = \mathsf{X}\neg\,\mathsf{X}p$. Their disjunction is not a tautology, because it is not satisfied on paths of length 1. Consider now two paths $\pi_1, \pi_2 \in \text{Pth}(\mathcal{K}, w)$ such that $(\pi_1)_1 = (\pi_2)_1$, $\mathcal{K}, (\pi_1)_2 \models p$, and $\mathcal{K}, (\pi_2)_2 \not\models p$. The two paths are distinct w.r.t. ψ_1 and ψ_2 because they have distinct third nodes, but they are identical w.r.t. $\psi_1 \vee \psi_2 \equiv \mathsf{X}\mathfrak{t}$. It is easy to believe that the hidden tautology may be found arbitrary deeper in the formula, that is why the disjunction-consistency cannot hold in its entirety.

Since it is not possible to define in general an easy property that relates the equivalence on a disjunction to the equivalence on the component formulas, we restrict our observations to a case where the tautology derived from the disjunction can appear only at the first node of paths. Hence, we consider only disjunctions between a state φ and a path formula ψ. In such a case, two paths equivalent w.r.t. the disjunction $\varphi \vee \psi \equiv \varphi \vee \neg\varphi \equiv \mathfrak{t}$ are equivalent w.r.t. one of the two state formulas, too. In the next section, we actually prove that this property does not contradict the previous ones.

Definition 1.12 (Local Disjunction Consistency) *An equivalence relation* $\equiv^{\cdot}_{\mathcal{K}}$ *on paths is said* local disjunction consistent *iff it holds that* $\pi_1 \equiv^{\varphi \vee \psi}_{\mathcal{K}} \pi_2$ *iff* $\pi_1 \equiv^{\varphi}_{\mathcal{K}} \pi_2$ *or* $\pi_1 \equiv^{\psi}_{\mathcal{K}} \pi_2$, *for all* $\pi_1, \pi_2 \in \text{Pth}(\mathcal{K})$, *where* φ *is a state formula.*

We further discuss an incidental property.

Consider a path formula ψ. Since in the semantics we only consider paths satisfying ψ when evaluating the truth nature of an existential or universal quantification, it is pointless to compare two paths if one of them does not satisfy ψ. However, suppose that there exist two paths π_1 and π_2 that do not satisfy a state formula φ, but that are equivalent w.r.t. φ. Also suppose that these paths satisfy a path formula ψ, but they are not equivalent w.r.t. ψ. Then, by local disjunction consistency the two paths would be equivalent w.r.t. $\varphi \vee \psi$, but it is unreasonable that there is only one path satisfying the disjunction while φ is not satisfied on them and there are two paths satisfying the formula ψ. In order to avoid such a problem, we may want to require that two paths are equivalent w.r.t. a formula only if they both satisfy it.

Definition 1.13 (Satisfiability Constraint) *An equivalence relation* $\equiv^{\cdot}_{\mathcal{K}}$ *on paths is said* satisfiability constrained *iff it holds that if* $\pi_1 \equiv^{\psi}_{\mathcal{K}} \pi_2$ *then* $\mathcal{K}, \pi_1 \models \psi$ *and* $\mathcal{K}, \pi_2 \models \psi$, *for all* $\pi_1, \pi_2 \in \mathrm{Pth}(\mathcal{K})$.

By the state-focus, local disjunction-consistency, and satisfiability-constraint properties, we can derive an equivalence on the quantification of a disjunction between a state and a path formula that allow to extract in a negated form the first one from the scope of the quantifier. Similarly, we can extract a negated state formula from a universal quantification of a conjunction between this and a path formula. Note that this property is not an extension of what we have in the case of ungraded quantifications.

Theorem 1.8 (Local Disjunction Quantification) *Let* $\equiv^{\cdot}_{\cdot}$ *be a state-focused, local disjunction-consistent, and satisfiability-constrained equivalence relation. Moreover, let* φ *and* ψ *be a state and a path formula, respectively, and* $g \in [2, \omega]$. *Then, the following holds:* (i) $\mathsf{E}^{\geq g}(\varphi \vee \psi) \equiv \neg\varphi \wedge \mathsf{E}^{\geq g}\psi$ *and* (ii) $\mathsf{A}^{< g}(\varphi \wedge \psi) \equiv \neg\varphi \vee \mathsf{A}^{< g}\psi$.

Proof [Only if]. If $\mathcal{K}, w_0 \models \mathsf{E}^{\geq g}\varphi \vee \psi$ then $|(\mathrm{Pth}(\mathcal{K}, w_0, \varphi \vee \psi)/\equiv^{\varphi\vee\psi}_{\mathcal{K}})| \geq g$, where w_0 is the initial world of $\mathcal{K}$. Suppose now by contradiction that $\mathcal{K}, w_0 \models \varphi$. Then, by the state-focus property, we have that $\pi_1 \equiv^{\varphi}_{\mathcal{K}} \pi_2$, for all paths $\pi_1, \pi_2 \in \mathrm{Pth}(\mathcal{K}, w_0)$. So, by the local disjunction-consistency property, we obtain that $\pi_1 \equiv^{\varphi\vee\psi}_{\mathcal{K}} \pi_2$ and then that $|(\mathrm{Pth}(\mathcal{K}, w_0, \varphi \vee \psi)/\equiv^{\varphi\vee\psi}_{\mathcal{K}})| = 1 < g$, but this contradict the hypothesis. Hence, $\mathcal{K}, w_0 \not\models \varphi$, i.e., $\mathcal{K}, w_0 \models \neg\varphi$. Then, by Item iv of Proposition 1.1, it is immediate to see that $\mathrm{Pth}(\mathcal{K}, w_0, \varphi \vee \psi) = \mathrm{Pth}(\mathcal{K}, w_0, \psi)$. Moreover, by the satisfiability-constraint property, we have that $\pi_1 \not\equiv^{\varphi}_{\mathcal{K}} \pi_2$, for all paths $\pi_1, \pi_2 \in \mathrm{Pth}(\mathcal{K}, w_0)$. Now, again by the local disjunction-consistency property, we obtain that $\pi_1 \equiv^{\varphi\vee\psi}_{\mathcal{K}} \pi_2$ iff $\pi_1 \equiv^{\psi}_{\mathcal{K}} \pi_2$. At this point, $(\mathrm{Pth}(\mathcal{K}, w_0, \varphi \vee \psi) /\equiv^{\varphi\vee\psi}_{\mathcal{K}}) = (\mathrm{Pth}(\mathcal{K}, w_0, \psi)/\equiv^{\varphi\vee\psi}_{\mathcal{K}}) = (\mathrm{Pth}(\mathcal{K}, w_0, \psi)/\equiv^{\psi}_{\mathcal{K}})$. Hence, $\mathcal{K}, w_0 \models \mathsf{E}^{\geq g}\psi$ and consequently $\mathcal{K}, w_0 \models \neg\varphi \wedge \mathsf{E}^{\geq g}\psi$.

[If]. If $\mathcal{K}, w_0 \models \neg\varphi \wedge \mathsf{E}^{\geq g}\psi$, we have that $\mathcal{K}, w_0 \not\models \varphi$ and $|(\mathrm{Pth}(\mathcal{K}, w_0, \psi)/\equiv^{\psi}_{\mathcal{K}})| \geq g$. Then, by Item iv of Proposition 1.1, it is immediate to see that $\mathrm{Pth}(\mathcal{K}, w_0, \psi) = \mathrm{Pth}(\mathcal{K}, w_0, \varphi \vee \psi)$. Moreover, by the satisfiability-constraint property, we have that $\pi_1 \not\equiv^{\varphi}_{\mathcal{K}} \pi_2$, for all paths $\pi_1, \pi_2 \in \mathrm{Pth}(\mathcal{K}, w_0)$. Now,

by the local disjunction-consistency property, we obtain that $\pi_1 \equiv_{\mathscr{K}}^{\varphi \vee \psi} \pi_2$ iff $\pi_1 \equiv_{\mathscr{K}}^{\psi} \pi_2$. At this point, $(\mathrm{Pth}(\mathscr{K}, w_0, \psi)/\equiv_{\mathscr{K}}^{\psi}) = (\mathrm{Pth}(\mathscr{K}, w_0, \varphi \vee \psi)/\equiv_{\mathscr{K}}^{\psi}) = (\mathrm{Pth}(\mathscr{K}, w_0, \varphi \vee \psi)/\equiv_{\mathscr{K}}^{\varphi \vee \psi})$. Hence, $\mathscr{K}, w_0 \models \mathsf{E}^{\geq g} \varphi \vee \psi$. □

1.4.4 Main Properties

We now summarize all the previous properties in the single concept of adequacy.

Definition 1.14 (Adequacy) *An equivalence relation* $\equiv_{\mathscr{K}}^{\cdot}$ *on paths is said* adequate *w.r.t. an equivalence structure* $\cong^{\cdot}_{\cdot}$ *iff it holds that it is* (i) *syntax independent,* (ii) *state focused,* (iii) *next consistent,* (iv) *weak next consistent w.r.t.* $\cong^{\cdot}_{\cdot}$, (v) *source dependent,* (vi) *conjunction consistent,* (vii) *local disjunction consistent, and* (viii) *satisfiability constrained.*

Next theorem shows four exponential fixpoint expressions that extend to graded formulas the corresponding well-known results for ungraded ones. These interesting equivalences among GCTL formulas are useful to describe important properties of its semantics.

Theorem 1.9 (GCTL **Fixpoint Equivalences**) *Let* $\equiv^{\cdot}_{\cdot}$ *be an adequate equivalence relation. Moreover, let* φ_1 *and* φ_2 *be two state formulas and* $g \in [2, \omega]$. *Then, the following equivalences hold:*

(1) $\mathsf{E}^{\geq g} \varphi_1 \mathsf{U} \varphi_2 \equiv \neg\varphi_2 \wedge \varphi_1 \wedge (\mathsf{EX}'(g, \varphi_1 \mathsf{U} \varphi_2) \vee \mathsf{E}^{\geq 1}\mathsf{X}\mathsf{E}^{\geq g} \varphi_1 \mathsf{U} \varphi_2)$;
(2) $\mathsf{E}^{\geq g} \varphi_1 \mathsf{R} \varphi_2 \equiv \varphi_2 \wedge \neg\varphi_1 \wedge (\mathsf{EX}'(g, \varphi_1 \mathsf{R} \varphi_2) \vee \mathsf{E}^{\geq 1}\mathsf{X}\mathsf{E}^{\geq g} \varphi_1 \mathsf{R} \varphi_2)$;
(3) $\mathsf{A}^{<g} \varphi_1 \tilde{\mathsf{U}} \varphi_2 \equiv \varphi_2 \vee \neg\varphi_1 \vee \mathsf{A}\tilde{\mathsf{X}}'(g, \varphi_1 \tilde{\mathsf{U}} \varphi_2) \wedge \mathsf{A}^{<1}\tilde{\mathsf{X}}\mathsf{A}^{<g} \varphi_1 \tilde{\mathsf{U}} \varphi_2$;
(4) $\mathsf{A}^{<g} \varphi_1 \tilde{\mathsf{R}} \varphi_2 \equiv \neg\varphi_2 \vee \varphi_1 \vee \mathsf{A}\tilde{\mathsf{X}}'(g, \varphi_1 \tilde{\mathsf{R}} \varphi_2) \wedge \mathsf{A}^{<1}\tilde{\mathsf{X}}\mathsf{A}^{<g} \varphi_1 \tilde{\mathsf{R}} \varphi_2$.

Proof To show Item 1 (resp., 2), it is possible to apply to the formula $\mathsf{E}^{\geq g} \varphi_1 \mathsf{U} \varphi_2$ (resp., $\mathsf{E}^{\geq g} \varphi_1 \mathsf{R} \varphi_2$) the following chain of equivalences: Item i (resp., ii) of Corollary 1.1 and Theorems 1.8 (resp., 1.7), 1.7 (resp., 1.8), 1.5, and 1.6. At the same way, to show Item 3 (resp., 4), it is possible to apply to the formula $\mathsf{A}^{<g} \varphi_1 \tilde{\mathsf{U}} \varphi_2$ (resp., $\mathsf{A}^{<g} \varphi_1 \tilde{\mathsf{R}} \varphi_2$) the following sequence of equivalences: Item vii (resp., viii) of Corollary 1.1, and Theorems 1.7 (resp., 1.8), 1.8 (resp., 1.7), 1.5, and 1.6. □

In the following, we use the four macros $\mathsf{EU}(g, \varphi_1, \varphi_2, Y)$, $\mathsf{ER}(g, \varphi_1, \varphi_2, Y)$, $\mathsf{A}\tilde{\mathsf{U}}(g, \varphi_1, \varphi_2, Y)$, and $\mathsf{A}\tilde{\mathsf{R}}(g, \varphi_1, \varphi_2, Y)$ defined below, to represent in short the expansion formulas for the existential U and R and the universal $\tilde{\mathsf{U}}$ and $\tilde{\mathsf{R}}$ temporal operators derived in the previous theorem and in Items 1, 2, 7, and 8 of Proposition 1.3.

- $\mathsf{EU}(g, \varphi_1, \varphi_2, Y) \triangleq \begin{cases} \varphi_2 \vee \varphi_1 \wedge \mathsf{E}^{\geq 1}\mathsf{X}Y, & \text{if } g = 1; \\ \neg\varphi_2 \wedge \varphi_1 \wedge (\mathsf{EX}'(g, \varphi_1 \mathsf{U} \varphi_2) \vee \mathsf{E}^{\geq 1}\mathsf{X}Y), & \text{otherwise.} \end{cases}$
- $\mathsf{ER}(g, \varphi_1, \varphi_2, Y) \triangleq \begin{cases} \varphi_2 \wedge (\varphi_1 \vee \mathsf{E}^{\geq 1}\mathsf{X}Y), & \text{if } g = 1; \\ \varphi_2 \wedge \neg\varphi_1 \wedge (\mathsf{EX}'(g, \varphi_1 \mathsf{R} \varphi_2) \vee \mathsf{E}^{\geq 1}\mathsf{X}Y), & \text{otherwise.} \end{cases}$

- $\mathsf{A\tilde{U}}(g,\varphi_1,\varphi_2,Y) \triangleq \begin{cases} \varphi_2 \vee \varphi_1 \wedge \mathsf{A}^{<1}\mathsf{X}Y, & \text{if } g=1; \\ \varphi_2 \vee \neg\varphi_1 \vee \mathsf{A\tilde{X}}'(g,\varphi_1\tilde{\mathsf{U}}\varphi_2) \wedge \mathsf{A}^{<1}\mathsf{X}Y, & \text{otherwise.} \end{cases}$
- $\mathsf{A\tilde{R}}(g,\varphi_1,\varphi_2,Y) \triangleq \begin{cases} \varphi_2 \wedge (\varphi_1 \vee \mathsf{A}^{<1}\mathsf{X}Y), & \text{if } g=1; \\ \neg\varphi_2 \vee \varphi_1 \vee \mathsf{A\tilde{X}}'(g,\varphi_1\tilde{\mathsf{R}}\varphi_2) \wedge \mathsf{A}^{<1}\mathsf{X}Y, & \text{otherwise.} \end{cases}$

It is immediate to see that $\mathsf{lng}(\mathsf{EU}(g,\varphi_1,\varphi_2,Y)) = \mathsf{lng}(\mathsf{ER}(g,\varphi_1,\varphi_2,Y)) = \Theta(\mathsf{lng}(Y)+(\mathsf{lng}(\varphi_1)+\mathsf{lng}(\varphi_2)+5)\cdot 2^{k\cdot\sqrt{g}})$ and $\mathsf{lng}(\mathsf{A\tilde{U}}(g,\varphi_1,\varphi_2,Y)) = \mathsf{lng}(\mathsf{A\tilde{R}}(g,\varphi_1,\varphi_2,Y)) = \Theta(\mathsf{lng}(Y)+(\mathsf{lng}(\varphi_1)+\mathsf{lng}(\varphi_2)+5)\cdot 2^{k\cdot\sqrt{g-1}})$, for a constant k. Moreover, for all $g \in [1,\omega]$, it holds that

- $\mathsf{E}^{\geq g}\varphi_1\mathsf{U}\varphi_2 \equiv \mathsf{EU}(g,\varphi_1,\varphi_2,\mathsf{E}^{\geq g}\varphi_1\mathsf{U}\varphi_2)$,
- $\mathsf{E}^{\geq g}\varphi_1\mathsf{R}\varphi_2 \equiv \mathsf{ER}(g,\varphi_1,\varphi_2,\mathsf{E}^{\geq g}\varphi_1\mathsf{R}\varphi_2)$,
- $\mathsf{A}^{<g}\varphi_1\tilde{\mathsf{U}}\varphi_2 \equiv \mathsf{A\tilde{U}}(g,\varphi_1,\varphi_2,\mathsf{A}^{<g}\varphi_1\tilde{\mathsf{U}}\varphi_2)$,
- $\mathsf{A}^{<g}\varphi_1\tilde{\mathsf{R}}\varphi_2 \equiv \mathsf{A\tilde{R}}(g,\varphi_1,\varphi_2,\mathsf{A}^{<g}\varphi_1\tilde{\mathsf{R}}\varphi_2)$.

Differently from the previous cases, we cannot hope to obtain similar general fixpoint equivalences for the existential $\tilde{\mathsf{U}}$ and $\tilde{\mathsf{R}}$ and the universal U and R temporal operators. This is due to the fact that we do not have general equivalences between the quantifications of $\mathsf{X}\psi$ and those of $\tilde{\mathsf{X}}\psi$. The next theorem shows the four exponential fixpoint properties we are able to derive for these cases.

Theorem 1.10 (GCTL **Almost Fixpoint Equivalences)** *Let $\equiv_{:}$ be an adequate equivalence relation w.r.t. the equivalence structure $\cong_{:}^{:}$. Moreover, let $\mathscr{K}$ be a* KS*, w_0 its initial world, φ_1 and φ_2 be two state formulas, and $g \in [2,\omega]$. Then, the following hold:*

(1) $\mathscr{K} \models \mathsf{E}^{\geq g}\varphi_1\tilde{\mathsf{U}}\varphi_2$ iff $\mathscr{K} \models \neg\varphi_2 \wedge \varphi_1 \wedge (\mathsf{EX}'(g,\varphi_1\tilde{\mathsf{U}}\varphi_2) \vee \mathsf{E}^{\geq 1}\mathsf{X}\mathsf{E}^{\geq g}\varphi_1\tilde{\mathsf{U}}\varphi_2)$ and $\tilde{\mathsf{X}}\varphi_1\tilde{\mathsf{U}}\varphi_2$ is not an $\cong_{\mathscr{K}}^{w_0}$-tautology;
(2) $\mathscr{K} \models \mathsf{E}^{\geq g}\varphi_1\tilde{\mathsf{R}}\varphi_2$ iff $\mathscr{K} \models \varphi_2 \wedge \neg\varphi_1 \wedge (\mathsf{EX}'(g,\varphi_1\tilde{\mathsf{R}}\varphi_2) \vee \mathsf{E}^{\geq 1}\mathsf{X}\mathsf{E}^{\geq g}\varphi_1\tilde{\mathsf{R}}\varphi_2)$ and $\tilde{\mathsf{X}}\varphi_1\tilde{\mathsf{R}}\varphi_2$ is not an $\cong_{\mathscr{K}}^{w_0}$-tautology;
(3) $\mathscr{K} \models \mathsf{A}^{<g}\varphi_1\mathsf{U}\varphi_2$ iff $\mathscr{K} \models \varphi_2 \vee \neg\varphi_1 \vee \mathsf{A\tilde{X}}'(g,\varphi_1\mathsf{U}\varphi_2) \wedge \mathsf{A}^{<1}\tilde{\mathsf{X}}\mathsf{A}^{<g}\varphi_1\mathsf{U}\varphi_2$ or $\neg\mathsf{X}\varphi_1\mathsf{U}\varphi_2$ is an $\cong_{\mathscr{K}}^{w_0}$-tautology;
(4) $\mathscr{K} \models \mathsf{A}^{<g}\varphi_1\mathsf{R}\varphi_2$ iff $\mathscr{K} \models \neg\varphi_2 \vee \varphi_1 \vee \mathsf{A\tilde{X}}'(g,\varphi_1\mathsf{R}\varphi_2) \wedge \mathsf{A}^{<1}\tilde{\mathsf{X}}\mathsf{A}^{<g}\varphi_1\mathsf{R}\varphi_2$ or $\neg\mathsf{X}\varphi_1\mathsf{R}\varphi_2$ is an $\cong_{\mathscr{K}}^{w_0}$-tautology.

Proof To show Item 1 (resp., 2), it is possible to apply to the formula $\mathsf{E}^{\geq g}\varphi_1\tilde{\mathsf{U}}\varphi_2$ (resp., $\mathsf{E}^{\geq g}\varphi_1\tilde{\mathsf{R}}\varphi_2$) the following chain of equivalences: Item iii (resp., iv) of Corollary 1.1, and Theorems 1.8 (resp., 1.7), 1.7 (resp., 1.8), 1.3, 1.5, and 1.6. At the same way, to show Item 3 (resp., 4), it is possible to apply to the formula $\mathsf{A}^{<g}\varphi_1\mathsf{U}\varphi_2$ (resp., $\mathsf{A}^{<g}\varphi_1\mathsf{R}\varphi_2$) the following sequence of equivalences: Item v (resp., vi) of Corollary 1.1, and Theorems 1.7 (resp., 1.8), 1.8 (resp., 1.7), 1.3, 1.5, and 1.6. □

As for the previous cases, in the following, we use the macros $\mathsf{E\tilde{U}}(g,\varphi_1,\varphi_2,Y,\varphi)$, $\mathsf{E\tilde{R}}(g,\varphi_1,\varphi_2,Y,\varphi)$, $\mathsf{AU}(g,\varphi_1,\varphi_2,Y,\varphi)$, and $\mathsf{AR}(g,\varphi_1,\varphi_2,Y,\varphi)$ defined below, to represent in short the expansion formulas for the existential $\tilde{\mathsf{U}}$ and $\tilde{\mathsf{R}}$ and the universal U and R temporal operators derived in the previous theorem and in Items 3, 4, 5, and 6 of Proposition 1.3.

- $\mathsf{E\tilde{U}}(g, \varphi_1, \varphi_2, Y, \varphi) \triangleq \begin{cases} \varphi_2 \vee \varphi_1 \wedge (\mathsf{E}^{\geq 1}\tilde{\mathsf{X}}\mathsf{f} \vee \mathsf{E}^{\geq 1}\mathsf{X}Y), & \text{if } g = 1; \\ \neg\varphi_2 \wedge \varphi_1 \wedge (\mathsf{EX}'(g, \varphi_1 \tilde{\mathsf{U}} \varphi_2) \vee \mathsf{E}^{\geq 1}\mathsf{X}Y) \wedge \varphi, & \text{otherwise.} \end{cases}$
- $\mathsf{E\tilde{R}}(g, \varphi_1, \varphi_2, Y, \varphi) \triangleq \begin{cases} \varphi_2 \wedge (\varphi_1 \vee \mathsf{E}^{\geq 1}\tilde{\mathsf{X}}\mathsf{f} \vee \mathsf{E}^{\geq 1}\mathsf{X}Y), & \text{if } g = 1; \\ \varphi_2 \wedge \neg\varphi_1 \wedge (\mathsf{EX}'(g, \varphi_1 \tilde{\mathsf{R}} \varphi_2) \vee \mathsf{E}^{\geq 1}\mathsf{X}Y) \wedge \varphi, & \text{otherwise.} \end{cases}$
- $\mathsf{AU}(g, \varphi_1, \varphi_2, Y, \varphi) \triangleq \begin{cases} \varphi_2 \vee \varphi_1 \wedge \mathsf{A}^{<1}\mathsf{X}\mathsf{t} \wedge \mathsf{A}^{<1}\mathsf{X}Y, & \text{if } g = 1; \\ \varphi_2 \vee \neg\varphi_1 \vee \mathsf{A\tilde{X}}'(g, \varphi_1 \mathsf{U} \varphi_2) \wedge \mathsf{A}^{<1}\tilde{\mathsf{X}}Y \vee \varphi, & \text{otherwise.} \end{cases}$
- $\mathsf{AR}(g, \varphi_1, \varphi_2, Y, \varphi) \triangleq \begin{cases} \varphi_2 \wedge (\varphi_1 \vee \mathsf{A}^{<1}\mathsf{X}\mathsf{t} \wedge \mathsf{A}^{<1}\mathsf{X}Y), & \text{if } g = 1; \\ \neg\varphi_2 \vee \varphi_1 \vee \mathsf{A\tilde{X}}'(g, \varphi_1 \mathsf{R} \varphi_2) \wedge \mathsf{A}^{<1}\tilde{\mathsf{X}}Y \vee \varphi, & \text{otherwise.} \end{cases}$

It is immediate to see that $\mathsf{lng}(\mathsf{E\tilde{U}}(g, \varphi_1, \varphi_2, Y, \varphi)) = \mathsf{lng}(\mathsf{E\tilde{R}}(g, \varphi_1, \varphi_2, Y, \varphi)) = \Theta(\mathsf{lng}(Y) + \mathsf{lng}(\varphi) + (\mathsf{lng}(\varphi_1) + \mathsf{lng}(\varphi_2) + 5) \cdot 2^{k \cdot \sqrt{g}})$ and $\mathsf{lng}(\mathsf{AU}(g, \varphi_1, \varphi_2, Y, \varphi)) = \mathsf{lng}(\mathsf{AR}(g, \varphi_1, \varphi_2, Y, \varphi)) = \Theta(\mathsf{lng}(Y) + \mathsf{lng}(\varphi) + (\mathsf{lng}(\varphi_1) + \mathsf{lng}(\varphi_2) + 5) \cdot 2^{k \cdot \sqrt{g-1}})$, for a constant k. As yet noted above, there are no general equivalences that directly link the formulas $\mathsf{E}^{\geq g}\varphi_1 \tilde{\mathsf{U}} \varphi_2$, $\mathsf{E}^{\geq g}\varphi_1 \tilde{\mathsf{R}} \varphi_2$, $\mathsf{A}^{<g}\varphi_1 \mathsf{U} \varphi_2$, and $\mathsf{A}^{<g}\varphi_1 \mathsf{R} \varphi_2$ with their expansions $\mathsf{E\tilde{U}}(g, \varphi_1, \varphi_2, \mathsf{E}^{\geq g}\varphi_1 \tilde{\mathsf{U}} \varphi_2, \varphi)$, $\mathsf{E\tilde{R}}(g, \varphi_1, \varphi_2, \mathsf{E}^{\geq g}\varphi_1 \tilde{\mathsf{R}} \varphi_2, \varphi)$, $\mathsf{AU}(g, \varphi_1, \varphi_2, \mathsf{A}^{<g}\varphi_1 \mathsf{U} \varphi_2, \varphi)$, and $\mathsf{AR}(g, \varphi_1, \varphi_2, \mathsf{A}^{<g}\varphi_1 \mathsf{R} \varphi_2, \varphi)$. Note that here the metavariable φ can be used at the same way of that of the macro $\mathsf{E\tilde{X}}(g, \psi, \varphi)$.

Finally, we show a fundamental equivalence that allows us to extract all state formulas from the scope of a quantification of a generic GCTL* path formula.

Theorem 1.11 (GCTL* **Path Expansion Equivalence**) *Let* $\equiv_{\cdot}^{\cdot}$ *be a syntax-in-dependent, state-focused, conjunction-consistent, local disjunction-consistent, and satisfiability-constrained equivalence relation. Moreover, let* φ_i *and* ψ_i *be, respectively, k state and path formulas,* $\mathsf{Op}_i \in \{\mathsf{X}, \tilde{\mathsf{X}}\}$, *and* $g \in [1, \omega]$. *Then, the following equivalences hold, where* $\psi = \bigwedge_{i=1}^{k}(\varphi_i \vee \mathsf{Op}_i \psi_i)$, $\varphi_I = \bigwedge_{i \in I} \varphi_i \wedge \bigwedge_{i \in [1,k] \setminus I} \neg\varphi_i$ *and* $\psi_I = \mathsf{Op} \bigwedge_{i \in [1,k] \setminus I} \psi_i$ *with* $\mathsf{Op} \in \{\mathsf{X}, \tilde{\mathsf{X}}\}$ *and* $\mathsf{Op} = \mathsf{X}$ *iff there is* $i \in [1, k] \setminus I$ *such that* $\mathsf{Op}_i = \mathsf{X}$.

(1) $\mathsf{E}^{\geq g}\psi \equiv \bigvee_{I \subseteq [1,k]} \varphi_I \wedge \mathsf{E}^{\geq g}\psi_I$;
(2) $\mathsf{A}^{<g}\neg\psi \equiv \bigvee_{I \subseteq [1,k]} \varphi_I \wedge \mathsf{A}^{<g}\neg\psi_I$.

Proof We have to prove that $\mathscr{K}, w_0 \models \mathsf{E}^{\geq g}\psi$ iff $\mathscr{K}, w_0 \models \bigvee_{I \subseteq [1,k]} \varphi_I \wedge \mathsf{E}^{\geq g}\psi_I$ (resp., $\mathscr{K}, w_0 \models \mathsf{A}^{<g}\neg\psi$ iff $\mathscr{K}, w_0 \models \bigvee_{I \subseteq [1,k]} \varphi_I \wedge \mathsf{A}^{<g}\neg\psi_I$), where w_0 is the initial world of $\mathscr{K}$, for all KS $\mathscr{K}$. First, let $I \subseteq [1, k]$ be the set of indexes of just the state formulas φ_i that are true on $\mathscr{K}$, i.e., such that *(i)* $\mathscr{K}, w_0 \models \varphi_i$, for all $i \in I$, and *(ii)* $\mathscr{K}, w_0 \not\models \varphi_i$, for all $i \in [1, k] \setminus I$. Thus, $\mathscr{K}, w_0 \models \varphi_I$. Note that such a set is uniquely determined by the KS $\mathscr{K}$.

By Items iii and iv of Proposition 1.1, it holds that $\mathrm{Pth}(\mathscr{K}, w_0, \psi) = \mathrm{Pth}(\mathscr{K}, w_0, \psi_I)$. What remains to prove is that $\pi_1 \equiv_{\mathscr{K}}^{\psi} \pi_2$ iff $\pi_1 \equiv_{\mathscr{K}}^{\psi_I} \pi_2$, for all $\pi_1, \pi_2 \in \mathrm{Pth}(\mathscr{K}, w_0)$. By the conjunction-consistency property, we have that $\pi_1 \equiv_{\mathscr{K}}^{\psi} \pi_2$ iff, for all $i \in [1, k]$, it holds that $\pi_1 \equiv_{\mathscr{K}}^{\varphi_i \vee \mathsf{Op}_i \psi_i} \pi_2$. Thus, by the local disjunction-consistency property, we obtain that $\pi_1 \equiv_{\mathscr{K}}^{\psi} \pi_2$ iff, for all $i \in [1, k]$, it holds that $\pi_1 \equiv_{\mathscr{K}}^{\varphi_i} \pi_2$ or $\pi_1 \equiv_{\mathscr{K}}^{\mathsf{Op}_i \psi_i} \pi_2$. Now, if $i \in I$, by the state-focus property, it holds that $\pi_1 \equiv^{\varphi_i}_{\mathcal{K}} \pi_2$. On the contrary, if $i \in [1,k] \setminus I$, by the satisfiability-constraint property, it holds that $\pi_1 \not\equiv^{\varphi_i}_{\mathcal{K}} \pi_2$. Hence, the previous coimplication between $\pi_1 \equiv^{\psi}_{\mathcal{K}} \pi_2$ and its expansion can be simplified as follows: $\pi_1 \equiv^{\psi}_{\mathcal{K}} \pi_2$ iff, for all $i \in [1,k] \setminus I$, it holds that $\pi_1 \equiv^{\mathsf{Op}_i \psi_i}_{\mathcal{K}} \pi_2$. At this point, again by the conjunction-consistency property, we have that $\pi_1 \equiv^{\psi}_{\mathcal{K}} \pi_2$ iff $\pi_1 \equiv^{\bigwedge_{i \in [1,k] \setminus I} \mathsf{Op}_i \psi_i}_{\mathcal{K}} \pi_2$. Now, it is easy to note that $\bigwedge_{i \in [1,k] \setminus I} \mathsf{Op}_i \psi_i \equiv \psi_I$. So, by the syntax-independence property, we can further simplify the previous coimplication in $\pi_1 \equiv^{\psi}_{\mathcal{K}} \pi_2$ iff $\pi_1 \equiv^{\psi_I}_{\mathcal{K}} \pi_2$, obtaining directly that $(\mathrm{Pth}(\mathcal{K}, w_0, \psi)/\equiv^{\psi}_{\mathcal{K}}) = (\mathrm{Pth}(\mathcal{K}, w_0, \psi_I)/\equiv^{\psi_I}_{\mathcal{K}})$. Thus, the assumption $\mathcal{K}, w_0 \models \varphi_I$ implies that $\mathcal{K}, w_0 \models \mathsf{E}^{\geq g} \psi$ iff $\mathcal{K}, w_0 \models \mathsf{E}^{\geq g} \psi_I$ (resp., $\mathcal{K}, w_0 \models \mathsf{A}^{<g} \neg \psi$ iff $\mathcal{K}, w_0 \models \mathsf{A}^{<g} \neg \psi_I$).

Now, on one hand, it is easy to see that, for each KS $\mathcal{K}$, there is a set $I \subseteq [1, k]$ such that $\mathcal{K}, w_0 \models \varphi_I$ and so $\mathsf{E}^{\geq g} \psi \Rightarrow \bigvee_{I \subseteq [1,k]} \varphi_I \wedge \mathsf{E}^{\geq g} \psi_I$ (resp., $\mathsf{A}^{<g} \neg \psi \Rightarrow \bigvee_{I \subseteq [1,k]} \varphi_I \wedge \mathsf{A}^{<g} \neg \psi_I$). On the other hand, the existence of a set $I \subseteq [1,k]$ such that $\mathcal{K}, w_0 \models \varphi_I$ and $\mathcal{K}, w_0 \models \mathsf{E}^{\geq g} \psi_I$ (resp., $\mathcal{K}, w_0 \models \mathsf{A}^{<g} \neg \psi_I$) implies $\mathsf{E}^{\geq g} \psi$ (resp., $\mathsf{A}^{<g} \neg \psi$), i.e., $\bigvee_{I \subseteq [1,k]} \varphi_I \wedge \mathsf{E}^{\geq g} \psi_I \Rightarrow \mathsf{E}^{\geq g} \psi$ (resp., $\bigvee_{I \subseteq [1,k]} \varphi_I \wedge \mathsf{A}^{<g} \neg \psi_I \Rightarrow \mathsf{A}^{<g} \neg \psi$). Hence, the thesis follows. □

It may be interesting to observe that the previous result is a generalization of Theorems 1.7 and 1.8 that can be obtained as the limit cases in which there are no conjunctions or disjunctions, respectively. Moreover, it is important to note that, differently from the case of the local conjunction quantification, here we need the full power of the conjunction-consistency property in order to prove this equivalence.

1.5 Prefix Path Equivalence

In this section, we introduce a suitable path equivalence relation that satisfies all the previously discussed properties. Hence, we show that those properties are not contradictory, by presenting one of the possible meaningful graded computation tree logics. In the sequel of the paper, we only refer to GCTL* under this specific equivalence relation.

1.5.1 Definition and Properties

In the definition of the GCTL* semantics, we use a generic equivalence relation $\equiv^{\cdot}_{\cdot}$ on paths that allows us to count how many ways a structure has to satisfy a path formula. So, two paths should be considered equivalent when they represent only one way to perform according to that formula. For many formulas, such a way results to be their common finite prefix. For example, all paths that satisfy $\mathsf{X}p$ and have the first two nodes in common may be regarded as equivalent because the first two nodes constitute the one sought way to satisfy the formula. For some other formula like $\tilde{\mathsf{X}}p$, the ways to satisfy it are less clear. For example, consider two paths π_1 and π_2

with only the starting node in common, such that the first satisfies $\mathsf{X}p$ while the latter $\mathsf{X}\neg p$. Then, the common node, if taken alone, i.e., without its successors, may be considered as a path satisfying $\tilde{\mathsf{X}}p$. So, the two paths would be equivalent. However, this looks unreasonable because π_2 does not satisfy $\tilde{\mathsf{X}}p$ and, thus, the common prefix failed to ensure the conservativeness of the satisfiability for this formula. Hence, a common prefix between two paths may be considered as a way to satisfy a path formula, if it satisfies the formula and somehow it allows us to deduce that this formula is true on all paths with that prefix in the structure. The following definition of the equivalence relation among paths formally captures the previous idea.

Definition 1.15 (Prefix Equivalence) *Two paths* $\pi_1, \pi_2 \in \mathrm{Pth}(\mathcal{K})$ *are* prefix equivalent *w.r.t. a path formula* ψ*, in symbols* $\pi_1 \equiv^{\psi}_{\mathcal{K}} \pi_2$*, iff either* $\pi_1 = \pi_2$ *or* (i) *the common track* $\rho = \mathsf{pfx}(\pi_1, \pi_2)$ *of* π_1 *and* π_2 *is not empty and* (ii) $\mathcal{K}, \rho \cdot \pi_{\geq 1} \models \psi$*, for every path/track* $\pi \in (\mathrm{Pth}(\mathcal{K}, \mathsf{lst}(\rho)) \cup \mathrm{Trk}(\mathcal{K}, \mathsf{lst}(\rho)))$.

Observe that when two paths are distinct w.r.t. $\equiv^{\psi}_{\mathcal{K}}$, there are always at least two successors of the last node of their common prefix. Hence, the KS $\mathcal{K}$ is never allowed to stop its computations at that node, i.e., the common prefix is a track but not a path in $\mathcal{K}$.

We now give few simple examples of the behavior of GCTL* under the use of the prefix equivalence.

Consider a finite KT $\mathcal{T}$ having just three nodes all labeled by p, the root and its two successors (see Fig. 1.2). Also, consider the formula $\varphi = \mathsf{E}^{\geq 2}\mathsf{F}p$. Because of the definition of the equivalence, the only two paths $\pi_1, \pi_2 \in \mathrm{Pth}(\mathcal{T}, \varepsilon)$ of length two satisfying $\mathsf{F}p$ are equivalent, since the common prefix $\rho = \mathsf{pfx}(\pi_1, \pi_2)$ containing just the root satisfies the formula too. Hence, $\mathcal{T} \not\models \varphi$. On the contrary, if we take a tree $\mathcal{T}'$ that is the same of $\mathcal{T}$, but with its root not labeled with p, we obtain that $\mathcal{T}' \models \varphi$, since $\mathcal{T}', \rho \not\models \mathsf{F}p$. This means that the particular equivalence allows us to count as different events only their first appearance along the paths.

Consider now the formula $\varphi = \mathsf{E}^{\geq 2}\mathsf{G}p$ and an infinite KT $\mathcal{T}$ having just two paths all labeled by p (see Fig. 1.3). Since $\mathsf{G}p$ cannot be satisfied on a track or finite path, we have that $\mathcal{T}, \rho \not\models \mathsf{G}p$, so the two infinite paths are not equivalent w.r.t. this formula, which implies that $\mathcal{T} \models \varphi'$. On the contrary, if we take $\varphi' = \mathsf{E}^{\geq 2}\tilde{\mathsf{G}}p$, then we obtain $\mathcal{T} \not\models \varphi'$, since each track and path completely labeled with p satisfies $\tilde{\mathsf{G}}p$.

We now define a new equivalence between path formulas that results to be compatible with the chosen prefix equivalence. Its definition, in particular, takes into

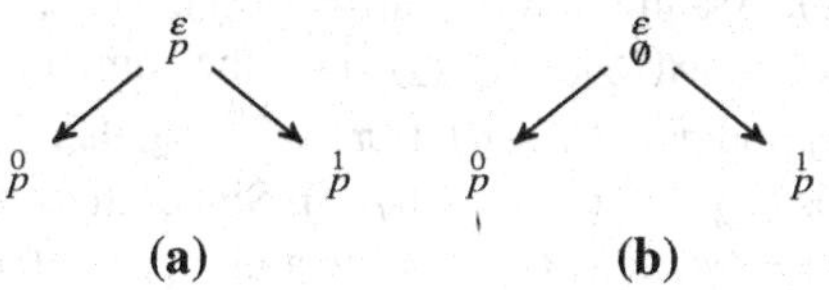

Fig. 1.2 Two finite KTs

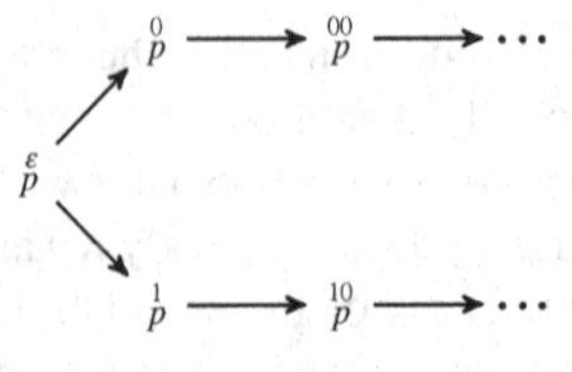

Fig. 1.3 An infinite KT

account a KS $\mathscr{K}$ and one of its worlds w in which we want to verify that the two formulas under exam are interchangeable for the logic.

Definition 1.16 (Structure Formula Equivalence) *Let $\mathscr{K}$ be a* KS, *w one of its worlds, and ψ_1 and ψ_2 be two path formulas. Then, ψ_1 is* structurally equivalent *to ψ_2 w.r.t. $\mathscr{K}$ and w, in symbols $\psi_1 \cong^{w}_{\mathscr{K}} \psi_2$, iff, for all paths/tracks $\pi \in (\mathrm{Pth}(\mathscr{K}, w) \cup \mathrm{Trk}(\mathscr{K}, w))$, it holds that $\mathscr{K}, \pi \models \psi_1$ iff $\mathscr{K}, \pi \models \psi_2$.*

Observe that $\cong^{\cdot}_{\cdot}$ is an equivalence structure according to Definition 1.6.

The following theorem shows that the prefix path relation satisfies the adequacy property defined in the previous section, if we consider the structure formula equivalence when we have to deal with the weak next operator.

Theorem 1.12 (Prefix Equivalence Adequacy) *The prefix equivalence relation is* adequate *w.r.t. $\cong^{\cdot}_{\cdot}$.*

Proof All the equivalence properties we want to show express that a given property on two paths implies a derived property on the same paths. So they are trivially satisfied when they concern two identical paths. For this reason in the following, we make the assumption that the two paths $\pi_1, \pi_2 \in \mathrm{Pth}(\mathscr{K})$ involved in the proof are distinct. Moreover, we use $\rho = \mathsf{pfx}(\pi_1, \pi_2)$ to indicate their common prefix.

(1) (Syntax independence). For $i \in \{1, 2\}$, if $\pi_1 \equiv^{\psi_i}_{\mathscr{K}} \pi_2$, then *(i)* $\rho \neq \varepsilon$ and *(ii)* $\mathscr{K}, \rho \cdot \pi_{\geq 1} \models \psi_i$, for all $\pi \in (\mathrm{Pth}(\mathscr{K}, \mathsf{lst}(\rho)) \cup \mathrm{Trk}(\mathscr{K}, \mathsf{lst}(\rho)))$. Since $\psi_1 \equiv \psi_2$, by Item i of Proposition 1.1, we obtain then that $\mathscr{K}, \rho \cdot \pi_{\geq 1} \models \psi_{3-i}$, for all $\pi \in (\mathrm{Pth}(\mathscr{K}, \mathsf{lst}(\rho)) \cup \mathrm{Trk}(\mathscr{K}, \mathsf{lst}(\rho)))$. Hence, $\pi_1 \equiv^{\psi_{3-i}}_{\mathscr{K}} \pi_2$.

(2) (State focus). Assume that $(\pi_1)_0 = (\pi_2)_0$, thus obtaining $\rho \neq \varepsilon$. Since φ is a state formula, by Item ii of Proposition 1.1, we have that $\mathscr{K}, (\rho)_0 \models \varphi$ implies $\mathscr{K}, \rho \cdot \pi_{\geq 1} \models \varphi$, for all $\pi \in (\mathrm{Pth}(\mathscr{K}, \mathsf{lst}(\rho)) \cup \mathrm{Trk}(\mathscr{K}, \mathsf{lst}(\rho)))$. Hence, $\pi_1 \equiv^{\varphi}_{\mathscr{K}} \pi_2$.

(3) (Next consistency). Assume that $(\pi_1)_0 = (\pi_2)_0$. Then, it is immediate to see that $\rho \neq \varepsilon$ and $\rho_{\geq 1} = \mathsf{pfx}((\pi_1)_{\geq 1}, (\pi_2)_{\geq 1})$ is the common prefix of the suffixes of the two paths π_1 and π_2. *[Only if].* If $\pi_1 \equiv^{\mathsf{X}\psi}_{\mathscr{K}} \pi_2$, then $\mathscr{K}, \rho \cdot \pi_{\geq 1} \models \mathsf{X}\psi$, for all $\pi \in (\mathrm{Pth}(\mathscr{K}, \mathsf{lst}(\rho)) \cup \mathrm{Trk}(\mathscr{K}, \mathsf{lst}(\rho)))$. Since $\mathsf{lst}(\rho) \in \mathrm{Trk}(\mathscr{K}, \mathsf{lst}(\rho))$, we have that $\mathscr{K}, \rho \cdot \varepsilon \models \mathsf{X}\psi$, i.e., $\mathscr{K}, \rho \models \mathsf{X}\psi$ and so, $\rho_{\geq 1} \neq \varepsilon$, by Item v of Proposition 1.1. Moreover, by the same item, one can note that $\mathscr{K}, (\rho \cdot \pi_{\geq 1})_{\geq 1} \models \psi$, i.e., $\mathscr{K}, \rho_{\geq 1} \cdot \pi_{\geq 1} \models \psi$, for all $\pi \in (\mathrm{Pth}(\mathscr{K}, \mathsf{lst}(\rho)) \cup \mathrm{Trk}(\mathscr{K}, \mathsf{lst}(\rho))) =$

$(\mathrm{Pth}(\mathcal{K}, \mathsf{lst}(\rho_{\geq 1})) \cup \mathrm{Trk}(\mathcal{K}, \mathsf{lst}(\rho_{\geq 1})))$. Hence, $(\pi_1)_{\geq 1} \equiv^{\psi}_{\mathcal{K}} (\pi_2)_{\geq 1}$. *[If].* If $(\pi_1)_{\geq 1} \equiv^{\psi}_{\mathcal{K}} (\pi_2)_{\geq 1}$, then $\mathcal{K}, \rho_{\geq 1} \cdot \pi_{\geq 1} \models \psi$, i.e., $\mathcal{K}, (\rho \cdot \pi_{\geq 1})_{\geq 1} \models \psi$, for all $\pi \in (\mathrm{Pth}(\mathcal{K}, \mathsf{lst}(\rho_{\geq 1})) \cup \mathrm{Trk}(\mathcal{K}, \mathsf{lst}(\rho_{\geq 1})))$. Now, by Item v of Proposition 1.1, one can note that $\mathcal{K}, \rho \cdot \pi_{\geq 1} \models \mathsf{X}\psi$, for all $\pi \in (\mathrm{Pth}(\mathcal{K}, \mathsf{lst}(\rho_{\geq 1})) \cup \mathrm{Trk}(\mathcal{K}, \mathsf{lst}(\rho_{\geq 1}))) = (\mathrm{Pth}(\mathcal{K}, \mathsf{lst}(\rho)) \cup \mathrm{Trk}(\mathcal{K}, \mathsf{lst}(\rho)))$. Hence, $\pi_1 \equiv^{\mathsf{X}\psi}_{\mathcal{K}} \pi_2$.

(4) (Weak next consistency). Assume that $(\pi_1)_0 = (\pi_2)_0$. As in the previous item, we have that $\rho \neq \varepsilon$ and $\rho_{\geq 1} = \mathsf{pfx}((\pi_1)_{\geq 1}, (\pi_2)_{\geq 1})$. *[Only if].* If $\pi_1 \equiv^{\tilde{\mathsf{X}}\psi}_{\mathcal{K}} \pi_2$, then $\mathcal{K}, \rho \cdot \pi_{\geq 1} \models \tilde{\mathsf{X}}\psi$, for all $\pi \in (\mathrm{Pth}(\mathcal{K}, \mathsf{lst}(\rho)) \cup \mathrm{Trk}(\mathcal{K}, \mathsf{lst}(\rho)))$. Now, suppose that $\tilde{\mathsf{X}}\psi$ is not an $\cong^{(\rho)_0}_{\mathcal{K}}$-tautology. Then, it is possible to see that $\rho_{\geq 1} \neq \varepsilon$. Indeed, suppose by contradiction that $\rho_{\geq 1} = \varepsilon$ and let $\pi \in (\mathrm{Pth}(\mathcal{K}, (\rho)_0) \cup \mathrm{Trk}(\mathcal{K}, (\rho_0))$ be the path/track not satisfying $\tilde{\mathsf{X}}\psi$, i.e., such that $\mathcal{K}, \pi \not\models \tilde{\mathsf{X}}\psi$. Since $(\rho)_0 = \mathsf{lst}(\rho)$, it is immediate to see that $\pi = \rho \cdot \pi_{\geq 1}$, so we have that $\mathcal{K}, \rho \cdot \pi_{\geq 1} \not\models \tilde{\mathsf{X}}\psi$, and this is in contradiction with the equivalence $\pi_1 \equiv^{\tilde{\mathsf{X}}\psi}_{\mathcal{K}} \pi_2$. At this point, by Item vi of Proposition 1.1, one can note that $\mathcal{K}, \rho_{\geq 1} \cdot \pi_{\geq 1} \models \psi$, for all $\pi \in (\mathrm{Pth}(\mathcal{K}, \mathsf{lst}(\rho)) \cup \mathrm{Trk}(\mathcal{K}, \mathsf{lst}(\rho))) = (\mathrm{Pth}(\mathcal{K}, \mathsf{lst}(\rho_{\geq 1})) \cup \mathrm{Trk}(\mathcal{K}, \mathsf{lst}(\rho_{\geq 1})))$. Hence, $(\pi_1)_{\geq 1} \equiv^{\psi}_{\mathcal{K}} (\pi_2)_{\geq 1}$. *[If].* On one hand, if $\tilde{\mathsf{X}}\psi$ is an $\cong^{(\rho)_0}_{\mathcal{K}}$-tautology, then all paths/tracks $\pi \in (\mathrm{Pth}(\mathcal{K}, (\rho)_0) \cup \mathrm{Trk}(\mathcal{K}, (\rho)_0))$ satisfy $\tilde{\mathsf{X}}\psi$, i.e., $\mathcal{K}, \pi \models \tilde{\mathsf{X}}\psi$. Thus, $\mathcal{K}, \rho \cdot \pi_{\geq 1} \models \tilde{\mathsf{X}}\psi$, for all $\pi \in (\mathrm{Pth}(\mathcal{K}, \mathsf{lst}(\rho)) \cup \mathrm{Trk}(\mathcal{K}, \mathsf{lst}(\rho)))$. Hence, $\pi_1 \equiv^{\tilde{\mathsf{X}}\psi}_{\mathcal{K}} \pi_2$. On the other hand, if $(\pi_1)_{\geq 1} \equiv^{\psi}_{\mathcal{K}} (\pi_2)_{\geq 1}$, then $\mathcal{K}, \rho_{\geq 1} \cdot \pi_{\geq 1} \models \psi$, for all $\pi \in (\mathrm{Pth}(\mathcal{K}, \mathsf{lst}(\rho_{\geq 1})) \cup \mathrm{Trk}(\mathcal{K}, \mathsf{lst}(\rho_{\geq 1})))$. Now, by Item vi of Proposition 1.1, one can note that $\mathcal{K}, \rho \cdot \pi_{\geq 1} \models \tilde{\mathsf{X}}\psi$, for all $\pi \in (\mathrm{Pth}(\mathcal{K}, \mathsf{lst}(\rho_{\geq 1})) \cup \mathrm{Trk}(\mathcal{K}, \mathsf{lst}(\rho_{\geq 1}))) = (\mathrm{Pth}(\mathcal{K}, \mathsf{lst}(\rho)) \cup \mathrm{Trk}(\mathcal{K}, \mathsf{lst}(\rho)))$. Hence, $\pi_1 \equiv^{\tilde{\mathsf{X}}\psi}_{\mathcal{K}} \pi_2$.

(5) (Source dependence). By definition, if the two paths π_1 and π_2 have no starting node in common, i.e., $(\pi_1)_0 \neq (\pi_2)_0$, they cannot be prefix equivalent because $\rho = \varepsilon$, i.e., they do not have any non-empty prefix in common at all.

(6) (Conjunction consistency). Let $\psi = \psi_1 \wedge \psi_2$. Then, it holds that $\pi_1 \equiv^{\psi}_{\mathcal{K}} \pi_2$ iff *(i)* $\rho \neq \varepsilon$ and *(ii)* $\mathcal{K}, \rho \cdot \pi_{\geq 1} \models \psi$, for all $\pi \in (\mathrm{Pth}(\mathcal{K}, \mathsf{lst}(\rho)) \cup \mathrm{Trk}(\mathcal{K}, \mathsf{lst}(\rho)))$. By Item iii of Proposition 1.1, the condition *(ii)* is equivalent to $\mathcal{K}, \rho \cdot \pi_{\geq 1} \models \psi_i$, for all $i \in \{1, 2\}$. Hence, $\pi_1 \equiv^{\psi}_{\mathcal{K}} \pi_2$ iff $\pi_1 \equiv^{\psi_1}_{\mathcal{K}} \pi_2$ and $\pi_1 \equiv^{\psi_2}_{\mathcal{K}} \pi_2$.

(7) (Local disjunction consistency). Let $\psi = \varphi \vee \psi'$, where φ is a state formula. *[Only if].* If $\pi_1 \equiv^{\psi}_{\mathcal{K}} \pi_2$, then *(i)* $\rho \neq \varepsilon$ and *(ii)* $\mathcal{K}, \rho \cdot \pi_{\geq 1} \models \psi$, for all $\pi \in (\mathrm{Pth}(\mathcal{K}, \mathsf{lst}(\rho)) \cup \mathrm{Trk}(\mathcal{K}, \mathsf{lst}(\rho)))$. First suppose that $\mathcal{K}, (\rho)_0 \models \varphi$. Then, by the state-focus property, we obtain that $\pi_1 \equiv^{\varphi}_{\mathcal{K}} \pi_2$. Suppose now that $\mathcal{K}, (\rho)_0 \not\models \varphi$. By Item ii of Proposition 1.1, we have that $\mathcal{K}, \rho \cdot \pi_{\geq 1} \not\models \varphi$, for all $\pi \in (\mathrm{Pth}(\mathcal{K}, \mathsf{lst}(\rho)) \cup \mathrm{Trk}(\mathcal{K}, \mathsf{lst}(\rho)))$, and so, by Item iv of Proposition 1.1, we obtain that $\mathcal{K}, \rho \cdot \pi_{\geq 1} \models \psi'$, for all $\pi \in (\mathrm{Pth}(\mathcal{K}, \mathsf{lst}(\rho)) \cup \mathrm{Trk}(\mathcal{K}, \mathsf{lst}(\rho)))$. Consequently, we obtain that $\pi_1 \equiv^{\psi'}_{\mathcal{K}} \pi_2$. *[If].* If $\pi_1 \equiv^{\varphi}_{\mathcal{K}} \pi_2$ (resp., $\pi_1 \equiv^{\psi'}_{\mathcal{K}} \pi_2$), then *(i)* $\rho \neq \varepsilon$ and *(ii)* $\mathcal{K}, \rho \cdot \pi_{\geq 1} \models \varphi$ (resp., $\mathcal{K}, \rho \cdot \pi_{\geq 1} \models \psi'$), for all

$\pi \in (\mathrm{Pth}(\mathcal{K}, \mathsf{lst}(\rho)) \cup \mathrm{Trk}(\mathcal{K}, \mathsf{lst}(\rho)))$. By Item iv of Proposition 1.1, we have that $\mathcal{K}, \rho \cdot \pi_{\geq 1} \models \psi$, for all $\pi \in (\mathrm{Pth}(\mathcal{K}, \mathsf{lst}(\rho)) \cup \mathrm{Trk}(\mathcal{K}, \mathsf{lst}(\rho)))$. Hence, $\pi_1 \equiv^{\psi}_{\mathcal{K}} \pi_2$.

(8) (Satisfiability constraint). If $\pi_1 \equiv^{\psi}_{\mathcal{K}} \pi_2$, then $\mathcal{K}, \rho \cdot \pi_{\geq 1} \models \psi$, for all $\pi \in (\mathrm{Pth}(\mathcal{K}, \mathsf{lst}(\rho)) \cup \mathrm{Trk}(\mathcal{K}, \mathsf{lst}(\rho)))$. Now, since there are two paths $\pi'_1, \pi'_2 \in \mathrm{Pth}(\mathcal{K}, \mathsf{lst}(\rho))$ such that $\pi_1 = \rho \cdot (\pi'_1)_{\geq 1}$ and $\pi_2 = \rho \cdot (\pi'_2)_{\geq 1}$, we obtain that $\mathcal{K}, \pi_1 \models \psi$ and $\mathcal{K}, \pi_2 \models \psi$. □

At this point, we are able to prove that we can express the concept of tautology in GCTL itself, due to the particular structure formula equivalence chosen for the logic.

Theorem 1.13 (Structure Formula Tautology) *Let $\mathcal{K} = \langle \mathrm{AP}, \mathrm{W}, R, \mathsf{L}, w_0 \rangle$ be a* KS *and $w \in \mathrm{W}$ be one of its worlds. Moreover, let φ, φ_1, and φ_2 be state formulas and ψ be a path formula. Then, the following holds:*

(1) φ is an $\cong^{w}_{\mathcal{K}}$-tautology iff $\mathcal{K}, w \models \varphi$;
(2) $\mathsf{X}\psi$ cannot be an $\cong^{w}_{\mathcal{K}}$-tautology;
(3) $\tilde{\mathsf{X}}\psi$ is an $\cong^{w}_{\mathcal{K}}$-tautology iff ψ is an $\cong^{w'}_{\mathcal{K}}$-tautology, for all $w' \in \mathrm{W}$ such that $(w, w') \in R$;
(4) $\varphi_1 \mathsf{U} \varphi_2$ is an $\cong^{w}_{\mathcal{K}}$-tautology iff $\mathcal{K}, w \models \varphi_2$;
(5) $\varphi_1 \mathsf{R} \varphi_2$ is an $\cong^{w}_{\mathcal{K}}$-tautology iff $\mathcal{K}, w \models \varphi_1 \wedge \varphi_2$;
(6) $\varphi_1 \tilde{\mathsf{U}} \varphi_2$ is an $\cong^{w}_{\mathcal{K}}$-tautology iff $\mathcal{K}, w \models \mathsf{A}^{<1} \varphi_1 \tilde{\mathsf{U}} \varphi_2$;
(7) $\varphi_1 \tilde{\mathsf{R}} \varphi_2$ is an $\cong^{w}_{\mathcal{K}}$-tautology iff $\mathcal{K}, w \models \mathsf{A}^{<1} \varphi_1 \tilde{\mathsf{R}} \varphi_2$.

Proof We prove the statements case by case. In particular, note that we implicitly make use of properties of Proposition 1.1. Moreover, for Items 6 and 7, we only prove the (if) direction, since the converse is immediate by the definition of $\cong^{w}_{\mathcal{K}}$-equivalence.

(1) The thesis directly derives from the definition of $\cong^{w}_{\mathcal{K}}$-tautology.
(2) The formula $\mathsf{X}\psi$ cannot be an $\cong^{w}_{\mathcal{K}}$-tautology, since $w \in \mathrm{Trk}(\mathcal{K}, w)$ and $\mathcal{K}, w \not\models \mathsf{X}\psi$, where we remind that w for the path formula satisfiability relation $\models$ is considered as the track built only by the world w itself.
(3) *[Only if]*. If $\tilde{\mathsf{X}}\psi$ is an $\cong^{w}_{\mathcal{K}}$-tautology, then $\mathcal{K}, \pi \models \tilde{\mathsf{X}}\psi$, for all $\pi \in (\mathrm{Pth}(\mathcal{K}, w) \cup \mathrm{Trk}(\mathcal{K}, w))$. Hence, we have that $\mathcal{K}, \pi_{\geq 1} \models \psi$, for all $\pi \in (\mathrm{Pth}(\mathcal{K}, w) \cup \mathrm{Trk}(\mathcal{K}, w))$ with $\pi_{\geq 1} \neq \varepsilon$, i.e., $\pi \neq w$, which implies that $\mathcal{K}, \pi \models \psi$, for all $\pi \in (\mathrm{Pth}(\mathcal{K}, w') \cup \mathrm{Trk}(\mathcal{K}, w'))$ with $(w, w') \in R$. Hence, the thesis follows. *[If]*. The converse direction is perfectly specular to the previous one.
(4) *[Only if]*. If $\varphi_1 \mathsf{U} \varphi_2$ is an $\cong^{w}_{\mathcal{K}}$-tautology, so is $\varphi_2 \vee \varphi_1 \wedge \mathsf{X} \varphi_1 \mathsf{U} \varphi_2$. Now, since $w \in \mathrm{Trk}(\mathcal{K}, w)$, we have that $\mathcal{K}, w \models \varphi_2 \vee \varphi_1 \wedge \mathsf{X} \varphi_1 \mathsf{U} \varphi_2$ and so, $\mathcal{K}, w \models \varphi_2$, since $\mathcal{K}, w \not\models \mathsf{X} \varphi_1 \mathsf{U} \varphi_2$. *[If]*. If $\mathcal{K}, w \models \varphi_2$, then $\mathcal{K}, \pi \models \varphi_2 \vee \varphi_1 \wedge \mathsf{X} \varphi_1 \mathsf{U} \varphi_2$ and so $\mathcal{K}, \pi \models \varphi_1 \mathsf{U} \varphi_2$, for all $\pi \in (\mathrm{Pth}(\mathcal{K}, w) \cup \mathrm{Trk}(\mathcal{K}, w))$. Hence, $\varphi_1 \mathsf{U} \varphi_2$ is an $\cong^{w}_{\mathcal{K}}$-tautology.
(5) *[Only if]*. If $\varphi_1 \mathsf{R} \varphi_2$ is an $\cong^{w}_{\mathcal{K}}$-tautology, so is $\varphi_2 \wedge (\varphi_1 \vee \mathsf{X} \varphi_1 \mathsf{R} \varphi_2)$. Now, since $w \in \mathrm{Trk}(\mathcal{K}, w)$, we have that $\mathcal{K}, w \models \varphi_2 \wedge (\varphi_1 \vee \mathsf{X} \varphi_1 \mathsf{R} \varphi_2)$ and so, $\mathcal{K}, w \models$

$\varphi_1 \wedge \varphi_2$, since $\mathscr{K}, w \not\models \mathsf{X}\varphi_1\mathsf{R}\varphi_2$. *[If]*. If $\mathscr{K}, w \models \varphi_1 \wedge \varphi_2$, then $\mathscr{K}, \pi \models \varphi_2 \wedge (\varphi_1 \vee \mathsf{X}\varphi_1\mathsf{R}\varphi_2)$ and so $\mathscr{K}, \pi \models \varphi_1\mathsf{R}\varphi_2$, for all $\pi \in (\mathrm{Pth}(\mathscr{K}, w) \cup \mathrm{Trk}(\mathscr{K}, w))$. Hence, $\varphi_1\mathsf{R}\varphi_2$ is an $\cong^w_{\mathscr{K}}$-tautology.

(6) By the hypothesis, we have that $\mathscr{K}, \pi \models \varphi_1\tilde{\mathsf{U}}\varphi_2$, for all $\pi \in \mathrm{Pth}(\mathscr{K}, w)$. Now, suppose by contradiction that $\varphi_1\tilde{\mathsf{U}}\varphi_2$ is not an $\cong^w_{\mathscr{K}}$-tautology, i.e., that there is a track $\rho \in \mathrm{Trk}(\mathscr{K}, w)$ such that $\mathscr{K}, \rho \not\models \varphi_1\tilde{\mathsf{U}}\varphi_2$. Then, we have that $\mathscr{K}, \rho \models (\neg\varphi_1)\mathsf{R}(\neg\varphi_2)$ and so $\mathscr{K}, \rho \models (\neg\varphi_2)\mathsf{U}(\neg\varphi_1 \wedge \neg\varphi_2)$, since ρ is necessarily finite. Now, consider a path $\pi \in \mathrm{Pth}(\mathscr{K}, w)$ having ρ as prefix, i.e., such that $\pi_{\leq(|\rho|-1)} = \rho$. Then, it is evident that $\mathscr{K}, \pi \models (\neg\varphi_2)\mathsf{U}(\neg\varphi_1 \wedge \neg\varphi_2)$ and this implies that $\mathscr{K}, \pi \not\models \varphi_1\tilde{\mathsf{U}}\varphi_2$, since there is no prefix in π satisfying φ_1 in all its positions before to reach a point in which φ_2 holds. Hence, we reached the contradiction.

(7) By the hypothesis, we have that $\mathscr{K}, \pi \models \varphi_1\tilde{\mathsf{R}}\varphi_2$, for all $\pi \in \mathrm{Pth}(\mathscr{K}, w)$. Now, suppose by contradiction that $\varphi_1\tilde{\mathsf{R}}\varphi_2$ is not an $\cong^w_{\mathscr{K}}$-tautology, i.e., that there exists a track $\rho \in \mathrm{Trk}(\mathscr{K}, w)$ such that $\mathscr{K}, \rho \not\models \varphi_1\tilde{\mathsf{R}}\varphi_2$. Then, we have that $\mathscr{K}, \rho \models (\neg\varphi_1)\mathsf{U}(\neg\varphi_2)$. Now, consider a path $\pi \in \mathrm{Pth}(\mathscr{K}, w)$ having ρ as prefix, i.e., such that $\pi_{\leq(|\rho|-1)} = \rho$. Then, it is evident that $\mathscr{K}, \pi \models (\neg\varphi_1)\mathsf{U}(\neg\varphi_2)$ and this implies that $\mathscr{K}, \pi \not\models \varphi_1\tilde{\mathsf{R}}\varphi_2$. Hence, we reached the contradiction. □

We now deduce two simple corollaries.

Corollary 1.2 (GCTL **Next Equivalences**) *Let* $\equiv_{:}$ *be the prefix path equivalence. Moreover, let* φ *be a state formula and* $g \in [1, \omega]$. *Then, it holds that* $\mathsf{E}^{\geq g}\tilde{\mathsf{X}}\varphi \equiv \mathsf{E}\tilde{\mathsf{X}}(g, \varphi, \mathsf{E}^{\geq 1}\mathsf{X}\neg\varphi)$ *and* $\mathsf{A}^{<g}\mathsf{X}\varphi \equiv \mathsf{AX}(g, \varphi, \mathsf{A}^{<1}\tilde{\mathsf{X}}\neg\varphi)$.

Proof By Theorem 1.3, $\equiv_{:}$ is adequate. Now, the thesis can be derived by Theorem 1.3 and Items 1 and 3 of Theorem 1.1. □

In the rest of the paper, we only consider formulas not containing any sub formula of the form $\mathsf{E}^{\geq g}\tilde{\mathsf{X}}\varphi$ with $\varphi \neq \mathfrak{f}$ and $\mathsf{A}^{<g}\mathsf{X}\varphi$ with $\varphi \neq \mathfrak{t}$. This can be done w.l.o.g. since each formula can be converted, with a linear blow-up only, into another one without the above quantifications, by using the equivalence of the previous corollary.

Corollary 1.3 (GCTL **Fixpoint Equivalences**) *Let* $\equiv_{:}$ *be the prefix path equivalence. Moreover, let* φ_1 *and* φ_2 *be two state formulas and* $g \in [1, \omega]$. *Then, the following holds:*

(1) $\mathsf{E}^{\geq g}\varphi_1\mathsf{U}\varphi_2 \equiv \mathsf{EU}(g, \varphi_1, \varphi_2, \mathsf{E}^{\geq g}\varphi_1\mathsf{U}\varphi_2)$;
(2) $\mathsf{E}^{\geq g}\varphi_1\mathsf{R}\varphi_2 \equiv \mathsf{ER}(g, \varphi_1, \varphi_2, \mathsf{E}^{\geq g}\varphi_1\mathsf{R}\varphi_2)$;
(3) $\mathsf{E}^{\geq g}\varphi_1\tilde{\mathsf{U}}\varphi_2 \equiv \mathsf{E}\tilde{\mathsf{U}}(g, \varphi_1, \varphi_2, \mathsf{E}^{\geq g}\varphi_1\tilde{\mathsf{U}}\varphi_2, \mathsf{E}^{\geq 1}\mathsf{X}\mathsf{E}^{\geq 1}\neg(\varphi_1\tilde{\mathsf{U}}\varphi_2))$;
(4) $\mathsf{E}^{\geq g}\varphi_1\tilde{\mathsf{R}}\varphi_2 \equiv \mathsf{E}\tilde{\mathsf{R}}(g, \varphi_1, \varphi_2, \mathsf{E}^{\geq g}\varphi_1\tilde{\mathsf{R}}\varphi_2, \mathsf{E}^{\geq 1}\mathsf{X}\mathsf{E}^{\geq 1}\neg(\varphi_1\tilde{\mathsf{R}}\varphi_2))$;
(5) $\mathsf{A}^{<g}\varphi_1\mathsf{U}\varphi_2 \equiv \mathsf{AU}(g, \varphi_1, \varphi_2, \mathsf{A}^{<g}\varphi_1\mathsf{U}\varphi_2, \mathsf{A}^{<1}\tilde{\mathsf{X}}\mathsf{A}^{<1}\neg(\varphi_1\mathsf{U}\varphi_2))$;
(6) $\mathsf{A}^{<g}\varphi_1\mathsf{R}\varphi_2 \equiv \mathsf{AR}(g, \varphi_1, \varphi_2, \mathsf{A}^{<g}\varphi_1\mathsf{R}\varphi_2, \mathsf{A}^{<1}\tilde{\mathsf{X}}\mathsf{A}^{<1}\neg(\varphi_1\mathsf{R}\varphi_2))$;
(7) $\mathsf{A}^{<g}\varphi_1\tilde{\mathsf{U}}\varphi_2 \equiv \mathsf{A}\tilde{\mathsf{U}}(g, \varphi_1, \varphi_2, \mathsf{A}^{<g}\varphi_1\tilde{\mathsf{U}}\varphi_2)$;
(8) $\mathsf{A}^{<g}\varphi_1\tilde{\mathsf{R}}\varphi_2 \equiv \mathsf{A}\tilde{\mathsf{R}}(g, \varphi_1, \varphi_2, \mathsf{A}^{<g}\varphi_1\tilde{\mathsf{R}}\varphi_2)$.

Proof By Theorem 1.12, $\equiv_{:}$ is adequate. Now, Items 1, 2, 7, and 8 follow directly by Theorem 1.9, while Items 3, 4, 5, and 6 can be derived by Theorem 1.10 and Items 3, 6, and 7 of Theorem 1.13. □

We now conclude this part of the section by showing two simple but fundamental properties of GCTL* that allow the application of the automata-theoretic approach to the solution of the satisfiability problem for GCTL.

By using a proof by induction, we prove that GCTL* is invariant under the unwinding of a model.

Theorem 1.14 **(GCTL* Unwinding Invariance)** *Let $\equiv_{:}$ be the prefix path equivalence. Then,* GCTL* *is invariant w.r.t. unwinding, i.e., $\mathcal{K} \models \varphi$ iff $\mathcal{K}_U \models \varphi$, for all state formulas φ.*

Proof Let $\mathcal{K} = \langle \mathrm{AP}, \mathrm{W}, R, \mathsf{L}, w_0 \rangle$ be a Ks and $\mathcal{K}_U = \langle \mathrm{AP}, \mathrm{W}', R', \mathsf{L}', \varepsilon \rangle$ be its unwinding. Then, we show that for each GCTL* state formula φ and world $w \in \mathrm{W}'$, it holds that $\mathcal{K}, \mathsf{unw}(w) \models \varphi$ iff $\mathcal{K}_U, w \models \varphi$, where $\mathsf{unw} : \mathrm{W}' \to \mathrm{W}$ is the unwinding function. As a side result, we also prove that $\mathcal{K}, \mathsf{unw}(\pi) \models \psi$ iff $\mathcal{K}_U, \pi \models \psi$, for all GCTL* path formulas ψ and paths/tracks $\pi \in (\mathrm{Pth}(\mathcal{K}_U, w) \cup \mathrm{Trk}(\mathcal{K}_U, w))$, where, in this case, $\mathsf{unw} : (\mathrm{Pth}(\mathcal{K}_U) \cup \mathrm{Trk}(\mathcal{K}_U)) \to (\mathrm{Pth}(\mathcal{K}) \cup \mathrm{Trk}(\mathcal{K}))$ is bijective function that extends the unwinding function on worlds to paths and tracks, i.e., $(\mathsf{unw}(\pi))_i = \mathsf{unw}((\pi)_i)$, for all $i \in [0, |\pi|[$.

The proof proceeds by induction on the structure of the formula φ. The basic case of atomic propositions and the inductive cases of Boolean combinations are immediate and left to the reader. Therefore, let us consider the inductive case where φ is an existential quantification of the form $\mathsf{E}^{\geq g}\psi$, with $g \in [1, \omega]$. The case of universal quantifications $\mathsf{A}^{<g}\psi$ can be treated similarly.

First observe that, by the inductive hypothesis, it holds that $\mathcal{K}, \mathsf{unw}(w) \models \varphi$ iff $\mathcal{K}_U, w \models \varphi$, for all $\varphi \in \mathsf{rcl}(\psi)$ and $w \in \mathrm{W}'$. Now, it is immediate to see that $\mathcal{K}, \mathsf{unw}(\pi) \models \psi$ iff $\mathcal{K}_U, \pi \models \psi$, for all paths $\pi \in (\mathrm{Pth}(\mathcal{K}_U, w) \cup \mathrm{Trk}(\mathcal{K}_U, w))$. Indeed, by the definition of semantics on paths, we have that $\mathcal{K}, \mathsf{unw}(\pi) \models \psi$ iff $\varpi_{\mathcal{K},\psi}(\mathsf{unw}(\pi)) \models \psi$ and $\mathcal{K}_U, \pi \models \psi$ iff $\varpi_{\mathcal{K}_U,\psi}(\pi) \models \psi$. Now, by the previous observation and the definition of the path transformation, we have that $\varpi_{\mathcal{K},\psi}(\mathsf{unw}(\pi)) = \varpi_{\mathcal{K}_U,\psi}(\pi)$. Consequently, it holds that $\mathsf{unw}(\pi) \in \mathrm{Pth}(\mathcal{K}, \mathsf{unw}(w), \psi)$ iff $\pi \in \mathrm{Pth}(\mathcal{K}_U, w, \psi)$, for all $\pi \in \mathrm{Pth}(\mathcal{K}_U, w)$.

At this point, in order to prove that $|(\mathrm{Pth}(\mathcal{K}, \mathsf{unw}(w), \psi)/\equiv^{\psi}_{\mathcal{K}})| \geq g$ iff $|(\mathrm{Pth}(\mathcal{K}_U, w, \psi)/\equiv^{\psi}_{\mathcal{K}_U})| \geq g$, it remains to shows that $\pi_1 \equiv^{\psi}_{\mathcal{K}_U} \pi_2$ iff $\mathsf{unw}(\pi_1) \equiv^{\psi}_{\mathcal{K}} \mathsf{unw}(\pi_2)$. The case $\pi_1 = \pi_2$ is trivial. Thus, consider the case $\pi_1 \neq \pi_2$, let $\rho = \mathsf{pfx}(\pi_1, \pi_2)$ be their common prefix, and observe that $\mathsf{unw}(\rho) = \mathsf{pfx}(\mathsf{unw}(\pi_1), \mathsf{unw}(\pi_2))$. Now, by definition of prefix path equivalence, we have that $\pi_1 \equiv^{\psi}_{\mathcal{K}_U} \pi_2$ iff $\rho \neq \varepsilon$ and $\mathcal{K}_U, \rho \cdot \pi_{\geq 1} \models \psi$, for all $\pi \in (\mathrm{Pth}(\mathcal{K}_U, \mathsf{lst}(\rho)) \cup \mathrm{Trk}(\mathcal{K}_U, \mathsf{lst}(\rho)))$, and $\mathsf{unw}(\pi_1) \equiv^{\psi}_{\mathcal{K}} \mathsf{unw}(\pi_2)$ iff $\mathsf{unw}(\rho) \neq \varepsilon$ and $\mathcal{K}, \mathsf{unw}(\rho) \cdot \pi'_{\geq 1} \models \psi$, for all $\pi' \in (\mathrm{Pth}(\mathcal{K}, \mathsf{lst}(\mathsf{unw}(\rho))) \cup \mathrm{Trk}(\mathcal{K}, \mathsf{lst}(\mathsf{unw}(\rho))))$. Now, using again the fact that $\mathcal{K}, \mathsf{unw}(\pi) \models \psi$ iff $\mathcal{K}_U, \pi \models \psi$, for all paths/tracks $\pi \in (\mathrm{Pth}(\mathcal{K}_U, w) \cup \mathrm{Trk}(\mathcal{K}_U, w))$, the thesis follows. □

Directly from the previous result, we obtain that GCTL* also enjoys the tree model property.

Corollary 1.4 (GCTL* **Tree Model Property**) *Let* $\equiv_{:}$ *be the prefix path equivalence. Then,* GCTL* *has the* tree model property.

Proof Consider a formula φ and suppose that it is satisfiable. Then, there is a KS $\mathcal{K}$ such that $\mathcal{K} \models \varphi$. By Theorem 1.14, φ is satisfied at the root of the unwinding $\mathcal{K}_U$ of $\mathcal{K}$. Thus, since $\mathcal{K}_U$ is a KT, we immediately have that φ is satisfied on a tree model. □

1.5.2 GCTL *Versus* GμCALCULUS *relationships*

The μCALCULUS (Kozen 1983) is a well-known modal logic augmented with fixed point operators, which subsumes the classical temporal logics such as LTL, CTL, and CTL*. The GμCALCULUS simply extends the μCALCULUS with graded state quantifiers (Kupferman et al. 2002; Bonatti et al. 2008).

In the next theorem, we show a double-exponential reduction from the significant fragment of GCTL without infinite-degree quantifications to GμCALCULUS.

Theorem 1.15 (GCTL-GμCALCULUS **Reduction**) *For each* GCTL *formula* φ *free of the* $\mathsf{E}^{\geq\omega}$ *and* $\mathsf{A}^{<\omega}$ *quantifications, it is possible to construct an equisatisfiable formula* χ *of* GμCALCULUS *with* $\mathsf{siz}(\chi) = \mathrm{O}(2^{k\cdot\sqrt{\mathsf{deg}(\varphi)}\cdot\mathsf{lng}(\varphi)})$, *for a constant k, i.e.,* φ *is satisfiable iff* χ *is satisfiable.*

Proof The reduction we now propose is almost a translation by equivalence. The only basic formulas that cannot be directly translated are the quantifications $\mathsf{E}^{\geq 1}\tilde{\mathsf{X}}\mathfrak{f}$ and $\mathsf{A}^{<1}\mathsf{X}\mathfrak{t}$ that are satisfied, respectively, only on worlds without and with successors. This is due to the fact that the μCALCULUS, and so the GμCALCULUS, is usually defined only on total KS, and $\mathsf{E}^{\geq 1}\tilde{\mathsf{X}}\mathfrak{f}$ and $\mathsf{A}^{<1}\mathsf{X}\mathfrak{t}$ are equivalent to $\mathfrak{f}$ and $\mathfrak{t}$, respectively, on such a kind of structures. To overcome this gap, we enrich each KS with a fresh atomic proposition *end*, representing the fact that a world has no successors, and translate $\mathsf{E}^{\geq 1}\tilde{\mathsf{X}}\mathfrak{f}$ in *end* and $\mathsf{A}^{<1}\mathsf{X}\mathfrak{t}$ in $\neg$*end*. Moreover, we force the translation of *(i)* $\mathsf{E}^{\geq g}\mathsf{X}\varphi$ to ensure that it is satisfied only on worlds not labeled with *end* and *(ii)* $\mathsf{A}^{<g}\tilde{\mathsf{X}}\varphi$ to allow that it is satisfied also on worlds labeled with *end*, where $g \in [1, \omega[$. Apart from the cases of the atomic propositions, the Boolean connectives, and the quantifiers $\mathsf{E}^{\geq 0}\psi$ and $\mathsf{A}^{<0}\psi$ that are equivalent to $\mathfrak{t}$ and $\mathfrak{f}$, respectively, the remaining case are solved using the equivalence showed in Corollary 1.3.

Formally, the translation $\chi = \overline{\varphi}$ of φ is inductively defined as follows, where $g \in [1, \omega[$:

(1) $\overline{p} \triangleq p$, for $p \in \mathrm{AP}$;
(2) $\overline{\neg\varphi} \triangleq \neg\,\overline{\varphi}$; $\overline{\varphi_1 \wedge \varphi_2} \triangleq \overline{\varphi_1} \wedge \overline{\varphi_2}$; $\overline{\varphi_1 \vee \varphi_2} \triangleq \overline{\varphi_1} \vee \overline{\varphi_2}$;
(3) $\overline{\mathsf{E}^{\geq 0}\psi} \triangleq \mathfrak{t}$; $\overline{\mathsf{A}^{<0}\psi} \triangleq \mathfrak{f}$;

(4) $\overline{\mathsf{E}^{\geq 1}\tilde{\mathsf{X}}\mathsf{f}} \triangleq end;\quad \overline{\mathsf{A}^{<1}\mathsf{X}\mathsf{t}} \triangleq \neg end;$
(5) $\overline{\mathsf{E}^{\geq g}\mathsf{X}\varphi} \triangleq \neg\, end \wedge \langle g-1\rangle\, \overline{\varphi};\quad \overline{\mathsf{A}^{<g}\tilde{\mathsf{X}}\varphi} \triangleq end \vee [g-1]\, \overline{\varphi};$
(6) $\overline{\mathsf{E}^{\geq g}(\varphi_1 \mathsf{U} \varphi_2)} \triangleq \mu Y.\overline{\mathsf{EU}(g, \varphi_1, \varphi_2, Y)};$
(7) $\overline{\mathsf{E}^{\geq g}(\varphi_1 \mathsf{R} \varphi_2)} \triangleq \nu Y.\overline{\mathsf{ER}(g, \varphi_1, \varphi_2, Y)};$
(8) $\overline{\mathsf{E}^{\geq g}(\varphi_1 \tilde{\mathsf{U}} \varphi_2)} \triangleq \mu Y.\overline{\mathsf{E}\tilde{\mathsf{U}}(g, \varphi_1, \varphi_2, Y, \mathsf{E}^{\geq 1}\mathsf{X}\mathsf{E}^{\geq 1}\neg(\varphi_1 \tilde{\mathsf{U}} \varphi_2))};$
(9) $\overline{\mathsf{E}^{\geq g}(\varphi_1 \tilde{\mathsf{R}} \varphi_2)} \triangleq \nu Y.\overline{\mathsf{E}\tilde{\mathsf{R}}(g, \varphi_1, \varphi_2, Y, \mathsf{E}^{\geq 1}\mathsf{X}\mathsf{E}^{\geq 1}\neg(\varphi_1 \tilde{\mathsf{R}} \varphi_2))};$
(10) $\overline{\mathsf{A}^{<g}(\varphi_1 \mathsf{U} \varphi_2)} \triangleq \mu Y.\overline{\mathsf{AU}(g, \varphi_1, \varphi_2, Y, \mathsf{A}^{<1}\tilde{\mathsf{X}}\mathsf{A}^{<1}\neg(\varphi_1 \mathsf{U} \varphi_2))};$
(11) $\overline{\mathsf{A}^{<g}(\varphi_1 \mathsf{R} \varphi_2)} \triangleq \nu Y.\overline{\mathsf{AR}(g, \varphi_1, \varphi_2, Y, \mathsf{A}^{<1}\tilde{\mathsf{X}}\mathsf{A}^{<1}\neg(\varphi_1 \mathsf{R} \varphi_2))};$
(12) $\overline{\mathsf{A}^{<g}(\varphi_1 \tilde{\mathsf{U}} \varphi_2)} \triangleq \mu Y.\overline{\mathsf{A}\tilde{\mathsf{U}}(g, \varphi_1, \varphi_2, Y)};$
(13) $\overline{\mathsf{A}^{<g}(\varphi_1 \tilde{\mathsf{R}} \varphi_2)} \triangleq \nu Y.\overline{\mathsf{A}\tilde{\mathsf{R}}(g, \varphi_1, \varphi_2, Y)}.$

By induction on the structure of the formula, it is not hard to see that, for each Ks $\mathcal{K} = \langle \mathrm{AP}, \mathrm{W}, R, \mathsf{L}, w_0\rangle$ model of φ, the Ks $\mathcal{K}' = \langle \mathrm{AP} \cup \{end\}, \mathrm{W}, R', \mathsf{L}', w_0\rangle$ is a model of $\overline{\varphi}$, where *(i)* $R' \cap (\mathrm{W} \setminus \mathrm{W}') \times \mathrm{W} = R$, *(ii)* $\mathsf{L}'(w) = \mathsf{L}(w)$, *(iii)* $\mathsf{L}'(w') = \mathsf{L}(w') \cup \{end\}$, and *(iv)* $(w', w') \in R'$, for all $w \in \mathrm{W} \setminus \mathrm{W}'$ and $w' \in \mathrm{W}'$, with $\mathrm{W}' = \{w \in \mathrm{W} : \nexists w' \in \mathrm{W}.(w, w') \in R\}$. Intuitively, we simply add to each world having no successors a self loop and the label *end*. Moreover, from a Ks $\mathcal{K} = \langle \mathrm{AP}, \mathrm{W}, R, \mathsf{L}, w_0\rangle$ model of $\overline{\varphi}$, it is possible to extract a Ks $\mathcal{K}' = \langle \mathrm{AP}, \mathrm{W}, R', \mathsf{L}, w_0\rangle$ model of φ, by simply substituting the transition relation R with a new relation R' defined as follows: $(w, w') \in R'$ iff $(w, w') \in R$ and $end \notin \mathsf{L}(w)$, for all $w, w' \in \mathrm{W}$. Intuitively, we simply cut out each edge exiting from a world labeled with *end*.

Finally, we turn to the size of χ. First, note that all points 1–5 are linear. Instead, points 6–13 are exponential in the degree of the original formula, and so, double exponential in its size, since they are based on the expansion formulas of Theorems 1.9 and 1.10. With more details, each of these transformations give a blow-up that is an $\mathsf{O}((\mathsf{lng}(\overline{\varphi_1}) + \mathsf{lng}(\overline{\varphi_2})) \cdot 2^{k \cdot \sqrt{g}})$. Now, by a simple calculation, since the nesting of such a kind of formulas is bounded by the length of φ, we obtain that $\mathsf{siz}(\chi) = \mathsf{O}(2^{k \cdot \sqrt{\mathsf{deg}(\varphi)} \cdot \mathsf{lng}(\varphi)})$. It is important to remark that the number of disjunctions in χ can be exponential in the degree $\mathsf{deg}(\varphi)$. Therefore, even using a DAG to represent χ, it would only reduce the overall size to $\mathsf{O}(\mathsf{lng}(\varphi) \cdot 2^{k \cdot \sqrt{\mathsf{deg}(\varphi)}})$. Hence, it remains double exponential in $\mathsf{siz}(\varphi)$. □

By the previous theorem and the fact that for GμCALCULUS the satisfiability problem is solvable in EXPTIME (Kupferman et al. 2002), we immediately get that the problem for the above fragment of GCTL is decidable and solvable in 3EXPTIME. However, in the next chapters we improve this result by showing that the problem for the whole GCTL is solvable in EXPTIME, by exploiting an automata-theoretic approach.

Finally, we show that GCTL is at least exponentially more succinct than GμCALCULUS, both with the binary coding of the degree. We prove the statement by showing a class of GCTL formulas φ_g, with $g \in [1, \omega[$, whose minimal equivalent GμCALCULUS formulas χ_g needs to be, in size, exponentially bigger than (the size of) φ_g. Classical techniques (Lange 2008; Lutz 2006; Wilke 1999) rely on the fact that in the more succinct logic there exists a formula having a *least finite model* whose size is double exponential in the size of the formula, while in the less succinct logic every satisfiable formula has finite models of size at most exponential in its size. Unfortunately, in our case we cannot apply this idea, since, as far as we know, both GCTL and the GμCALCULUS satisfy the small model property, i.e., all their satisfiable formulas have always a model at most exponential in their size. Hence, to prove the succinctness of GCTL, we explore a technique based on a characteristic property of our logic. Specifically, it is based on the fact that, using GCTL, we can write a set of formulas φ_g each one having a number of "characterizing models" that is exponential in the degree g of φ_g, while every GμCALCULUS formula has at most a polynomial number of those models in its degree.

Consider the property "in a tree, there are exactly g grandchildren of the root having only one path leading from them, and these grandchildren are all and only the nodes labeled with p". Such a property can be easily described by the GCTL formula $\varphi_g = \varphi' \wedge \varphi''_g$, where $\varphi' = \neg p \wedge \mathsf{A}^{<1}\mathsf{X}(\neg p \wedge \mathsf{A}^{<1}\mathsf{X}(p \wedge \mathsf{A}^{<1}\mathsf{X}\mathsf{A}^{<1}\mathsf{G}(\neg p \wedge \mathsf{A}^{<2}\tilde{\mathsf{X}}\mathsf{f})))$ and $\varphi''_g = \mathsf{E}^{=g}\mathsf{F}p$. By simple a calculation, we can see that $\mathsf{lng}(\varphi_g) = 31$, $\mathsf{deg}(\varphi_g) = g$, and $\mathsf{siz}(\varphi_g) = 32 + \lceil \log(g) \rceil + \lceil \log(g+1) \rceil$. So, its size is $\Theta(\lceil \log(g) \rceil)$. We claim that a G$\mu$CALCULUS formula χ_g requires exponential size to express the same property. More formally, our aim is to prove the following theorem.

Theorem 1.16 (GCTL **Exponential Succinctness**) *Let* $\varphi_g = \varphi' \wedge \varphi''_g$, *with* $\varphi' = \neg p \wedge \mathsf{A}^{<1}\mathsf{X}(\neg p \wedge \mathsf{A}^{<1}\mathsf{X}(p \wedge \mathsf{A}^{<1}\mathsf{X}\mathsf{A}^{<1}\mathsf{G}(\neg p \wedge \mathsf{A}^{<2}\tilde{\mathsf{X}}\mathsf{f})))$, $\varphi''_g = \mathsf{E}^{=g}\mathsf{F}p$, *and* $g \in [1, \omega[$. *Then, each* GμCALCULUS *formula* χ_g *equivalent to* φ_g *has size* $\Omega(2^{\mathsf{siz}(\varphi_g)})$.

The proof of this theorem proceeds directly by proving the following lemma and observing that, since $\mathsf{siz}(\varphi_g) = \Theta(\lceil \log(g) \rceil)$, we can easily derive that $\mathsf{siz}(\chi_g) = \Omega(2^{\mathsf{siz}(\varphi_g)})$.

Lemma 1.4 (GμCALCULUS **Polynomial Degree Lower Bound**) *For all the* GμCALCULUS *formulas* χ_g *equivalent to* φ_g, *it holds that they have size* $\Omega(g)$.

Proof To prove this, we use an automata-theoretic approach. We first recall that the automata model developed in Kupferman et al. (2002), used to accept all and only the tree models of a GμCALCULUS formula χ, has as set of states the closure set of χ. On every accepting run, when the automaton is in a state q on a node x of the input tree, the subtree rooted at that node is a model of q. Our aim now is to prove that the automaton $\mathscr{A}_{\chi_g}$ for χ_g can accept all and only the models of χ_g, and so of φ_g, only if its state space contains either a formula $\langle i \rangle \phi$ or a formula $[i]\phi$, for all $i \in [0, g[$. Recall that the GμCALCULUS formulas $\langle i \rangle \phi$ and $[i]\phi$ mean that there are at least $i+1$ successor satisfying ϕ and all but at most i successors satisfy ϕ, respectively. Suppose by contradiction that there is no formula ϕ such that $\langle i \rangle \phi$ or $[i]\phi$ are in the state space of $\mathscr{A}_{\chi_g}$, for a given index i. Since $\mathscr{A}_{\chi_g}$ accepts all the models of φ_g, it accepts the input tree $\mathscr{T} = \langle \mathrm{T}, \mathsf{v} \rangle$, where $\mathrm{T} = \{\varepsilon\} \cup \{0, 1\} \cup \{0 \cdot 0 \cdot 0^*, \ldots, 0 \cdot (i-1) \cdot 0^*, 1 \cdot 0 \cdot 0^*, \ldots, 1 \cdot (g-i-1) \cdot 0^*\}$, every node x, with $|x| = 2$, i.e., of level equal to 2, is labeled with $\mathsf{v}(x) = \{p\}$, and every other node y is labeled with $\mathsf{v}(y) = \emptyset$. Informally, node 0 has i successors labeled with p, while node 1 has $g-i$ successors labeled in the same way. Now, on the accepting run $\mathscr{R}$ of $\mathscr{A}_{\chi_g}$ on $\mathscr{T}$ in the node 0, the active states represent what are needed to be satisfied in the current node and such requirements do no contain any existential $\langle i \rangle \phi$ or universal $[i]\phi$. Hence, if we substitute $\mathcal{T}$ with a new tree $\mathcal{T}'$ having only $i - 1$ successor of 0 (labeled with p), then we obtain that also $\mathcal{T}'$ is accepted, reaching in this way the contradiction. This is due to the fact that, we can easily modify the run $\mathcal{R}$ to construct an accepting run $\mathcal{R}'$ for $\mathcal{T}'$, by removing all its subtrees rooted at a node whose label contains the node $0 \cdot l$, with $l \in [0, g[$, not in $\mathcal{T}'$. Indeed, when $\mathcal{A}_{\chi_g}$ is on the node 0, every non-quantified formula is already satisfied. A formula $\langle j \rangle \phi$ with $j > i$ could not be required on $\mathcal{T}$, and so on $\mathcal{T}'$, since it would be trivially false anyway. A formula $[j]\phi$ with $j > i$ is trivially true on both the trees. Finally, formulas $\langle j \rangle \phi$ or $[j]\phi$, with $j < i$, are satisfied on $\mathcal{T}$ by hypothesis. Now, since the subtrees rooted at the successor nodes of 0 are all equal, they all satisfy ϕ. Thus, by removing one of them, the quantifier formula is still satisfied. This reasoning shows that the closure of χ_g contains at least an existential or universal formula for each degree $i \in [0, g[$. Hence, the formula χ_g must have at least size $\Omega(g)$. □

Note that, as far as we know, the size of the smallest GμCALCULUS formula χ equivalent to φ has size double exponential in the binary coding of the degree g. In particular, χ can be obtained by using the translation $\overline{\varphi}$ described in Theorem 1.15. So, there is an exponential gap between upper and lower bound for the translation from GCTL to GμCALCULUS. Actually, we conjecture that the succinctness is tight for double exponential, but the technique used in the previous lemma does not seem to be adaptable for a double exponential lower bound.

1.6 Alternating Tree Automata

In this section, we briefly introduce an automaton model used to solve efficiently the satisfiability problems for GCTL in EXPTIME w.r.t. the size of the formula, by reducing this problem to the emptiness of the automaton. We recall that, in general, an approach with tree automata to the solution of the satisfiability problem is only possible once the logic satisfies the tree model property. In fact, this property holds for GCTL*, and consequently for GCTL, as we have proved in Corollary 1.4.

1.6.1 Classic Automata

Nondeterministic tree automata are a generalization to infinite trees of the classical *nondeterministic word automata* (see Thomas 1990, for an introduction). *Alternating tree automata* are a further generalization of nondeterministic tree automata (Muller and Schupp 1987). Intuitively, on visiting a node of the input tree, while the latter sends exactly one copy of itself to each of the successors of the node, the first can send several copies of itself to the same successor.

We now give the formal definition of alternating tree automata.

Definition 1.17 (Alternating Tree Automata) *An* alternating tree automaton *(*ATA*, for short) is a tuple* $\mathscr{A} \triangleq \langle \Sigma, \Delta, \mathrm{Q}, \delta, q_0, \mathrm{F} \rangle$*, where* Σ*,* Δ*, and* Q *are non-empty finite sets of* input symbols, directions, *and* states, *respectively,* $q_0 \in \mathrm{Q}$ *is an* initial state, F *is an* acceptance condition *to be defined later, and* $\delta : \mathrm{Q} \times \Sigma \rightarrow \mathrm{B}^+(\Delta \times \mathrm{Q})$ *is an* alternating transition function *that maps each pair of states and input symbols to a positive Boolean combination on the set of propositions of the form* $(d, q) \in \Delta \times \mathrm{Q}$*, a.k.a.* moves.

A *nondeterministic tree automaton* (NTA, for short) is a special ATA in which each conjunction in the transition function δ has exactly one move (d, q) associated with each direction d. In addition, a *universal tree automaton* (UTA, for short) is a special ATA in which all the Boolean combinations that appear in δ are only conjunctions of moves.

The semantics of the ATAs is now given through the following concept of run.

Definition 1.18 (ATA **Run)** *A* run *of an* ATA $\mathscr{A} = \langle \Sigma, \Delta, \mathrm{Q}, \delta, q_0, \mathrm{F} \rangle$ *on a* Σ*-labeled* Δ*-tree* $\mathscr{T} = \langle \mathrm{T}, \mathsf{v} \rangle$ *is a* $(\mathrm{Q} \times \mathrm{T})$*-labeled* $\mathbb{N}$*-tree* $\mathscr{R} \triangleq \langle \mathrm{R}, \mathsf{r} \rangle$ *such that* (i) $\mathsf{r}(\varepsilon) = (q_0, \varepsilon)$ *and* (ii) *for all nodes* $y \in \mathrm{R}$ *with* $\mathsf{r}(y) = (q, x)$*, there is a set of moves* $\mathrm{S} \subseteq \Delta \times \mathrm{Q}$ *with* $\mathrm{S} \models \delta(q, \mathsf{v}(x))$ *such that, for all* $(d, q') \in \mathrm{S}$*, there is an index* $j \in [0, |\mathrm{S}|[$ *for which it holds that* $y \cdot j \in \mathrm{R}$ *and* $\mathsf{r}(y \cdot j) = (q', x \cdot d)$.

In the following, we only consider ATAs along with the *parity* acceptance condition (APT, for short) $\mathrm{F} = (\mathrm{F}_1, \ldots, \mathrm{F}_k) \in (2^{\mathrm{Q}})^+$ with $\mathrm{F}_1 \subseteq \cdots \subseteq \mathrm{F}_k = \mathrm{Q}$ (see Kupferman et al. 2000, for more). The number k of sets in F is called the *index* of the automaton.

Let $\mathscr{R} = \langle \mathrm{R}, \mathsf{r} \rangle$ be a run of an ATA $\mathscr{A}$ on a tree $\mathscr{T} = \langle \mathrm{T}, \mathsf{v} \rangle$ and $\mathrm{R}' \subseteq \mathrm{R}$ one of its branches. Then, by $\mathsf{inf}(\mathrm{R}') \triangleq \{q \in \mathrm{Q} : |\{y \in \mathrm{R}' : \exists x \in \mathrm{T}. \mathsf{r}(y) = (q, x)\}| = \omega\}$ we denote the set of states that occur infinitely often as labeling of the nodes in the branch R'. We say that a branch R' of $\mathscr{T}$ satisfies the parity acceptance condition $\mathrm{F} = (\mathrm{F}_1, \ldots, \mathrm{F}_k)$ iff the least index $i \in [1, k]$ for which $\mathsf{inf}(\mathrm{R}') \cap \mathrm{F}_i \neq \emptyset$ is even.

At this point, we can define the concept of language accepted by an ATA.

Definition 1.19 (ATA **Acceptance)** *An* ATA $\mathscr{A} = \langle \Sigma, \Delta, \mathrm{Q}, \delta, q_0, \mathrm{F} \rangle$ accepts *a* Σ*-labeled* Δ*-tree* $\mathscr{T}$ *iff there exists a run* $\mathscr{R}$ *of* $\mathscr{A}$ *on* $\mathscr{T}$ *such that all its infinite branches satisfy the acceptance condition* F*, where the concept of satisfaction is dependent from the definition of* F.

By $\mathrm{L}(\mathscr{A})$ we denote the language accepted by the ATA $\mathscr{A}$, i.e., the set of trees $\mathscr{T}$ accepted by $\mathscr{A}$. Moreover, $\mathscr{A}$ is said to be *empty* if $\mathrm{L}(\mathscr{A}) = \emptyset$. The *emptiness problem* for $\mathscr{A}$ is to decide whether $\mathrm{L}(\mathscr{A}) = \emptyset$ or not.

1.6.2 Automata with Satellite

As a generalization of ATA, here we consider *alternating tree automata with satellites* (ATAS, for short), in a similar way it has been done in Kupferman and Vardi (2006), with the main difference that our satellites are nondeterministic and can work on trees and not only on words. The satellite is used to ensure that the input tree satisfies some structural properties and it is kept apart from the main automaton as it allows to show a tight complexity for the satisfiability problems.

We now formally define this new fundamental concept of automaton.

Definition 1.20 (Alternating Tree Automata with Satellite) *An* alternating tree automaton with satellite *(*ATAS, *for short) is a tuple* $\langle \mathscr{A}, \mathscr{S} \rangle$*, where* $\mathscr{A} \triangleq \langle \Sigma \times \mathrm{P}_E, \Delta, \mathrm{Q}, \delta, q_0, \mathrm{F} \rangle$ *is an* ATA *and* $\mathscr{S} \triangleq \langle \Sigma, \Delta, \mathrm{P}, \zeta, \mathrm{P}_0 \rangle$ *is a* nondeterministic safety automaton, *a.k.a.* satellite, *where* $\mathrm{P} = \mathrm{P}_E \times \mathrm{P}_I$ *is a non-empty finite set of* states *split in two components,* external P_E *and* internal P_I *states,* $\mathrm{P}_0 \subseteq \mathrm{P}$ *is a set of* initial states, *and* $\zeta : \mathrm{P} \times \Sigma \rightarrow 2^{\mathrm{P}^\Delta}$ *is a* nondeterministic transition function *that maps a state and an input symbol to a set of functions from directions to states. The set* Σ *is the* alphabet *of the* ATAS $\langle \mathscr{A}, \mathscr{S} \rangle$.

The semantics of satellites is given through the following concepts of run, acceptance, and building. It is possible to note a similarity with the concept of cascade product automata that can be found in literature.

Definition 1.21 (Satellite Run) *A* run *of a satellite* $\mathscr{S} = \langle \Sigma, \Delta, \mathrm{P}, \zeta, \mathrm{P}_0 \rangle$ *on a* Σ*-labeled* Δ*-tree* $\mathscr{T} = \langle \mathrm{T}, \mathsf{v} \rangle$ *is a* P*-labeled* Δ*-tree* $\mathscr{R} \triangleq \langle \mathrm{T}, \mathsf{r} \rangle$ *such that* (i) $\mathsf{r}(\varepsilon) \in \mathrm{P}_0$ *and* (ii) *for all nodes* $x \in \mathrm{T}$ *with* $\mathsf{r}(x) = p$*, there is a function* $g \in \zeta(p, \mathsf{v}(x))$ *such that, for all* $d \in \Delta$ *with* $x \cdot d \in \mathrm{T}$*, it holds that* $\mathsf{r}(x \cdot d) = g(d)$.

Definition 1.22 (Satellite Acceptance) *A satellite* $\mathscr{S} = \langle \Sigma, \Delta, \mathrm{P}, \zeta, \mathrm{P}_0 \rangle$ accepts *a* Σ*-labeled* Δ*-tree* $\mathscr{T}$ *iff there exists a run* $\mathscr{R}$ *of* $\mathscr{S}$ *on* $\mathscr{T}$.

For the coming definition we have to introduce an extra notation. Given a $(\Sigma' \times \Sigma'')$-labeled Δ-tree $\mathscr{T} = \langle \mathrm{T}, \mathsf{v} \rangle$, we define the *projection* of $\mathscr{T}$ on Σ' as the Σ'-labeled Δ-tree $\mathscr{T}_{\downarrow \Sigma'} \triangleq \langle \mathrm{T}, \mathsf{v}' \rangle$ such that, for all nodes $x \in \mathrm{T}$, we have $\mathsf{v}(x) = (\mathsf{v}'(x), \sigma)$, for some $\sigma \in \Sigma''$. Moreover, given a Σ'-labeled Δ-tree $\mathscr{T}' = \langle \mathrm{T}, \mathsf{v}' \rangle$ and a Σ''-labeled Δ-tree $\mathscr{T}'' = \langle \mathrm{T}, \mathsf{v}'' \rangle$, we define the *combination* of $\mathscr{T}'$ with $\mathscr{T}''$ as the $(\Sigma' \times \Sigma'')$-labeled Δ-tree $\mathscr{T}' \otimes \mathscr{T}'' \triangleq \langle \mathrm{T}, \mathsf{v} \rangle$ such that, for all nodes $x \in \mathrm{T}$, we have $\mathsf{v}(x) = (\mathsf{v}'(x), \mathsf{v}''(x))$.

Definition 1.23 (Satellite Building) *A satellite* $\mathscr{S} = \langle \Sigma, \Delta, \mathrm{P}, \zeta, \mathrm{P}_0 \rangle$ *with* $\mathrm{P} = \mathrm{P}_E \times \mathrm{P}_I$ builds *a* $\Sigma \times \mathrm{P}_E$*-labeled* Δ*-tree* $\mathscr{T}_{\mathscr{S}}$ *over a* Σ*-labeled* Δ*-tree* $\mathscr{T}$ *iff there exists a run* $\mathscr{R}$ *of* $\mathscr{S}$ *on* $\mathscr{T}$ *such that* $\mathscr{T}_{\mathscr{S}}$ *is the combination* $\mathscr{T} \otimes \mathscr{R}_{\downarrow \mathrm{P}_E}$ *of* $\mathscr{T}$ *with the projection of* $\mathscr{R}$ *on* P_E.

At this point, we can define the language accepted by an ATAS.

Definition 1.24 (ATAS **Acceptance**) *A* Σ*-labeled* Δ*-tree* $\mathscr{T}$ *is accepted by an* ATAS $\langle\mathscr{A},\mathscr{S}\rangle$, *where* $\mathscr{A} = \langle\Sigma \times P_E, \Delta, Q, \delta, q_0, F\rangle$, $\mathscr{S} = \langle\Sigma, \Delta, P, \zeta, P_0\rangle$, *and* $P = P_E \times P_I$, *iff* $\mathscr{S}$ *builds a tree* $\mathscr{T}_{\mathscr{S}}$ *over* $\mathscr{T}$ *such that* $\mathscr{T}_{\mathscr{S}}$ *is accepted by the* ATA $\mathscr{A}$.

In words, first the satellite $\mathscr{S}$ guesses and adds to the input tree $\mathscr{T}$ an additional labeling over the set P_E, thus returning the built tree $\mathscr{T}_{\mathscr{S}}$. Then, the main automaton $\mathscr{A}$ computes a new run on $\mathscr{T}_{\mathscr{S}}$ taken as input. By $L(\langle\mathscr{A},\mathscr{S}\rangle)$ we denote the language accepted by the ATAS $\langle\mathscr{A},\mathscr{S}\rangle$.

In the following, we consider, in particular, ATAS along with the parity acceptance condition (APTS, for short).

Note that satellites are just a convenient way to describe an ATA in which the state space can be partitioned into two components, one of which is nondeterministic, independent from the other, and that has no influence on the acceptance. Indeed, it is just a matter of technicality to see that automata with satellites inherit all the closure properties of alternating automata. In particular, the following theorem, directly derived by a proof idea of (Kupferman and Vardi 2006), shows how the separation between $\mathscr{A}$ and $\mathscr{S}$ gives a tight analysis of the complexity of the relative emptiness problem.

Theorem 1.17 (APTS **Emptiness**) *The* emptiness problem *for an* APTS $\langle\mathscr{A},\mathscr{S}\rangle$ *with alphabet size h, where the main automaton* $\mathscr{A}$ *has n states and index k and the satellite* $\mathscr{S}$ *has m states, can be decided in time* $2^{O(\log(h)+(n\cdot k)\cdot((n\cdot k)\cdot\log(n\cdot k)+\log(m)))}$.

Proof As first thing, we use the Muller-Schupp exponential-time nondeterminization procedure (Muller and Schupp 1995) that leads from the APT $\mathscr{A}$ to an NPT $\mathscr{N}$, with $2^{O((n\cdot k)\cdot\log(n\cdot k))}$ states and index $O(n\cdot k)$, such that $L(\mathscr{A}) = L(\mathscr{N})$. Since an NPT is a particular APT, we immediately have that $L(\langle\mathscr{N},\mathscr{S}\rangle) = L(\langle\mathscr{A},\mathscr{S}\rangle)$. At this point, by taking the product-automaton between $\mathscr{N}$ and the satellite $\mathscr{S}$, we obtain another NPT $\mathscr{N}^\star$, with $2^{O((n\cdot k)\cdot\log(n\cdot k)+\log(m))}$ states and index $O(n\cdot k)$, such that $L(\mathscr{N}^\star) = L(\langle\mathscr{N},\mathscr{S}\rangle)$. With more details, if $\mathscr{N} = \langle\Sigma \times P_E, \Delta, Q, \delta, Q_0, F\rangle$ and $\mathscr{S} = \langle\Sigma, \Delta, P, \zeta, P_0\rangle$ with $P = P_E \times P_I$ and $F = (F_1, \ldots, F_k)$, we have that $\mathscr{N}^\star \triangleq \langle\Sigma, \Delta, Q \times P, \delta^\star, Q_0 \times P_0, F^\star\rangle$ with $F^\star \triangleq (F_1 \times P, \ldots, F_k \times P)$ and $\delta^\star((q,(p_E,p_I)),\sigma) \triangleq (\bigvee_{g\in\zeta((p_E,p_I),\sigma)} \delta(q,(\sigma,p_E)))[(d,q') \in \Delta \times Q/(d,(q',g(d)))]$, where by $f[x/y]$ we denote the formula in which all occurrences of a proposition x in f are replaced by the proposition y. In words, $\delta^\star((q,(p_E,p_I)),\sigma)$ is obtained by guessing what is the choice g of the satellite in the state (p_E,p_I) when it reads σ and then by substituting in $\delta(q,(\sigma,p_E))$ each occurrence of a move (d,q') with a new move of the form $(d,(q',p'))$, where $p' = g(d)$ represents the new state sent by the satellite in the direction d. Hence, it is evident that $L(\mathscr{N}^\star) = L(\langle\mathscr{A},\mathscr{S}\rangle)$ by definition of ATAS. Now, the emptiness of $\mathscr{N}^\star$ can be checked in polynomial running-time in its number of states, exponential in its index, and linear in the alphabet size (see Theorem 5.1 of Kupferman and Vardi (1998)). Overall, with this procedure, we obtain that the emptiness problem for an APTS is solvable in time $2^{O(\log(h)+(n\cdot k)\cdot((n\cdot k)\cdot\log(n\cdot k)+\log(m)))}$. □

1.7 GCTL Model Transformations

At this point, we can start to describe the decision procedure for the satisfiability problem of GCTL. As we discussed in the introduction, we exploit an automata-theoretic approach by using satellites that are able to accept binary tree-encodings of tree models of a formula. So, we first introduce the binary tree encoding and then, in the next section, we show how to build the automaton accepting all tree-model encodings satisfying the formula of interest.

The tree encoding works as follows. Given a tree model $\mathscr{T}$ of φ, we first build its *widening* $\mathscr{T}_W$, obtaining in this way a full tree with infinite branching. Then, from $\mathscr{T}_W$, we derive a *delayed generation* tree $\mathscr{T}_D$ that embeds $\mathscr{T}_W$ in a binary tree. Finally, we enrich the labeling of $\mathscr{T}_W$ with degree functions that allow to propagate the information related to the degree g of the formula along the paths. This is done by using a set B of elements, called bases, that are used in the domain of the degree functions. The obtained tree $\mathscr{T}_{D_{B,g}}$ is named B-*based g-degree delayed generation*. Intuitively, a base is used to represent a subformula of φ to which we associate, by means of the degree functions, the related number of paths required to be satisfied. This turns to be a key step in the whole satisfiability procedure we show in the next section.

In the following, to simplify the technical reasoning, we use as unwinding of a KS $\mathscr{K}$, not the KT $\mathscr{K}_U$ itself, but one of the complete 2^{AP}-labeled $\mathbb{N}$-tree $\mathscr{T}$ isomorphic to $\mathscr{K}_U$.

1.7.1 Binary Tree Model Encoding

As first step in our binary encoding construction, we define the widening of a 2^{AP}-labeled $\mathbb{N}$-tree $\mathscr{T}$, i.e., a transformation that, taken $\mathscr{T}$, returns a full infinite tree $\mathscr{T}_W$ having infinite branching degree and embedding $\mathscr{T}$ itself (see Figure 1.4). This transformation ensures that in $\mathscr{T}_W$ all nodes have the same branching degree and all

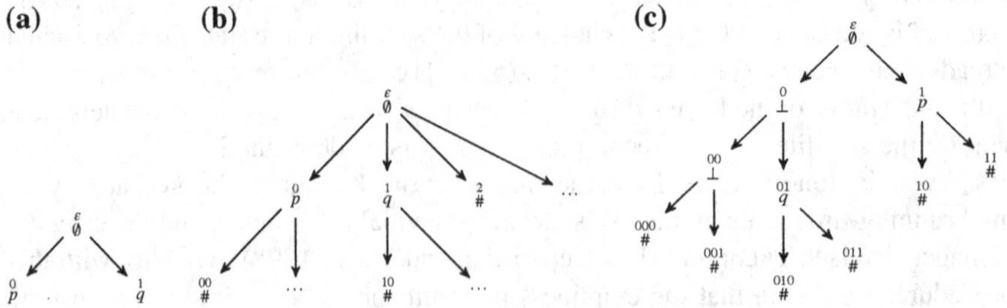

Fig. 1.4 A tree $\mathscr{T}$ **(a)**, its widening $\mathscr{T}_W$ **(b)**, and the related delayed generation $\mathscr{T}_D$ **(c)**

branches are infinite. To this aim, we use a fresh label # to denote fake nodes, as described in the following definition.

Definition 1.25 (Widening) *Let* $\mathcal{T} = \langle \mathrm{T}, \mathsf{v} \rangle$ *be a* Σ*-labeled* $\mathbb{N}$*-tree such that* $\# \notin \Sigma$. *Then, the* widening *of* $\mathcal{T}$ *is the* Σ_W*-labeled* $\mathbb{N}$*-tree* $\mathcal{T}_W \triangleq \langle \mathbb{N}^*, \mathsf{v}_W \rangle$ *such that* (i) $\Sigma_W \triangleq \Sigma \cup \{\#\}$, (ii) *for* $x \in \mathrm{T}$, $\mathsf{v}_W(x) \triangleq \mathsf{v}(x)$, *and* (iii) *for* $y \in \mathbb{N}^* \setminus \mathrm{T}$, $\mathsf{v}_W(y) \triangleq \#$.

Now, we define a sharp transformation of $\mathcal{T}_W$ in a full binary tree $\mathcal{T}_D$. This is inspired but different from that used to embed the logic $S\omega S$ into $S2S$ (Rabin 1969). Intuitively, the transformation allows to delay n abstract decisions, to be taken at a node y in $\mathcal{T}_W$ and corresponding to its n successors $y \cdot i$, along some corresponding nodes $x, x \cdot 0, x \cdot 00, \ldots$ in $\mathcal{T}_D$. In particular, when we are on a node $x \cdot 0^i$, we are able to split the decision on $y \cdot i$ into an immediate action, which is sent to the right (effective) successor $x \cdot 0^i \cdot 1$, while the remaining actions are sent to its copy $x \cdot 0^{i+1}$. To differentiate the meaning of left and right successors of a node in $\mathcal{T}_D$, we use the fresh symbol $\perp$ (see Fig. 1.4).

Definition 1.26 (Delayed Generation) *Let* $\mathcal{T}_W = \langle \mathbb{N}^*, \mathsf{v}_W \rangle$ *be the widening of a* Σ*-labeled tree* $\mathcal{T}$ *such that* $\perp \notin \Sigma$. *Then, the* delayed generation *of* $\mathcal{T}$ *is the* Σ_D*-labeled* $\{0, 1\}$*-tree* $\mathcal{T}_D \triangleq \langle \{0, 1\}^*, \mathsf{v}_D \rangle$ *such that* (i) $\Sigma_D \triangleq \Sigma_W \cup \{\perp\}$ *and* (ii) *there exists a surjective function* $\mathsf{s} : \{0, 1\}^* \to \mathbb{N}^*$, *with* $\mathsf{s}(\varepsilon) \triangleq \varepsilon$, $\mathsf{s}(x \cdot 0^i) \triangleq \mathsf{s}(x)$, *and* $\mathsf{s}(x \cdot 0^i \cdot 1) \triangleq \mathsf{s}(x) \cdot i$, *where* $x \in \{0, 1\}^*$ *and* $i \in \mathbb{N}$, *such that* (ii.i) $\mathsf{v}_D(x) \triangleq \mathsf{v}_W(\mathsf{s}(x))$, *for all* $x \in \{\varepsilon\} \cup \{0, 1\}^* \cdot \{1\}$, *and* (ii.ii) *if* $\mathsf{v}_D(x \cdot 1) = \#$ *then* $\mathsf{v}_D(x \cdot 0) \triangleq \#$ *else* $\mathsf{v}_D(x \cdot 0) \triangleq \perp$, *for all* $x \in \{0, 1\}^*$.

To complete the tree encoding, we have also to delay the degree associated to each node in the input tree model. We recall that, an original tree model of a graded formula may require a fixed number of paths satisfying the formula going through the same node. Such a number is the degree associated to that node and which we need to delay. To this aim, we enrich the label of a node with a function mapping a set of elements, named *bases*, into triples of numbers representing the splitting of the node degree into two components. The first is the delayed degree, while the second is the degree associated to one of the effective successors of the node. Such a splitting is the delayed abstract action mentioned above, when it is customized to the need of having information on the degrees. We further use a flag with values in $\{\flat, \not\flat\}$ to indicate if the labeling is or not active, i.e., if it actually represents the splitting of the degree of a given base that needs to be propagated in the two tree directions. Note that, for a formula with degree g, it is not important to monitor the presence of a finite number of paths of cardinality greater than g. To this purpose, we use the symbol $\not\omega$ to efficiently represent the infinite set $]g, \omega[$. We relate ω and $\not\omega$ to the finite number in $[0, g]$ in the expected way: *(i)* $i < \not\omega < \omega$, for all $i \in [0, g]$; *(ii)* $i + j \triangleq \not\omega$, for all $i, j \in [0, g]$ such that $i + j > g$; *(iii)* $i + j = j + i \triangleq i$, for all $i \in \{\not\omega, \omega\}$ and $j \in [0, g] \cup \{\not\omega, \omega\}$ such that $j \leq i$. The whole idea of the degree encoding is formalized through the following four definitions.

Definition 1.27 (**(Σ, B)-Enriched** g**-Degree Tree**) *Let* Σ *and* B *be two sets,* $g \in \mathbb{N}$, *and* $\mathrm{H}(g) \triangleq \{(d, d_1, d_2) \in ([0, g] \cup \{/\omega, \omega\})^3 : d = d_1 + d_2\} \times \{\flat, /\flat\}$.

Then, a (Σ, B)-enriched g-degree tree *is a* $(\Sigma \times \mathrm{H}(g)^{\mathrm{B}})$*-labeled* $\{0, 1\}$*-tree* $\mathcal{T} = \langle \{0, 1\}^*, \mathsf{v} \rangle$.

We now introduce a (Σ_D, B)-enriched g-degree tree $\mathcal{T}_{D_{\mathrm{B},g}}$ as the extension of the delayed generation $\mathcal{T}_D$ of $\mathcal{T}$ with degree functions in its labeling. Intuitively, each function in a node represents how to distribute and propagate an information on the degrees along its successors.

Definition 1.28 **(B-Based g-Degree Delayed Generation)** *Let* B *be a set,* $g \in \mathbb{N}$, *and* $\mathcal{T}_D = \langle \{0, 1\}^*, \mathsf{v}_D \rangle$ *be the delayed generation of a* Σ*-labeled tree* $\mathcal{T}$*. Then, a* B-based g-degree delayed generation *of* $\mathcal{T}$ *is a* (Σ_D, B)*-enriched* g*-degree tree* $\mathcal{T}_{D_{\mathrm{B},g}} = \langle \{0, 1\}^*, \mathsf{v}_{D_{\mathrm{B},g}} \rangle$ *such that there is an* $\mathsf{h} \in \mathrm{H}(g)^{\mathrm{B}}$ *with* $\mathsf{v}_{D_{\mathrm{B},g}}(x) = (\mathsf{v}_D(x), \mathsf{h})$, *for all* $x \in \{0, 1\}^*$.

In order to have a sound construction for $\mathcal{T}_{D_{\mathrm{B},g}}$, we need to impose a coherence property on the information between a node and its two successors. In particular, whenever we enter a node x labeled with # in its first part, as it represents that the node is fictitious, we have to take no splitting of the degree by sending to x the value 0. In addition, we force children labeled with # to have necessarily the flag set to $\not\flat$. On the other nodes, we need to match the value of the first component of the splitting with the degree of the left successor. Moreover, in dependence of the flag in $\{\flat, \not\flat\}$, we may have also to match the value of the second component with the degree of the right successor. With more details, we require a coherence that is not punctual ($=$) but rather, depending on the particular kind of bases we are analyzing, it has to be either superior ($\geq$) or inferior ($\leq$) to the value given by the parent of the node. Specifically, to distinguish between these kinds of bases, we split them into the two subsets $\mathrm{B}_{\mathrm{sup}}$ and $\mathrm{B}_{\mathrm{inf}}$. So, a tree has to be superiorly coherent w.r.t. $\mathrm{B}_{\mathrm{sup}}$ and inferiorly coherent w.r.t. $\mathrm{B}_{\mathrm{inf}}$.

Definition 1.29 **(GCTL Sup/Inf Coherence)** *Let* $\mathcal{T} = \langle \{0, 1\}^*, \mathsf{v} \rangle$ *be a* $(\Sigma \cup \{\#\}, \mathrm{B})$*-enriched* g*-degree tree. Then,* $\mathcal{T}$ *is* superiorly *(resp.,* inferiorly*)* coherent *w.r.t. a base* $b \in \mathrm{B}$ *iff, for* $x \in \{0, 1\}^*$ *and* $i \in \{0, 1\}$ *with* $\mathsf{v}(x) = (\sigma, \mathsf{h})$, $\mathsf{h}(b) = (d, d_0, d_1, \beta)$, $\mathsf{v}(x \cdot i) = (\sigma_i, \mathsf{h}_i)$, *and* $\mathsf{h}_i(b) = (d^i, d_0^i, d_1^i, \beta^i)$, *it holds that* (i) *if* $\sigma_i = \#$ *then* $d_i = 0$ *and* $\beta^i = \not\flat$ *and* (ii) *if* $i = 0$ *or* $\beta = \flat$ *then* $d_i \leq d^i$ *(resp.,* $d_i \geq d^i$*).*

Finally, with the following definition, we extend the local concept of sup/inf coherence of a particular base to a pair of sets of bases $\mathrm{B}_{\mathrm{sup}}, \mathrm{B}_{\mathrm{inf}} \subseteq \mathrm{B}$.

Definition 1.30 **(GCTL Full Coherence)** *A* $(\Sigma \cup \{\#\}, \mathrm{B})$*-enriched* g*-degree tree* $\mathcal{T}$ *is* full coherent *w.r.t. a pair* $(\mathrm{B}_{\mathrm{sup}}, \mathrm{B}_{\mathrm{inf}})$*, where* $\mathrm{B}_{\mathrm{sup}} \cup \mathrm{B}_{\mathrm{inf}} \subseteq \mathrm{B}$*, iff it is superiorly and inferiorly coherent w.r.t. all bases* $b \in \mathrm{B}_{\mathrm{sup}}$ *and* $b \in \mathrm{B}_{\mathrm{inf}}$*, respectively.*

Note that the sets $\mathrm{B}_{\mathrm{sup}}$ and $\mathrm{B}_{\mathrm{inf}}$ turn out to be useful, in the satisfiability algorithm we give, to deal with the degree of existential and universal path quantifications, respectively. In particular, the whole construction ensures that the degrees of all formulas are correctly propagated along the tree, i.e., in other words, that the model is full coherent.

1.7.2 The Coherence Structure Satellites

We now define the satellites we use to verify that the tree encoding the model of the formula has a correct shape w.r.t. the whole transformation described in the previous paragraph. In particular, we first introduce a satellite that checks if the "enriched degree tree" in input is the result of a "based degree delayed generation" of the unraveled model of the formula. Then, we show how to create the additional labeling of the tree that satisfies the coherence properties on the degrees required by the semantics of the logic. The following automaton checks if the # and $\perp$ labels of the input tree are correct w.r.t. Definitions 1.25 and 1.26.

Definition 1.31 (Structure Satellite) *The* structure satellite *is the satellite* $\mathscr{S}^\star \triangleq \langle \Sigma_D, \{0,1\}, \{\#, \perp, @\}, \zeta, \{@\}\rangle$ *on binary trees, where* ζ *is set as follows: if* $p = \sigma = \#$ *then* $\zeta(p,\sigma) \triangleq \{(\#,\#)\}$ *else if either* $p = \sigma = \perp$ *or* $p = @$ *and* $\sigma \in \Sigma$ *then* $\zeta(p,\sigma) \triangleq \{(\perp, @), (\#,\#)\}$, *otherwise* $\zeta(p,\sigma) \triangleq \emptyset$.

The satellite $\mathscr{S}^\star$ has constant size 3. Its transition function ζ is defined to directly represent the constraints on the # and $\perp$ labels and, in particular, the state @ is used to represents a real node of the original tree with values in Σ. So, next lemma easily follows.

Lemma 1.5 (Structure Satellite) *The* $\mathscr{S}^\star$ *satellite accepts all and only the* Σ_D*-labeled* $\{0,1\}$*-trees* $\mathscr{T}_D$ *that can be obtained as the delayed generation of* Σ*-labeled* $\mathbb{N}$*-trees* $\mathscr{T}$.

The next satellite creates the additional labeling of the input tree, for the main automaton, in such a way that it is full coherent w.r.t. the pair of sets $(\mathrm{B}_{\mathrm{sup}}, \mathrm{B}_{\mathrm{inf}})$. Precisely, if the satellite accepts the input tree, the additional labeling of the built tree is given by its states.

Definition 1.32 (GCTL Coherence Satellite) *The* (Σ, B)-enriched g-degree $(\mathrm{B}_{\mathrm{sup}}, \mathrm{B}_{\mathrm{inf}})$-coherence satellite *with* $\mathrm{B}_{\mathrm{sup}} \cup \mathrm{B}_{\mathrm{inf}} \subseteq \mathrm{B}$ *is the binary satellite* $\mathscr{S}_{\mathrm{B},g}^{\Sigma,(\mathrm{B}_{\mathrm{sup}},\mathrm{B}_{\mathrm{inf}})} \triangleq \langle \Sigma \cup \{\#\}, \{0,1\}, \mathrm{H}(g)^{\mathrm{B}}, \zeta, \mathrm{H}(g)^{\mathrm{B}}\rangle$, *where* ζ *is set as follows:* (i) *if* $\sigma = \#$, *then* $\zeta(p,\sigma) \triangleq \{(p,p)\}$, *if for all* $b \in \mathrm{B}$ *it holds* $p(b) = (0,0,0,\not{b})$, *and* $\zeta(p,\sigma) \triangleq \emptyset$, *otherwise;* (ii) *if* $\sigma \neq \#$ *then* $\zeta(p,\sigma)$ *contains all and only the pairs of states* $(p_0, p_1) \in (\mathrm{H}(g)^{\mathrm{B}})^{\{0,1\}}$ *such that, for all* $b \in \mathrm{B}_\alpha$ *with* $\alpha = \sup$ *(resp.,* $\alpha = \inf$*),* $p(b) = (d, d_0, d_1, \beta)$, *and* $p_i(b) = (d^i, d_0^i, d_1^i, \beta^i)$, *it holds that if* $i = 0$ *or* $\beta = b$ *then* $d_i \leq d^i$ *(resp.,* $d_i \geq d^i$*), for all* $i \in \{0,1\}$.

The transition function is structured to directly represent the constraints of Definitions 1.29 and 1.30. Note that the satellite $\mathscr{S}_{\mathrm{B},g}^{\Sigma,(\mathrm{B}_{\mathrm{sup}},\mathrm{B}_{\mathrm{inf}})}$ is polynomial in g and exponential in $|\mathrm{B}|$, since its number of states is equal to $(2 \cdot (g+3)^2)^{|\mathrm{B}|}$. Next lemma follows by construction.

Lemma 1.6 (GCTL Coherence Satellite) *The* $\mathscr{S}_{\mathrm{B},g}^{\Sigma,(\mathrm{B}_{\mathrm{sup}},\mathrm{B}_{\mathrm{inf}})}$ *satellite builds all and only the* $(\Sigma \cup \{\#\}, \mathrm{B})$*-enriched* g*-degree trees* $\mathscr{T}'$ *over* $\Sigma \cup \{\#\}$*-labeled* $\{0,1\}$*-tree* $\mathscr{T}$ *that are full coherent w.r.t. the pair* $(\mathrm{B}_{\mathrm{sup}}, \mathrm{B}_{\mathrm{inf}})$.

Finally, we introduce the satellite that checks if the tree in input is coherent or not by merging the behavior of the two previous described satellites.

Definition 1.33 (GCTL **Coherence Structure Satellite**) *The B-based g-degree* structure (B_{sup}, B_{inf})-coherence satellite *with* $B_{sup} \cup B_{inf} \subseteq B$ *is the binary satellite* $\mathscr{S}_{B,g}^{B_{sup},B_{inf}} = \langle \Sigma_D, \{0,1\}, P_E \times P_I, \zeta, P_{E_0} \times P_{I_0} \rangle$, *where* $P_E = P_{E_0} \triangleq H(g)^B$, $P_I \triangleq \{\#, \bot, @\}$, *and* $P_{I_0} \triangleq \{@\}$, *obtained as the product of the* $(\Sigma \cup \{\bot\}, B)$*-enriched g-degree* (B_{sup}, B_{inf})*-full coherent satellite* $\mathscr{S}_{B,g}^{\Sigma\cup\{\bot\},(B_{sup},B_{inf})}$ *with the structure satellite* $\mathscr{S}^{\star}$.

Clearly, the size of $\mathscr{S}_{B,g}^{B_{sup},B_{inf}}$ is polynomial in g and exponential in $|B|$, since its number of states is equal to $3 \cdot (2 \cdot (g+3)^2)^{|B|}$. Due to the product structure of the automaton, next result directly follows from Lemmas 1.5 and 1.6.

Theorem 1.18 (GCTL **Coherence Structure Satellite**) *The* $\mathscr{S}_{B,g}^{B_{sup},B_{inf}}$ *satellite builds all and only the B-based g-degree delayed generations* $\mathscr{T}_{D_{B,g}}$ *of* Σ*-labeled* $\mathbb{N}$*-trees* $\mathscr{T}$ *over their delayed generation* $\mathscr{T}_D$ *that are full coherent w.r.t. the pair* (B_{sup}, B_{inf}).

1.8 GCTL Satisfiability

In this section, we finally introduce an APT $\mathscr{A}_\varphi$ that checks whether a complete 2^{AP}-labeled $\mathbb{N}$-tree $\mathscr{T}$ satisfies a given formula φ by evaluating all B-based g-degree delayed generation trees $\mathscr{T}_{D_{B,g}}$ associated with $\mathscr{T}$, where $g \triangleq \mathsf{deg}(\varphi)$ is the maximum finite degree of φ and $B \triangleq \mathsf{qcl}(\varphi)$ is the *quantification closure* of φ, i.e., the set of all the quantification formulas, contained in the closure, deprived of the degree. To formally define this concept, we have first to introduce the *extended closure* $\mathsf{ecl}(\varphi)$ of a GCTL formula φ that is construct in the same way of $\mathsf{cl}(\varphi)$, by also asserting that *(i)* if $\mathsf{E}^{\geq g}\varphi_1 \mathsf{Op}\ \varphi_2 \in \mathsf{ecl}(\varphi)$ then $\mathsf{E}^{\geq 1}\varphi_1 \mathsf{Op}\ \varphi_2 \in \mathsf{ecl}(\varphi)$, *(ii)* if $\mathsf{E}^{\geq g}\varphi_1 \widetilde{\mathsf{Op}}\ \varphi_2 \in \mathsf{ecl}(\varphi)$ then $\mathsf{E}^{\geq 1}\neg(\varphi_1 \widetilde{\mathsf{Op}}\ \varphi_2) \in \mathsf{ecl}(\varphi)$, *(iii)* if $\mathsf{A}^{<g}\varphi_1 \mathsf{Op}\ \varphi_2 \in \mathsf{ecl}(\varphi)$ then $\mathsf{A}^{<1}\neg(\varphi_1 \mathsf{Op}\ \varphi_2) \in \mathsf{ecl}(\varphi)$, and *(iv)* if $\mathsf{A}^{<g}\varphi_1 \widetilde{\mathsf{Op}}\ \varphi_2 \in \mathsf{ecl}(\varphi)$ then $\mathsf{A}^{<1}\varphi_1 \widetilde{\mathsf{Op}}\ \varphi_2 \in \mathsf{ecl}(\varphi)$, for all $\mathsf{Op} \in \{\mathsf{U}, \mathsf{R}\}$, and $g \in [2, \omega]$. Intuitively, the difference between $\mathsf{cl}(\varphi)$ and $\mathsf{ecl}(\varphi)$ resides in the fact that, in the latter, we also include the formulas with degree 1 used to deal with the $\equiv_{\mathscr{T}}^{x}$-tautologies and their negations. Note that $|\mathsf{ecl}(\varphi)| = \mathrm{O}(|\mathsf{cl}(\varphi)|)$. The quantification closure is consequently defined as follows: $\mathsf{qcl}_{\mathsf{E}}(\varphi) \triangleq \{\mathsf{E}\psi : \mathsf{E}^{\geq g}\psi \in \mathsf{ecl}(\varphi)\} \setminus \{\mathsf{E}\tilde{\mathsf{X}}\mathsf{f}\}$, $\mathsf{qcl}_{\mathsf{A}}(\varphi) \triangleq \{\mathsf{A}\psi : \mathsf{A}^{<g}\psi \in \mathsf{ecl}(\varphi)\} \setminus \{\mathsf{AXt}\}$, and $\mathsf{qcl}(\varphi) \triangleq \mathsf{qcl}_{\mathsf{E}}(\varphi) \cup \mathsf{qcl}_{\mathsf{A}}(\varphi)$. In particular, observe that we do not need any base for the formulas checking whether there is or not a successor of a node.

The automaton runs on every B-based g-degree generation tree, even those that are not associated with a complete tree. However, we make the assumptions that the trees in input are really associated with this kind of trees and that they are coherent with respect to (B_{sup}, B_{inf}), where $B_{sup} \triangleq \mathsf{qcl}_{\mathsf{E}}(\varphi)$ and $B_{inf} \triangleq \mathsf{qcl}_{\mathsf{A}}(\varphi)$. By Theorem 1.18, we are able to enforce such properties by using $\mathscr{A}_\varphi$ as the main part of an APTS having the B-based g-degree structure $(\mathrm{B_{sup}}, \mathrm{B_{inf}})$-coherence satellite $\mathcal{S}_{\mathrm{B},g}^{\mathrm{B_{sup}},\mathrm{B_{inf}}}$ as second component.

In order to understand how the formula automaton works, it is useful to gain more insights into the meaning of the tree $\mathcal{T}_{D_{\mathrm{B},g}}$ associated with $\mathcal{T}$. First of all, the widening operation has the purpose to make the tree full by adding fake nodes labeled with #. Through this, we obtain the tree $\mathcal{T}_W$. Then, the delaying operation transforms $\mathcal{T}_W$ into a binary tree $\mathcal{T}_D$, such that at every level a node x associated to a node y in $\mathcal{T}$ generates only one of the successor of y at a time in the direction 1, meanwhile it sends a duplicate of itself on the direction 0 labeled with $\perp$. The following duplicates have to generate the remaining successors in a recursive way. However, if there are no more successors to generate, the node x does not send in the direction 0 a duplicate of itself anymore, but just a fake node labeled with #. At this point, to obtain the tree $\mathcal{T}_{D_{\mathrm{B},g}}$, we enrich the labeling of the delayed generation tree, by adding a degree function $\mathsf{h} : \mathrm{B} \to \mathrm{H}(g)$. In the hypothesis that $\mathcal{T}$ satisfies φ, for every formula $\varphi' \in \mathrm{B}$ and node $x \in \{0,1\}^*$ with $\mathsf{v}_{D_{\mathrm{B},g}}(x) = (\sigma, \mathsf{h})$, we have that $\mathsf{h}(\varphi') = (d, d_0, d_1, \beta)$ describes the degree with which the formula φ' is supposed to be satisfied on x. In particular d is the degree in the current node, the decomposition $d = d_0 + d_1$ explains how this degree is partitioned in the following left and right children, and the β flag represents whether this splitting of degrees is meaningful or not. More precisely, β is set to $\not\beta$ iff the inner formula of φ' or its negation is a structure formula tautology in x. Hence, there is no point in splitting the degree, since the formula is already verified or falsified. Moreover, d_1 represents the degree sent to the direction 1, which usually corresponds to a concrete node in $\mathcal{T}$. Hence, it is the degree sent to that node. Meanwhile, d_0 represents the degree sent to the direction 0, which usually corresponds to a duplicate of the previous node. Hence, d_0 represents the degree that have yet to be partitioned among the remaining successors of the node y associated to x. To this aim, the coherence requirement asks: *(i)* for an existential formula, the degree found in a successor node is not lower than the degree the father sent to that node (it may be higher as the node may satisfy the formula by finding more paths with a certain property, so it surely satisfies what the formula requires); *(ii)* for a universal formula, the degree found in a successor node is not greater than the degree the father sent to that node (it may be smaller as the node may satisfy the formula by finding less paths with a certain negated property, so it surely satisfies what the formula requires).

In the hypothesis of coherence, the formula automaton needs only to check that *(i)* the degree of every existential and universal formula is correctly initiated on the node in which the formula first appears in (e.g., for an existential formula it needs to check that the degree in the label of the node is not lower than the degree required by the formula), and *(ii)* that every node of the tree satisfies the existential or universal formula with the degree specified in the node labeling. To do this, the automaton $\mathcal{A}_\varphi$ has as state space the set $\mathsf{ecl}(\varphi) \cup \mathsf{mcl}(\varphi) \cup \mathsf{qcl}(\varphi) \cup \{\#, \neg\#\}$, where $\mathsf{mcl}(\varphi)$ is the *modified closure* of φ defied as follows: $\mathsf{mcl}(\varphi) \triangleq \mathsf{mcl}_1(\varphi) \cup \mathsf{mcl}_\omega(\varphi)$, $\mathsf{mcl}_1(\varphi) \triangleq \mathsf{mcl}_{\mathsf{E}^1}(\varphi) \cup \mathsf{mcl}_{\mathsf{A}^1}(\varphi)$, $\mathsf{mcl}_{\mathsf{E}^1}(\varphi) \triangleq \bigcup_{\mathsf{Op} \in \{\mathsf{U},\mathsf{R}\}}^{i \in \{0,1\}} \mathsf{mcl}_{\mathsf{EOp},i}(\varphi)$, $\mathsf{mcl}_{\mathsf{A}^1}(\varphi) \triangleq \bigcup_{\mathsf{Op} \in \{\mathsf{U},\mathsf{R}\}}^{i \in \{0,1\}} \mathsf{mcl}_{\mathsf{AOp},i}(\varphi)$, $\mathsf{mcl}_{\mathsf{EOp},i}(\varphi) \triangleq \{\mathsf{E}_i^{\geq 1} \psi : \mathsf{E}\psi \in \mathsf{qcl}_{\mathsf{E}}(\varphi) \wedge \psi \in \{\varphi_1 \mathsf{Op}\ \varphi_2, \varphi_1 \widetilde{\mathsf{Op}}\ \varphi_2\}\}$, $\mathsf{mcl}_{\mathsf{AOp},i}(\varphi) \triangleq$

$\{\mathsf{A}_i^{<1}\psi : \mathsf{A}\psi \in \mathsf{qcl}_\mathsf{A}(\varphi) \wedge \psi \in \{\varphi_1 \mathsf{Op}\ \varphi_2, \varphi_1 \widetilde{\mathsf{Op}}\ \varphi_2\}\}$, $\mathsf{mcl}_\omega(\varphi) \triangleq \mathsf{mcl}_{\mathsf{E}^\omega}(\varphi) \cup \mathsf{mcl}_{\mathsf{A}^\omega}(\varphi)$, $\mathsf{mcl}_{\mathsf{E}^\omega}(\varphi) \triangleq \{\mathsf{E}^{\geq\omega}\psi : \mathsf{E}\psi \in \mathsf{qcl}_\mathsf{E}(\varphi)\}$, and $\mathsf{mcl}_{\mathsf{A}^\omega}(\varphi) \triangleq \{\mathsf{A}^{<\omega}\psi : \mathsf{A}\psi \in \mathsf{qcl}_\mathsf{A}(\varphi)\}$. On one hand, the formulas in $\mathsf{qcl}(\varphi)$ ask the automaton to verify them completely relying on the degree of the labeling. On the other hand, the existential and universal formulas in $\mathsf{ecl}(\varphi) \cup \mathsf{mcl}(\varphi)$ ask the automaton even to check that their degrees agree with that contained in the labeling. The states # and ¬# are used to verify the existence or not of a successor of a node when we have to deal with the formulas $\mathsf{E}^{\geq 1}\tilde{\mathsf{X}}\mathfrak{f}$ and $\mathsf{A}^{<1}\mathsf{X}\mathfrak{t}$. Finally, states in $\mathsf{mcl}(\varphi) \cup \mathsf{qcl}(\varphi)$ are also used for the parity acceptance condition.

Definition 1.34 (GCTL **Formula Automaton)** *The formula automaton for φ is the binary* APT $\mathscr{A}_\varphi \triangleq \langle \Sigma_\varphi \times \mathrm{P}_{E_\varphi}, \{0,1\}, \mathrm{Q}_\varphi, \delta, \varphi, \mathrm{F}_\varphi \rangle$, *where* $\Sigma_\varphi \triangleq 2^{\mathrm{AP}} \cup \{\#, \bot\}$, $\mathrm{P}_{E_\varphi} \triangleq \mathsf{H}(\mathsf{deg}(\varphi))^{\mathsf{qcl}(\varphi)}$, $\mathrm{Q}_\varphi \triangleq \mathsf{ecl}(\varphi) \cup \mathsf{mcl}(\varphi) \cup \mathsf{qcl}(\varphi) \cup \{\#, \neg\#\}$, $\mathrm{F}_\varphi \triangleq (\mathrm{F}_1, \mathrm{F}_2, \mathrm{Q}_\varphi)$ *with* $\mathrm{F}_1 \triangleq \mathsf{mcl}_{\mathsf{AU},1}(\varphi) \cup \mathsf{mcl}_{\mathsf{A}^\omega}(\varphi)$ *and* $\mathrm{F}_2 \triangleq \mathsf{qcl}_\mathsf{A}(\varphi) \cup \mathsf{mcl}_{\mathsf{A}^1}(\varphi) \cup \mathsf{mcl}_\omega(\varphi) \cup \mathsf{mcl}_{\mathsf{ER},1}(\varphi)$, *and* $\delta : \mathrm{Q}_\varphi \times (\Sigma_\varphi \times \mathrm{P}_{E_\varphi}) \rightarrow \mathrm{B}^+(\{0,1\} \times \mathrm{Q}_\varphi)$ *is defined in the body of the article.*

We now describe the structure of the whole transition function $\delta(q, (\sigma, \mathsf{h}))$ through a case analysis on the state space.

As first thing, when $\sigma = \#$, the automaton is on a fake node $x = x' \cdot i$ of the input tree $\mathscr{T}_{\mathrm{DB},g}$, so no formula should be checked on it. However, in the instant the automaton reaches such a node, by passing through its antecedent x', it is not asking to verify the formula represented by the state q, due to the fact that it is sent by another state q' on x' which corresponds to a universal formula. In this case, indeed, we are checking that its "core" is satisfied on all successors (but a given number of them). Hence, since x does not exist in the original tree $\mathscr{T}$, we do not have to verify the property of q on it. Moreover, we are sure that q' does not represent any existential property. This is due to the fact that *(i)* the degree d_i related to the state q' in the labeling of x' needs to be 0 by the coherence requirements of Definition 1.29 and *(ii)*, as we show later, the transition on existential formulas do not send any state to a direction $j \in \{0,1\}$ having $d_j = 0$. For this reason, we set $\delta(q, (\#, \mathsf{h})) \triangleq \mathfrak{t}$, for all $q \in \mathrm{Q}_\varphi$ and $\mathsf{h} \in \mathrm{P}_{E_\varphi}$.

Furthermore, the structure of the transition function does not send a state q belonging to the set $(\mathsf{ecl}(\varphi) \setminus \mathsf{mcl}_\omega(\varphi)) \cup \bigcup_{\mathsf{Op} \in \{\mathsf{U},\mathsf{R}\}} (\mathsf{mcl}_{\mathsf{EOp},1}(\varphi) \cup \mathsf{mcl}_{\mathsf{AOp},1}(\varphi))$ to a node labeled with $\sigma = \bot$ and a state q belonging to the set $\bigcup_{\mathsf{Op} \in \{\mathsf{U},\mathsf{R}\}} (\mathsf{mcl}_{\mathsf{EOp},0}(\varphi) \cup \mathsf{mcl}_{\mathsf{AOp},0}(\varphi))$ to a node labeled with $\sigma \neq \bot$. For this reason, w.l.o.g., we can set $\delta(q, (\sigma, \mathsf{h})) \triangleq \mathfrak{f}$, for all these cases.

Now, we describe the remaining part of the definition of $\delta(q, (\sigma, \mathsf{h}))$ with the proviso that *(i)* $\sigma \neq \#$, *(ii)* if $q \in (\mathsf{ecl}(\varphi) \setminus \mathsf{mcl}_\omega(\varphi)) \cup \bigcup_{\mathsf{Op} \in \{\mathsf{U},\mathsf{R}\}} (\mathsf{mcl}_{\mathsf{EOp},1}(\varphi) \cup \mathsf{mcl}_{\mathsf{AOp},1}(\varphi))$ then $\sigma \neq \bot$, and *(iii)* if $q \in \bigcup_{\mathsf{Op} \in \{\mathsf{U},\mathsf{R}\}} (\mathsf{mcl}_{\mathsf{EOp},0}(\varphi) \cup \mathsf{mcl}_{\mathsf{AOp},0}(\varphi))$ then $\sigma = \bot$.

(1) If $q \in \mathrm{Lit} \triangleq \mathrm{AP} \cup \neg\mathrm{AP}$, where $\neg\mathrm{AP} \triangleq \{\neg p : p \in \mathrm{AP}\}$, the automaton has to verify if the literal is locally satisfied or not. To do this, we set $\delta(q, (\sigma, \mathsf{h})) \triangleq \mathfrak{t}$, if either $q \in \mathrm{AP}$ and $q \in \sigma$ or $q \in \neg\mathrm{AP}$ and $q \notin \sigma$, and $\delta(q, (\sigma, \mathsf{h})) \triangleq \mathfrak{f}$, otherwise.
(2) The boolean cases are treated in the classical way: $\delta(\varphi_1 \wedge \varphi_2, (\sigma, \mathsf{h})) \triangleq \delta(\varphi_1, (\sigma, \mathsf{h})) \wedge \delta(\varphi_2, (\sigma, \mathsf{h}))$ and $\delta(\varphi_1 \vee \varphi_2, (\sigma, \mathsf{h})) \triangleq \delta(\varphi_1, (\sigma, \mathsf{h})) \vee \delta(\varphi_2, (\sigma, \mathsf{h}))$.

(3) The case $\mathsf{E}^{\geq 1}\tilde{\mathsf{X}}\mathfrak{f}$ (resp., $\mathsf{A}^{<1}\mathsf{X}\mathfrak{t}$) is simply solved by setting $\delta(\mathsf{E}^{\geq 1}\tilde{\mathsf{X}}\mathfrak{f}, (\sigma, \mathsf{h})) \triangleq (1, \#)$ (resp., $\delta(\mathsf{A}^{<1}\mathsf{X}\mathfrak{t}, (\sigma, \mathsf{h})) \triangleq (1, \neg\#)$) and $\delta(\#, (\sigma, \mathsf{h})) \triangleq \mathfrak{t}$ (resp., $\delta(\neg\#, (\sigma, \mathsf{h})) \triangleq \mathfrak{f}$), if $\sigma = \#$, and $\delta(\#, (\sigma, \mathsf{h})) \triangleq \mathfrak{f}$ (resp., $\delta(\neg\#, (\sigma, \mathsf{h})) \triangleq \mathfrak{t}$), otherwise.

(4) Let $\mathsf{h}(\mathsf{EX}\varphi) = (d, d_0, d_1, \beta)$ (resp., $\mathsf{h}(\mathsf{A}\tilde{\mathsf{X}}\varphi) = (d, d_0, d_1, \beta)$). For a state of the form $\mathsf{EX}\varphi$ (resp., $\mathsf{A}\tilde{\mathsf{X}}\varphi$) we verify that this formula holds with degree d. The flag β needs to be $\not\beta$, since a next formula on a successor node is not related to one in the current node, due to the fact that this kind of formula never propagate itself. Recall that in the input tree the pair of degrees (d_0, d_1) describe the distribution of the degree d on the nodes, which need to (resp., are allowed to not) satisfy φ, among the successors of the current node. Since the nodes on the direction 1 are real successors of the node in the original input tree $\mathscr{T}$ we need to ask that the state formula φ holds on them iff $d_1 = 1$ (resp., $d_1 = 0$). However, we cannot ask that a state formula holds more than one time, so, if $d_1 > 1$, the input tree cannot be accepted, since $\mathsf{E}^{\geq d_1}\varphi \equiv \mathfrak{f}$ (resp., we do not make any difference in dependence of a value $d_1 > 0$, since $\mathsf{A}^{\leq d_1}\varphi \equiv \mathfrak{t}$). Finally, on direction 0, we send the same state $\mathsf{EX}\varphi$ (resp., $\mathsf{A}\tilde{\mathsf{X}}\varphi$) if $0 < d_0 < \omega$ (resp., $0 \leq d_0 < \not\omega$), in order to ask that the residual degree d_0 is distributed on the remaining successors. When we deal with the infinite degree ω (resp., finite but unbounded degree $\not\omega$) we have to ensure that the formula φ is verified infinitely often (resp. falsified finitely often) on the successors of the current node. To this aim, every time a non-null degree is sent to direction 1, we sent the state $\mathsf{E}^{\geq\omega}\mathsf{X}\varphi$ (resp. $\mathsf{A}^{<\omega}\mathsf{X}\varphi$) to direction 0. Formally, $\delta(\mathsf{EX}\varphi, (\sigma, h))$ (resp., $\delta(\mathsf{A}\tilde{\mathsf{X}}\varphi, (\sigma, h))$) is set to $\mathfrak{f}$, if $\beta = \flat$, and to the following conjunction, otherwise:

- $\begin{cases} \mathfrak{t}, & \text{if } d_0 = 0; \\ (0, \mathsf{EX}\varphi), & \text{if } d_0 < \omega; \\ (0, \mathsf{EX}\varphi), & \text{if } d_0 = \omega \text{ and } d_1 = 0; \\ (0, \mathsf{E}^{\geq\omega}\mathsf{X}\varphi), & \text{if } d_0 = \omega \text{ and } d_1 \neq 0; \end{cases} \wedge \begin{cases} \mathfrak{t}, & \text{if } d_1 = 0; \\ (1, \varphi), & \text{if } d_1 = 1; \\ \mathfrak{f}, & \text{if } d_1 > 1. \end{cases}$
- $\begin{cases} (0, \mathsf{A}\tilde{\mathsf{X}}\varphi), & \text{if } d_0 < \not\omega; \\ (0, \mathsf{A}\tilde{\mathsf{X}}\varphi), & \text{if } d_0 = \not\omega \text{ and } d_1 \neq 0; \\ (0, \mathsf{A}^{<\omega}\tilde{\mathsf{X}}\varphi), & \text{if } d_0 = \not\omega \text{ and } d_1 = 0; \\ \mathfrak{f}, & \text{if } d_0 = \omega; \end{cases} \wedge \begin{cases} (1, \varphi), & \text{if } d_1 = 0; \\ \mathfrak{t}, & \text{if } d_1 > 0. \end{cases}$

For a state of the form $\mathsf{E}^{\geq g}\mathsf{X}\varphi$ (resp., $\mathsf{A}^{<g}\tilde{\mathsf{X}}\varphi$) we have only to further verify that the degree g agrees with the value d, i.e., $d \geq g$ (resp., $d < g$). Formally, $\delta(\mathsf{E}^{\geq g}\mathsf{X}\varphi, (\sigma, \mathsf{h}))$ (resp., $\delta(\mathsf{A}^{<g}\tilde{\mathsf{X}}\varphi, (\sigma, \mathsf{h}))$) is set to $\mathfrak{f}$, if $d < g$ (resp., $d \geq g$), and to $\delta(\mathsf{EX}\varphi, (\sigma, h))$ (resp., $\delta(\mathsf{A}\tilde{\mathsf{X}}\varphi, (\sigma, h))$), otherwise.

(5) A state $\mathsf{E}_i^{\geq 1}\psi$ (resp., $\mathsf{A}_i^{<1}\psi$) in $\mathsf{mcl}(\varphi)$ is used to verify that there is a branch satisfying (resp., all branch satisfy) the inner path formula $\psi = \varphi_1 \mathsf{Op}\, \varphi_2$, regardless the precise value of the added degree labels. What is important is only to follow paths in which the degrees are not null (for the existential case only). The related transition function simply reflects the one-step unfolding of the CTL formulas, shown in Proposition 1.3. When this requirement needs to be propagated on

some successor node, we send different states in the two tree directions, with the sole purpose to distinguish these ones for acceptance reasons.

- $\delta(\mathsf{E}_i^{\geq 1}\varphi_1 \mathsf{U}\varphi_2, (\sigma, \mathsf{h})) \triangleq \delta(\varphi_2, (\sigma, \mathsf{h})) \vee \delta(\varphi_1, (\sigma, \mathsf{h})) \wedge \bigvee_{j \in \{0,1\}}^{d_j > 0} (j, \mathsf{E}_j^{\geq 1}\varphi_1 \mathsf{U}\varphi_2)$;
- $\delta(\mathsf{A}_i^{<1}\varphi_1 \mathsf{U}\varphi_2, (\sigma, \mathsf{h})) \triangleq \delta(\varphi_2, (\sigma, \mathsf{h})) \vee \delta(\varphi_1, (\sigma, \mathsf{h})) \wedge \bigwedge_{j \in \{0,1\}} (j, \mathsf{A}_j^{<1}\varphi_1 \mathsf{U}\varphi_2) \wedge \delta(\mathsf{A}^{<1}\mathsf{X}\mathfrak{t}, (\sigma, \mathsf{h}))$;
- $\delta(\mathsf{E}_i^{\geq 1}\varphi_1 \mathsf{R}\varphi_2, (\sigma, \mathsf{h})) \triangleq \delta(\varphi_2, (\sigma, \mathsf{h})) \wedge (\delta(\varphi_1, (\sigma, \mathsf{h})) \vee \bigvee_{j \in \{0,1\}}^{d_j > 0} (j, \mathsf{E}_j^{\geq 1}\varphi_1 \mathsf{R} \varphi_2))$;
- $\delta(\mathsf{A}_i^{<1}\varphi_1 \mathsf{R}\varphi_2, (\sigma, \mathsf{h})) \triangleq \delta(\varphi_2, (\sigma, \mathsf{h})) \wedge (\delta(\varphi_1, (\sigma, \mathsf{h})) \vee \bigwedge_{j \in \{0,1\}} (j, \mathsf{A}_j^{<1}\varphi_1 \mathsf{R} \varphi_2) \wedge \delta(\mathsf{A}^{<1}\mathsf{X}\mathfrak{t}, (\sigma, \mathsf{h})))$;
- $\delta(\mathsf{E}_i^{\geq 1}\varphi_1 \tilde{\mathsf{U}}\varphi_2, (\sigma, \mathsf{h})) \triangleq \delta(\varphi_2, (\sigma, \mathsf{h})) \vee \delta(\varphi_1, (\sigma, \mathsf{h})) \wedge (\bigvee_{j \in \{0,1\}}^{d_j > 0} (j, \mathsf{E}_j^{\geq 1}\varphi_1 \tilde{\mathsf{U}}\varphi_2) \vee \delta(\mathsf{E}^{\geq 1}\tilde{\mathsf{X}}\mathfrak{f}, (\sigma, \mathsf{h})))$;
- $\delta(\mathsf{A}_i^{<1}\varphi_1 \tilde{\mathsf{U}}\varphi_2, (\sigma, \mathsf{h})) \triangleq \delta(\varphi_2, (\sigma, \mathsf{h})) \vee \delta(\varphi_1, (\sigma, \mathsf{h})) \wedge \bigwedge_{j \in \{0,1\}} (j, \mathsf{A}_j^{<1}\varphi_1 \tilde{\mathsf{U}}\varphi_2)$;
- $\delta(\mathsf{E}_i^{\geq 1}\varphi_1 \tilde{\mathsf{R}}\varphi_2, (\sigma, \mathsf{h})) \triangleq \delta(\varphi_2, (\sigma, \mathsf{h})) \wedge (\delta(\varphi_1, (\sigma, \mathsf{h})) \vee \bigvee_{j \in \{0,1\}}^{d_j > 0} (j, \mathsf{E}_j^{\geq 1}\varphi_1 \tilde{\mathsf{R}} \varphi_2) \vee \delta(\mathsf{E}^{\geq 1}\tilde{\mathsf{X}}\mathfrak{f}, (\sigma, \mathsf{h})))$;
- $\delta(\mathsf{A}_i^{<1}\varphi_1 \tilde{\mathsf{R}}\varphi_2, (\sigma, \mathsf{h})) \triangleq \delta(\varphi_2, (\sigma, \mathsf{h})) \wedge (\delta(\varphi_1, (\sigma, \mathsf{h})) \vee \bigwedge_{j \in \{0,1\}} (j, \mathsf{A}_j^{<1}\varphi_1 \tilde{\mathsf{R}} \varphi_2))$.

For a state of the form $\mathsf{E}^{\geq 1}\psi$ (resp., $\mathsf{A}^{<1}\psi$) we have only to further verify that $d \geq 1$ (resp., $d < 1$). Formally, $\delta(\mathsf{E}^{\geq 1}\psi, (\sigma, \mathsf{h}))$ (resp., $\delta(\mathsf{A}^{<1}\psi, (\sigma, \mathsf{h}))$) is set to $\mathfrak{f}$, if $d < 1$ (resp., $d \geq 1$), and to $\delta(\mathsf{E}_1^{\geq 1}\psi, (\sigma, h))$ (resp., $\delta(\mathsf{A}_1^{<1}\psi, (\sigma, h))$), otherwise.

(6) Let $\mathsf{h}(\mathsf{E}\psi) = (d, d_0, d_1, \beta)$ (resp., $\mathsf{h}(\mathsf{A}\psi) = (d, d_0, d_1, \beta)$), where $\psi = \varphi_1 \mathsf{Op}\ \varphi_2$. For a state of the form $\mathsf{E}\psi$ (resp., $\mathsf{A}\psi$) we verify that this formula holds with degree d. If the node is not a duplicate of a previous node, i.e., $\sigma \neq \perp$, we have to check the formula, which should hold in the current node, by applying the one-step unfolding property derived by the semantics and reported in Corollary 1.3. At this point, we may need to propagate the formula in the two directions of the tree, by taking into account the requirements established by the degree in those directions. If such degree d_i is 0 (resp., ω) then the existential (resp., universal) formula is immediately true (resp., false). If $d_i = 1$ (resp., $d_i = 0$), we propagate a particular requirement with the meaning that we are looking for a path (resp., all paths) satisfying the internal path formula $\varphi_1 \mathsf{Op}\ \varphi_2$. Precisely, in order to differentiate between the two directions, we send the state $\mathsf{E}_i^{\geq 1}\varphi_1 \mathsf{Op}\ \varphi_2$ (resp., $\mathsf{A}_i^{<1}\varphi_1 \mathsf{Op}\ \varphi_2$) to direction $i \in \{0, 1\}$. If $d_i > 1$ (resp., $0 < d_i < \omega$) we propagate the original requirement by leaving to the degree of the successor nodes the task to specify how many paths (resp., do not) satisfy the inner formula. However, when we deal with the infinite degree ω (resp., finite but unbounded degree $\not{\omega}$) we have to ensure that the formula $\varphi_1 \mathsf{Op}\ \varphi_2$ is verified on infinitely (resp. falsified on finitely) many paths. To this aim, we use the apposite state $\mathsf{E}^{\geq \omega}\varphi_1 \mathsf{Op}\ \varphi_2$ (resp., $\mathsf{A}^{<\omega}\varphi_1 \mathsf{Op}\ \varphi_2$), which is sent on one direction iff on the other one there is a non null (resp., null) degree. In this way, we can keep track of a possible infinite splitting of the degree which is required (resp., forbidden) by an infinite (resp., finite) number of paths. In the following we describe such a propagation of the states

by means of the following macro: $\gamma_{\mathsf{EOp}}(d_0, d_1) \triangleq \gamma^0_{\mathsf{EOp}}(d_0, d_1) \wedge \gamma^1_{\mathsf{EOp}}(d_0, d_1)$ (resp., $\gamma_{\mathsf{AOp}}(d_0, d_1) \triangleq \gamma^0_{\mathsf{AOp}}(d_0, d_1) \wedge \gamma^1_{\mathsf{AOp}}(d_0, d_1)$), where

- $$\gamma^i_{\mathsf{EOp}}(d_0, d_1) \triangleq \begin{cases} \mathfrak{t}, & \text{if } d_i = 0; \\ (i, \mathsf{E}^{\geq 1}_i \varphi_1 \mathsf{Op}\ \varphi_2), & \text{if } d_i = 1; \\ (i, \mathsf{E}\varphi_1 \mathsf{Op}\ \varphi_2), & \text{if } d_i < \omega; \\ (i, \mathsf{E}\varphi_1 \mathsf{Op}\ \varphi_2), & \text{if } d_i = \omega \text{ and } d_{1-i} = 0; \\ (i, \mathsf{E}^{\geq \omega}\varphi_1 \mathsf{Op}\ \varphi_2) \wedge & \\ \wedge (1-i, \mathsf{E}^{\geq 1}_{1-i}\varphi_1 \mathsf{Op}\ \varphi_2), & \text{if } d_i = \omega \text{ and } d_{1-i} \neq 0. \end{cases}$$
- $$\gamma^i_{\mathsf{AOp}}(d_0, d_1) \triangleq \begin{cases} (i, \mathsf{A}^{<1}_i \varphi_1 \mathsf{Op}\ \varphi_2), & \text{if } d_i = 0; \\ (i, \mathsf{A}\varphi_1 \mathsf{Op}\ \varphi_2), & \text{if } d_i < \not\omega; \\ (i, \mathsf{A}\varphi_1 \mathsf{Op}\ \varphi_2), & \text{if } d_i = \not\omega \text{ and } d_{1-i} = 0; \\ (i, \mathsf{A}^{<\omega}\varphi_1 \mathsf{Op}\ \varphi_2), & \text{if } d_i = \not\omega \text{ and } d_{1-i} \neq 0; \\ \mathfrak{f}, & \text{if } d_i = \omega. \end{cases}$$

Observe that the last case requires the existence of a path satisfying the inner formula ψ in the direction $1-i$. This is due to the fact that, when we verify the existential formula with infinite degree, we risk that the latter is always regenerated without actually completing a real path satisfying ψ. By coupling this condition with that about the infinite generation, we ensure that we actually find infinitely many paths satisfying ψ. (Resp., the first to last case may also require that in the direction $1-i$ there is no path falsifying the inner formula ψ. However, this requirement is implicit in the whole structure of $\gamma_{\mathsf{AOp}}(d_0, d_1)$.)

(7) Let $\mathsf{h}(\mathsf{Qn}\ \psi) = (d, d_0, d_1, \beta)$ with $\psi = \varphi_1 \mathsf{Op}\ \varphi_2$. Due to the meaning of the flag β, when $\beta = \not\flat$, the automaton has to verify that either ψ or $\neg\psi$ is a tautology. On the contrary, when $\beta = \flat$, it has to verify that no one of them is a tautology. Thus, we need two components of the transition function, $\eta_\psi(\sigma, \mathsf{h})$ and $\overline{\eta}_\psi(\sigma, \mathsf{h})$, to ensure, respectively, that ψ is or not a tautology on a node labeled with σ. These components have to require the automaton to check the truth of the formulas equivalent to the tautologies, as described in Theorem 1.13.

- $\eta_{\varphi_1 \mathsf{U} \varphi_2}(\sigma, \mathsf{h}) \triangleq \delta(\varphi_2, (\sigma, \mathsf{h}))$;
- $\overline{\eta}_{\varphi_1 \mathsf{U} \varphi_2}(\sigma, \mathsf{h}) \triangleq \delta(\neg\varphi_2, (\sigma, \mathsf{h}))$;
- $\eta_{\varphi_1 \mathsf{R} \varphi_2}(\sigma, \mathsf{h}) \triangleq \delta(\varphi_1, (\sigma, \mathsf{h})) \wedge \delta(\varphi_2, (\sigma, \mathsf{h}))$;
- $\overline{\eta}_{\varphi_1 \mathsf{R} \varphi_2}(\sigma, \mathsf{h}) \triangleq \delta(\neg\varphi_1, (\sigma, \mathsf{h})) \vee \delta(\neg\varphi_2, (\sigma, \mathsf{h}))$;
- $\eta_{\varphi_1 \tilde{\mathsf{U}} \varphi_2}(\sigma, \mathsf{h}) \triangleq \delta(\mathsf{A}^{<1}\varphi_1 \tilde{\mathsf{U}} \varphi_2, (\sigma, \mathsf{h}))$;
- $\overline{\eta}_{\varphi_1 \tilde{\mathsf{U}} \varphi_2}(\sigma, \mathsf{h}) \triangleq \delta(\mathsf{E}^{\geq 1}\neg(\varphi_1 \tilde{\mathsf{U}} \varphi_2), (\sigma, \mathsf{h}))$;
- $\eta_{\varphi_1 \tilde{\mathsf{R}} \varphi_2}(\sigma, \mathsf{h}) \triangleq \delta(\mathsf{A}^{<1}\varphi_1 \tilde{\mathsf{R}} \varphi_2, (\sigma, \mathsf{h}))$;
- $\overline{\eta}_{\varphi_1 \tilde{\mathsf{R}} \varphi_2}(\sigma, \mathsf{h}) \triangleq \delta(\mathsf{E}^{\geq 1}\neg(\varphi_1 \tilde{\mathsf{R}} \varphi_2), (\sigma, \mathsf{h}))$.

(8) Now, we discuss the general structure of a transition function for a state of the form $\mathsf{E}\psi$ (resp., $\mathsf{A}\psi$) with $\psi = \varphi_1 \mathsf{Op}\ \varphi_2$. Let $\mathsf{h}(\mathsf{E}\psi) = (d, d_0, d_1, \beta)$ (resp., $\mathsf{h}(\mathsf{A}\psi) = (d, d_0, d_1, \beta)$). Note that the degree d is never equal to 0 or 1 (resp.

0 or ω), because the requirement $\gamma_{\mathsf{EOp}}(d_0, d_1)$ (resp., $\gamma_{\mathsf{AOp}}(d_0, d_1)$) discussed above never propagates an existential (resp., universal) state without degree on a direction i when $d_i = 0$ or $d_i = 1$ (resp. $d_i = 0$ or $d_i = \omega$). If the node is not a duplicate of a previous node, i.e., $\sigma \neq \bot$, we verify that the formula holds in the current node by applying the one-step unfolding property derived by the semantics, as reported in Corollary 1.3. Precisely, since $d > 1$ (resp. $0 < d < \omega$) ψ cannot (resp., can) be a tautology, otherwise (resp., since) we would find only one minimal path satisfying ψ. On the other hand, $\neg\psi$ cannot (resp., can) be a tautology, otherwise (resp., since) we would find only one minimal path non satisfying ψ. So, the automaton has to verify that ψ and $\neg\psi$ are not tautologies in the current node and has to propagate the existential state on the successors through the $\gamma_{\mathsf{EOp}}(d_0, d_1)$ requirement (resp., the automaton has to verify either that ψ or $\neg\psi$ is a tautology or that both are not tautologies and that the universal requirement $\gamma_{\mathsf{AOp}}(d_0, d_1)$ is propagated on the successors). Due to the non-tautological nature of ψ and $\neg\psi$, the automaton has to reject the input tree when $\beta = \not\flat$ (resp. the automaton has to verify that ψ or $\neg\psi$ is a tautology iff $\beta = \not\flat$). If $\sigma = \bot$ then the current node is simply a replica of a previous node with $\sigma \neq \bot$. Since the existential (resp., universal) state have been propagated on direction 0, we already know that ψ and $\neg\psi$ are not tautologies, hence we need just to propagate the state through the relative $\gamma_{\mathsf{EOp}}(d_0, d_1)$ (resp., $\gamma_{\mathsf{AOp}}(d_0, d_1)$) requirement. Due to the fact that, when $\sigma = \bot$, both ψ and $\neg\psi$ are not tautologies, the automaton has to reject the tree when $\beta = \not\flat$.

- $\delta(\mathsf{E}\psi, (\sigma, \mathsf{h})) \triangleq \begin{cases} \mathfrak{f}, & \text{if } \beta = \not\flat; \\ \gamma_{\mathsf{EOp}}(d_0, d_1), & \text{if } \sigma = \bot \text{ and } \beta = \flat; \\ \overline{\eta}_{\psi}(\sigma, \mathsf{h}) \wedge \overline{\eta}_{\neg\psi}(\sigma, \mathsf{h}) & \text{if } \sigma \neq \bot \text{ and } \beta = \flat. \\ \wedge \gamma_{\mathsf{EOp}}(d_0, d_1), & \end{cases}$
- $\delta(\mathsf{A}\psi, (\sigma, \mathsf{h})) \triangleq \begin{cases} \mathfrak{f}, & \text{if } \sigma = \bot \text{ and } \beta = \not\flat; \\ \gamma_{\mathsf{AOp}}(d_0, d_1), & \text{if } \sigma = \bot \text{ and } \beta = \flat; \\ \eta_{\psi}(\sigma, \mathsf{h}) \vee \eta_{\neg\psi}(\sigma, \mathsf{h}), & \text{if } \sigma \neq \bot \text{ and } \beta = \not\flat; \\ \overline{\eta}_{\psi}(\sigma, \mathsf{h}) \wedge \overline{\eta}_{\neg\psi}(\sigma, \mathsf{h}) & \text{if } \sigma \neq \bot \text{ and } \beta = \flat. \\ \wedge \gamma_{\mathsf{AOp}}(d_0, d_1), & \end{cases}$

Note that, the whole transition function can be simplified, case by case, because of the redundancy of some of its components. For example, consider the case EU when $\sigma \neq \bot$ and $\beta = \flat$. By definition, we obtain that $\delta(\mathsf{E}\varphi_1\mathsf{U}\varphi_2, (\sigma, \mathsf{h})) = \delta(\neg\varphi_2, (\sigma, \mathsf{h})) \wedge \delta(\mathsf{E}^{\geq 1}\varphi_1\mathsf{U}\varphi_2, (\sigma, \mathsf{h})) \wedge \gamma_{\mathsf{EU}}(d_0, d_1)$, which can be equivalently written as follows: $\delta(\neg\varphi_2, (\sigma, \mathsf{h})) \wedge \delta(\varphi_1, (\sigma, \mathsf{h})) \wedge \delta(\mathsf{E}^{\geq 1}\mathsf{X}\mathsf{E}^{\geq 1}\varphi_1\mathsf{U}\varphi_2, (\sigma, \mathsf{h})) \wedge \gamma_{\mathsf{EU}}(d_0, d_1)$. Now, since the requirement $\gamma_{\mathsf{EU}}(d_0, d_1)$ ensure the existence of $d = d_0 + d_1 > 1$ non equivalent paths starting on the successors of the current node, we have that the $\delta(\mathsf{E}^{\geq 1}\mathsf{X}\mathsf{E}^{\geq 1}\varphi_1\mathsf{U}\varphi_2, (\sigma, \mathsf{h}))$ component is surely verified. So, this piece is redundant. The remaining expression $\delta(\neg\varphi_2, (\sigma, \mathsf{h})) \wedge \delta(\varphi_1, (\sigma, \mathsf{h})) \wedge \gamma_{\mathsf{EU}}(d_0, d_1)$ simply reflects what is required by Item 1 of Corollary 1.3. Now, for a state of the form $\mathsf{E}^{\geq g}\psi$ (resp., $\mathsf{A}^{<g}\psi$), with $g \in [2, \omega]$, we have only

to further verify that $d \geq g$ (resp., $d < g$). Formally, $\delta(\mathsf{E}^{\geq g}\psi, (\sigma, \mathsf{h}))$ (resp., $\delta(\mathsf{A}^{<g}\psi, (\sigma, \mathsf{h}))$) is set to $\mathfrak{f}$, if $d < g$ (resp., $d \geq g$), and to $\delta(\mathsf{E}\psi, (\sigma, h))$ (resp., $\delta(\mathsf{A}\psi, (\sigma, h))$), otherwise.

We now briefly discuss the parity acceptance condition for $\mathscr{A}_\varphi$. Note that, in our reasonings, we assume $\mathrm{F}_\varphi = (\mathrm{F}_1, \mathrm{F}_2, \mathrm{F}_3)$ with $\mathrm{F}_3 = \mathrm{Q}_\varphi$.

Let $\mathscr{T}$ be a complete $\mathbb{N}$-tree, $\mathscr{T}_{D_{\mathrm{B},g}}$ be one of its B-based g-degree delayed generation in input to $\mathscr{A}_\varphi$, and $\mathscr{R}$ be a related run. It is easy to see that states in $\mathsf{cl}(\varphi) \setminus \mathsf{mcl}_\omega(\varphi)$ represents literals, ands, ors, and quantified formulas with finite degree that never generate themselves, so, they never progress infinitely often. On the other hand, formulas in $\mathsf{mcl}(\varphi) \cup \mathsf{qcl}(\varphi)$ may be generated infinitely often, but only some of them should be allowed to do so (due to their intrinsic semantics).

(1) Existential next states $\mathsf{EX}\varphi$ and $\mathsf{E}^{\geq\omega}\mathsf{X}\varphi$ are never sent to direction 1 and they can only progress indefinitely along direction 0. The propagation of an existential formula without degree represents a delay of the choice of the particular successors of the replicated node on which it is needed to verify φ. When the associated degree is finite, the formula needs to be satisfied on a finite number of successors. So, the choice of the successors must be eventually made, and the formula cannot be propagated indefinitely. When the degree is infinite, instead, the formula is allowed to progress under the condition that successors satisfying ψ are found infinitely often. Hence, we use two states: a ω-grade version is generated every time a successor satisfying φ is found and a grade-less version is used when the successor is skipped. Hence, the existential next formulas $\mathsf{EX}\varphi$ is not allowed to progress indefinitely and, thus, it belongs to F_3 but not to F_2. On the other hand the formulas $\mathsf{E}^{\geq\omega}\mathsf{X}\varphi$ are allowed to occur infinitely often and, thus, they belong to F_2 but not F_1.
(2) Universal next states $\mathsf{AX}\varphi$ and $\mathsf{A}^{\geq\omega}\mathsf{X}\varphi$ are never sent to direction 1 and they can only progress indefinitely along direction 0. An infinite generation of an universal next formulas represents the propagation of a requirement demanded on infinitely many successors of the replicated node with the aim to check that only a finite number of them does not satisfy it. This should be allowed, however, when the associated degree is finite but not a priori determined, i.e., if it is $\not\omega$. Generally, this degree can be split infinitely many times without decreasing, so, we risk to allow infinitely many successors to not satisfy φ. In order to avoid such a problem, we use two states: a ω-grade version is generated every time a successor is allowed to not satisfy $\neg\varphi$ and a grade-less version is used when the successor satisfies φ. Hence, the universal formulas $\mathsf{AX}\varphi$ is allowed to progress indefinitely on such branches and, thus, it belongs to F_2 but not to F_1. On the other hand the universal formula $\mathsf{A}^{<\omega}\mathsf{X}\varphi$ is not allowed to occur infinitely often, even when $\mathsf{AX}\varphi$ does, thus, it belongs to F_1.
(3) Existential non-next formulas $\mathsf{E}_{\bar{i}}^{\geq 1}\psi$, with degree 1, have to trace a path satisfying the inner path formula $\psi \in \{\varphi_1 \mathsf{U} \varphi_2, \varphi_1 \mathsf{R} \varphi_2, \varphi_1 \tilde{\mathsf{U}} \varphi_2, \varphi_1 \tilde{\mathsf{R}} \varphi_2\}$. When ψ is an until or weak until formula, the path has to eventually reach a point in which the formula is locally satisfied. So, the relative states $\mathsf{E}_{\bar{i}}^{\geq 1}\psi$ are not allowed to progress indefinitely and, thus, they belong to F_3 but not to F_2. When ψ is a

release or weak release formula, it may happened that there are no points in which the formula is locally satisfied. However, only paths that progress infinitely often along direction 1 are real paths of the input tree (following the replica indefinitely would yield no path). Hence, states $\mathsf{E}_{\bar{0}}^{\geq 1}\psi$ belong to F_3 but not to F_2, and states $\mathsf{E}_{\bar{1}}^{\geq 1}\psi$ belong to F_2 but not F_1.

(4) Universal non-next formulas $\mathsf{A}_i^{<1}\psi$, with degree 1, have to trace all paths and prove that they satisfy the inner path formula $\psi \in \{\varphi_1 \mathsf{U} \varphi_2, \varphi_1 \mathsf{R} \varphi_2, \varphi_1 \tilde{\mathsf{U}} \varphi_2, \varphi_1 \tilde{\mathsf{R}} \varphi_2\}$. When ψ is a release or weak release formula, it may happened that there are no points in which the formula is locally satisfied. So, the relative states $\mathsf{A}_i^{<1}\psi$ are allowed to progress indefinitely and, thus, they belong to F_2 but not to F_1. When ψ is an until or weak until formula, the path has to eventually reach a point in which the formula is locally satisfied. However, we need to propagate it infinitely often along direction 0, in order to ask it on all successor of the replicated node. Now, since on paths that progress infinitely often along direction 1 it is possible to generate both the states $\mathsf{A}_{\bar{0}}^{\geq 1}\psi$ and $\mathsf{A}_{\bar{1}}^{\geq 1}\psi$, the infinite generation of $\mathsf{A}_{\bar{1}}^{\geq 1}\psi$ has an higher non-acceptance priority with respect to that of $\mathsf{A}_{\bar{0}}^{\geq 1}\psi$. This is due to the fact that those paths represent real branches of the input tree where ψ need to eventually hold. Hence, states $\mathsf{A}_{\bar{0}}^{\geq 1}\psi$ belong to F_2 but not to F_1, and states $\mathsf{A}_{\bar{1}}^{\geq 1}\psi$ belong to F_1.

(5) Existential non-next formulas with infinite degree $\mathsf{E}^{\geq \omega}\psi$ or without degree $\mathsf{E}\psi$ have to trace a non singleton set of paths satisfying the inner path formula $\psi \in \{\varphi_1 \mathsf{U} \varphi_2, \varphi_1 \mathsf{R} \varphi_2, \varphi_1 \tilde{\mathsf{U}} \varphi_2, \varphi_1 \tilde{\mathsf{R}} \varphi_2\}$. One one hand, if the number of such paths is finite, the automaton will eventually reach a node from which there in only one outgoing path model of ψ, since all the paths have to eventually split. When this happens, the automaton verify the existence of such a path with the relative 1-grade version $\mathsf{E}_{\bar{i}}^{\geq 1}\psi$. Hence, when a grade-less formula is accompanied by a finite degree it must not progress infinitely often. On the other hand, when the number of paths the automaton needs to follow is infinite, we should allow the existential formula to progress infinitely often. However, by doing so, we risk to trace just one path in the input tree along which we propagate the existential formula and, obviously, it cannot provide the infinite number of paths we need in order to verify the formula itself. Thus, when we propagate the existential requirement on direction i, we have to use the two versions of the requirement itself. The ω-grade formula is sent on direction i when on direction $1-i$ is ensured the existence of a path satisfying ψ. Instead, the grade-less version is used when such an existence is not verified. Consequently, when the ω-grade version is generated infinitely often along the path, there are infinite branches coming out from this and satisfying ψ. On the contrary, when the grade-less version is definitively propagated, we are just following a unique path which cannot provide the infinite paths we need. Hence, all grade-less non-next existential formula belong to F_3 but not to F_2 and their ω-grade versions belong to F_2 but not to F_1.

(6) Universal non-next formulas with infinite degree $\mathsf{A}^{<\omega}\psi$ or without degree $\mathsf{A}\psi$ have to trace a set of paths that are allowed to not satisfy the inner path formula

$\psi \in \{\varphi_1 \mathsf{U} \varphi_2, \varphi_1 \mathsf{R} \varphi_2, \varphi_1 \tilde{\mathsf{U}} \varphi_2, \varphi_1 \tilde{\mathsf{R}} \varphi_2\}$. There may be cases in which the automaton eventually reach a node from which there are no outgoing paths model of $\neg\psi$. When this happens, the automaton needs to verify the universal validity of ψ with the relative 1-grade version $\mathsf{A}_i^{<1}\psi$. Also, the automaton may reach a point where the ψ or $\neg\psi$ are tautologies and, thus, it stops by verifying one of them. However, it is also possible that the universal requirement progress infinitely often. In such a case, we have that it is tracing one path that may not satisfy ψ, even if it would be allowed to trace more paths. Since the accompanying degree is greater than 0, this does not result to be a problem and, hence, we allow the infinite propagation. Moreover, every time we meet an universal formula with finite but non a priori determined degree, i.e., if such degree is $\not\omega$, the formula may split in the two direction and allow paths to not satisfy the ψ formula on both of them. If this happens infinitely often along the single path on which we are propagating the requirement, we would allow an infinite numbers of path to not satisfy ψ, contradicting what we want to verify. Thus, when we propagate the universal requirement on direction i, we have to use the two versions of the requirement itself. The ω-grade formula is sent on direction i when on direction $1-i$ is allowed the existence of a path non-satisfying ψ. Instead, the grade-less version is used when such an existence is forbidden. Consequently, when the ω-grade version is generated infinitely often along the path, there may be infinite branches coming out from this and non-satisfying ψ. On the contrary, when the grade-less version is definitively propagated, we are just following a unique path which does not allow the existence of the infinite number of paths we want to avoid. Hence, all grade-less non-next universal formula belong to F_2 but not to F_1 and their ω-grade versions belong to F_1.

We now prove the following main result about the decidability of GCTL satisfiability.

Theorem 1.19 (GCTL **Satisfiability**) *Let φ be a* GCTL *formula, with $g = \mathsf{deg}(\varphi)$, $\mathrm{B} = \mathsf{qcl}(\varphi)$, $\mathrm{B}_{\text{sup}} = \mathsf{qcl}_{\mathsf{E}}(\varphi)$, and $\mathrm{B}_{\text{inf}} = \mathsf{qcl}_{\mathsf{A}}(\varphi)$. Then, φ is satisfiable iff $\mathrm{L}(\langle \mathscr{A}_\varphi, \mathscr{S}_{\mathrm{B},g}^{\mathrm{B}_{\text{sup}},\mathrm{B}_{\text{inf}}} \rangle) \neq \emptyset$.*

Proof [Only if]. Given a 2^{AP}-labeled $\mathbb{N}$-tree $\mathscr{T} = \langle \mathrm{T}, \mathsf{v} \rangle$ model of φ, we first show how to recursively construct one of its B-based g-degree delayed generation trees $\mathscr{T}_{D_{\mathrm{B},g}} = \langle \{0,1\}^*, \mathsf{v}_{D_{\mathrm{B},g}} \rangle$, necessarily full coherent w.r.t. the pair $(\mathrm{B}_{\text{sup}}, \mathrm{B}_{\text{inf}})$, along with a partial map $\mathsf{t} : \{0,1\}^* \rightharpoonup \mathrm{T}$ that links each node $x \in \{0,1\}^*$ of $\mathscr{T}_{D_{\mathrm{B},g}}$, with $\mathsf{v}_{D_{\mathrm{B},g}}(x) = (\sigma, \mathsf{h})$ and $\sigma \neq \#$, to the corresponding one $\mathsf{t}(x) \in \mathrm{T}$ in $\mathscr{T}$. This function, is simply the restriction to real nodes, i.e., nodes not labeled with #, of the s function introduced in Definition 1.26 of the delayed generation.

To each subtree $\mathscr{T}_{D_{\mathrm{B},g}}^x$ of $\mathscr{T}_{D_{\mathrm{B},g}}$ rooted in $x = x' \cdot 0^j$, with $x' \in \{\varepsilon\} \cup 0^* \cdot 1$, $\mathsf{v}_{D_{\mathrm{B},g}}(x) = (\sigma, \mathsf{h})$ such that $\sigma \neq \#$, we associate the subtree $\mathscr{T}^x$ of $\mathscr{T}$ rooted in $y = \mathsf{t}(x)$. Observe that $\mathscr{T}^{x \cdot 1}$ is the subtree of $\mathscr{T}$ rooted at the $(j+1)$-th successor of y and that $\mathscr{T}^{x \cdot 0} = \mathscr{T}^x$. Moreover, by $\mathscr{T}'^x$ we denote the subtree of $\mathscr{T}^x$ in which the first j successors of the root are deleted. Note that $\mathcal{T}'^{x\cdot 1} = \mathcal{T}^{x\cdot 1}$ and $\mathcal{T}'^{x\cdot 0}$ is the subtree of $\mathcal{T}'^{x}$ with the first successor of the root deleted.

In the rest of the proof, we say that a path formula ψ is *locally determined* on a node x iff either ψ or $\neg\psi$ is an $\equiv^{\varepsilon}_{\mathcal{T}x}$-tautology.

For each node $x \in \{0,1\}^*$ and base $b \in \mathrm{B}$ with $\mathsf{v}_{D_{\mathrm{B},g}}(x) = (\sigma, \mathsf{h})$, $\mathsf{h}(b) = (d, d_0, d_1, \beta)$, $\mathsf{v}_{D_{\mathrm{B},g}}(x \cdot 0) = (\sigma_0, \mathsf{h}_0)$, and $\mathsf{v}_{D_{\mathrm{B},g}}(x \cdot 1) = (\sigma_1, \mathsf{h}_1)$ we set: if $\sigma = \#$ then $d = d_0 = d_1 \triangleq 0$ and $\beta \triangleq \not\flat$, if $\sigma_0 = \#$ then $d_0 \triangleq 0$, if $\sigma_1 = \#$ then $d_1 \triangleq 0$. For the other cases, we set the values of the degrees as follows, where we recall that $\not\omega$ is in place of any finite number greater than g.

(1) $b = \mathsf{EX}\varphi$. Then, $\beta \triangleq \not\flat$ and d (resp., d_0) is set to the maximum degree $l \in [0, g] \cup \{\not\omega, \omega\}$ with which the formula $\mathsf{E}^{\geq l}\mathsf{X}\varphi$ is satisfied on $\mathcal{T}'^{x}$ (resp., $\mathcal{T}'^{x\cdot 0}$ if $\sigma_0 \neq \#$). Moreover, d_1 is set to 1, if φ is satisfied on $\mathcal{T}'^{x\cdot 1}$, and to 0 otherwise.
(2) $b = \mathsf{A}\tilde{\mathsf{X}}\varphi$. Then, $\beta \triangleq \not\flat$ and d (resp., d_0) is set to the minimum degree $l \in [0, g] \cup \{\not\omega, \omega\}$ with which the formula $\mathsf{A}^{<l+1}\tilde{\mathsf{X}}\varphi$ is satisfied on $\mathcal{T}'^{x}$ (resp., $\mathcal{T}'^{x\cdot 0}$ if $\sigma_0 \neq \#$). Moreover, d_1 is set to 1, if φ is not satisfied on $\mathcal{T}'^{x\cdot 1}$, and to 0 otherwise.
(3) $b = \mathsf{E}\psi$ is a non-next formula. Then, $\beta \triangleq \not\flat$ if ψ is locally determined on x. If $\beta = \flat$, then d (resp., d_0, d_1) is set to the maximum degree $l \in [0, g] \cup \{\not\omega, \omega\}$ with which the formula $\mathsf{E}^{\geq l}\mathsf{X}\psi$ (resp., $\mathsf{E}^{\geq l}\mathsf{X}\psi$, $\mathsf{E}^{\geq l}\psi$) is satisfied on $\mathcal{T}'^{x}$ (resp., $\mathcal{T}'^{x\cdot 0}$ if $\sigma_0 \neq \#$, $\mathcal{T}'^{x\cdot 1}$ if $\sigma_1 \neq \#$). If $\beta = \not\flat$, only d and d_0 are set as stated before, while d_1 is arbitrary.
(4) $b = \mathsf{A}\psi$ is a non-next formula. Then, $\beta \triangleq \not\flat$ if ψ is locally determined on x. If $\beta = \flat$, then d (resp., d_0, d_1) is set to the minimum degree $l \in [0, g] \cup \{\not\omega, \omega\}$ with which the formula $\mathsf{A}^{<l+1}\mathsf{X}\psi$ (resp., $\mathsf{A}^{<l+1}\mathsf{X}\psi$, $\mathsf{A}^{<l+1}\psi$) is satisfied on $\mathcal{T}'^{x}$ (resp., $\mathcal{T}'^{x\cdot 0}$ if $\sigma_0 \neq \#$, $\mathcal{T}'^{x\cdot 1}$ if $\sigma_1 \neq \#$). If $\beta = \not\flat$, only d and d_0 are set as stated before, while d_1 is arbitrary.

It is immediate to see that, if $\beta = \flat$ then $d = d_0 + d_1$. Moreover, let $\mathsf{h}_0(b) = (d^0, d^0_0, d^0_1, \beta^0)$ and $\mathsf{h}_1(b) = (d^1, d^1_0, d^1_1, \beta^1)$, we have that $d^0 = d_0$ and if $\beta = \flat$ then $d^1 = d_1$. Now, by Definition 1.30, we can derive that the tree $\mathcal{T}_{D_{\mathrm{B},g}}$ is actually full coherent w.r.t. the pair $(\mathrm{B}_{\mathrm{sup}}, \mathrm{B}_{\mathrm{inf}})$. Hence, by Theorem 1.18, we have that it can be obtained as a building of the satellite $\mathcal{S}^{\mathrm{B}_{\mathrm{sup}},\mathrm{B}_{\mathrm{inf}}}_{\mathrm{B},g}$ over the delayed generation $\mathcal{T}_D$ of $\mathcal{T}$ itself.

It remains to prove that $\mathcal{T}_{D_{\mathrm{B},g}}$ is accepted by $\mathcal{A}_\varphi$. The proof proceeds by induction on the structure of the set of states derived by the formula φ and on the degree d associated to the state. In particular, we use the following ordering $\prec \subseteq \mathrm{Q} \times \mathrm{Q}$ between states: *(i)* for all formulas $\varphi', \varphi'' \in \mathrm{Q}$ with $\varphi'' \in \mathsf{ecl}(\varphi')$ and $\varphi'' \neq \varphi'$, we set $\varphi'' \prec \varphi'$; *(ii)* $\mathsf{E}\psi \prec \mathsf{E}^{\geq l}\psi$ (resp., $\mathsf{A}\psi \prec \mathsf{A}^{<l}\psi$) and $\mathsf{E}^{\geq\omega}\psi \prec \mathsf{E}^{\geq l}\psi$ (resp., $\mathsf{A}^{<\omega}\psi \prec \mathsf{A}^{<l}\psi$), for all $l \in [2, \omega[$; *(iii)* $\mathsf{E}^{\geq 1}\psi \prec \mathsf{E}\psi$ (resp., $\mathsf{A}^{<1}\psi \prec \mathsf{A}\psi$) and $\mathsf{E}^{\geq 1}\psi \prec \mathsf{E}^{\geq\omega}\psi$ (resp., $\mathsf{A}^{<1}\psi \prec \mathsf{A}^{<\omega}\psi$); *(iv)* $\mathsf{E}^{\geq 1}_i\psi \prec \mathsf{E}^{\geq 1}\psi$ (resp., $\mathsf{A}^{<1}_i\psi \prec \mathsf{A}^{<1}\psi$), for all $i \in \{0, 1\}$. We also use the following inductive hypotheses: *(i)* each state $q = \mathsf{E}\psi$ is sent to a node x with the related degree greater than 1, i.e., with $\mathsf{v}_{D_{\mathrm{B},g}}(x) = (\sigma, \mathsf{h})$, $\mathsf{h}(q) = (d, d_0, d_1, \beta)$, and $d > 1$; *(ii)* each state $q = \mathsf{E}^{\geq\omega}\psi$ is sent to a node x

with infinite related degree, i.e., with $\mathsf{v}_{D_{B,g}}(x) = (\sigma, \mathsf{h})$, $\mathsf{h}(\mathsf{E}\psi) = (d, d_0, d_1, \beta)$, and $d = \omega$.

Intuitively, if the automaton $\mathscr{A}_\varphi$ is on a state $q = \mathsf{Qn}\ \psi$ (resp. $q = \mathsf{Qn}\ \mathsf{X}\psi$), where Qn is a quantification, on a node x of the tree $\mathscr{T}_{D_{B,g}}$, with label $\mathsf{v}_{D_{B,g}}(x) = (\sigma, \mathsf{h})$ and $\sigma \neq \#$, then it accepts the subtree $\mathscr{T}^x_{D_{B,g}}$ if either it is able to check the truth of formulas of lower order than q w.r.t. $\prec$, implying already the validity of q itself, or it checks other formulas lower than ψ w.r.t. $\prec$, implying the non-validity of the negation of the formula represented by q, and verifies that the subtree $\mathscr{T}'^x$ satisfies the formula represented by $\mathsf{Qn}\ \mathsf{X}\psi$ (resp., $\mathsf{Qn}\ \psi$) with degree given either by the formula q itself or, if such degree is not present in it, by the d component of the function h valuated on the relative base.

We now give a detailed explanation only for the inductive case of $q = \mathsf{E}\psi$ with $\psi = \varphi_1 \mathsf{Op}\ \varphi_2$, when we are on a node $x = x' \cdot 1$. The other cases are a variation on theme.

Let $\mathsf{h}(q) = (d, d_0, d_1, \beta)$. By the inductive hypothesis, the degree d is greater than 1. Hence ψ is not a tautology (otherwise, we would find only one path satisfying ψ). So, we have $\beta = \flat$. Consequently, the related path formulas $\mathsf{X}\psi$ and ψ are true on some of the successors of $\mathsf{t}(x)$ partitioned between $\mathscr{T}'^{x\cdot 0}$ and $\mathscr{T}'^{x\cdot 1}$. Precisely, we have $\mathsf{E}^{=d_0}\mathsf{X}\psi$ is satisfied on $\mathscr{T}'^{x\cdot 0}$ and $\mathsf{E}^{=d_1}\psi$ is satisfied on $\mathscr{T}^{x\cdot 1}$. The transition function checks that ψ and $\neg\psi$ are not tautologies, by verifying formulas of lower order than ψ w.r.t. $\prec$, through the use of the components $\overline{\eta}_\psi(\sigma, \mathsf{h})$ and $\overline{\eta}_{\neg\psi}(\sigma, \mathsf{h})$. Moreover, the transition function verifies the same state q on $\mathscr{T}'^{x\cdot 0}$ and $\mathscr{T}^{x\cdot 1}$, through the component $\gamma_{\mathsf{EOp}}(d_0, d_1)$. Observe that this formula sends the states $\mathsf{E}\psi$ and $\mathsf{E}^{\geq\omega}\psi$ on direction i only if $d_i > 1$ and $d_i = \omega$, respectively.

At this point, we have to distinguish between the two cases $d < \omega$ and $d = \omega$.

In the first, it is possible that the automaton needs to check only states of lower order w.r.t. $\prec$, so the acceptance is deduced by the inductive hypothesis. On the contrary, it may also happen that the state propagates itself with the same degree on one direction. But, this propagation cannot happen indefinitely, since the degree eventually splits, and so, it eventually incurs in the first possibility.

In the second case, instead, the state q surely propagates on one direction q itself or its ω-degree version $\mathsf{E}^{\geq\omega}\psi$. So, the induction does not reach a lower case. Let $t = x_0 \cdot x_1 \cdot \dots$ with $x_0 = x$ be the branch on which the infinite degree d is propagated: formally, for each $k \in \mathbb{N}$ with $\mathsf{v}_{D_{B,g}}(x_k) = (\sigma, \mathsf{h})$ and $\mathsf{h}(\mathsf{E}\psi) = (d^k, d_0^k, d_1^k, \beta^k)$, we have $d^k = \omega$. Moreover, let $\mathsf{f} : \mathbb{N} \to \{0, 1\}$ be the direction function that associates to each index $k \in \mathbb{N}$ the direction of the successor of x_k, i.e., $x_{k+1} = x_k \cdot \mathsf{f}(k)$. Then, we distinguish the two following cases, where only the first one can actually happen, meanwhile the second one yield a contradiction.

(1) $d^k_{1-\mathsf{f}(k)} > 0$, for infinitely many $k \in \mathbb{N}$. In this case, the automaton passes, on the branch t, through the state $\mathsf{E}^{\geq\omega}\psi$ infinitely often, so it accepts the branch t.
(2) $d^k_{1-\mathsf{f}(k)} = 0$, for all $k \in \mathbb{N}$. We distinguish two sub-cases: t progresses definitively on the direction 0 and t progresses infinitely often through direction 1.

(a) $\mathsf{f}(k) = 0$ so, $x_k = x_0 \cdot 0^k$, for all $k \in \mathbb{N}$. By construction of $\mathcal{T}^{x_0}$, we have that the tree $\mathcal{T}^{x_k \cdot 1}$ does not contain a path that satisfies the formula, for all $k \in \mathbb{N}$. This means that there is no path satisfying the formula through any successor of $\mathsf{t}(x_0)$. But this contradicts the hypothesis that $\mathcal{T}^x$ satisfies the q with infinite degree.

(b) $\mathsf{f}(k) = 1$, for infinitely many $k \in \mathbb{N}$. Than, there is an infinite set of indexes $\{j_0, j_1, \ldots\} \subseteq \mathbb{N}$ with $j_0 = 0$ such that, for all $l \in \mathbb{N}$ and $k \in [j_l, j_{l+1}[$, it holds that $x_k = x_{j_l} \cdot 0^{k-j_l}$, and $x_{j_{l+1}} = x_{j_l - 1} \cdot 1$. Let $y_l = \mathsf{t}(x_{j_l})$, for all $l \in \mathbb{N}$. Then, the branch $r = y_0 \cdot y_1 \cdot \cdots$ is an infinite path in $\mathcal{T}^{x_0}$ on which there are infinite non-equivalent paths that starting in y_l and satisfying ψ, for all $l \in \mathbb{N}$. Now, since $d^k_{1-\mathsf{f}(k)} = 0$, all these paths have to pass through y_{l+1}. By induction, we obtain that all the paths that start from y_0 and satisfy ψ must pass through all the nodes of r. But this is a contradiction, since it means that they are actually one unique path.

[If]. The converse direction is specular. Since a tree $\mathcal{T}_D$ is accepted by $\langle \mathcal{A}_\varphi, \mathcal{S}^{\mathrm{B_{sup}},\mathrm{B_{inf}}}_{\mathrm{B},g} \rangle$, we can assert that *(i)* it is actually a delayed generation of a 2^{AP}-labeled tree $\mathcal{T}$ and *(ii)* the B-based g-degree delayed generation tree $\mathcal{T}_{D_{\mathrm{B},g}}$ built by the satellite $\mathcal{S}^{\mathrm{B_{sup}},\mathrm{B_{inf}}}_{\mathrm{B},g}$ on $\mathcal{T}_D$ is full coherent w.r.t. $(\mathrm{B_{sup}}, \mathrm{B_{inf}})$ and it is accepted by $\mathcal{A}_\varphi$. Using these facts, by induction on the structure of the formula, we can prove that every time $\mathcal{A}_\varphi$ is in a state q on a node x of the tree $\mathcal{T}_{D_{\mathrm{B},g}}$ with label (σ, h), $\mathcal{T}^x$ satisfies the formula represented by q with the related degree iff the automaton accepts the subtree $\mathcal{T}^x_{D_{\mathrm{B},g}}$. Actually, this fact happens if x is a right node, i.e., when x does not terminate with 0. When x is a left node, the transition function only requires that $\mathcal{T}^x$ satisfies the next formulas in the one-step unfolding of q. However, since the formulas not in the scope of the next are yet verified on a previous right node, we also obtain that $\mathcal{T}^x$ satisfies the whole q. Finally, since $\mathcal{A}_\varphi$ accepts $\mathcal{T}^\varepsilon_{D_{B,g}}$ by hypothesis, we have that the tree $\mathcal{T}$ is a model of φ. □

By a matter of calculation, it holds that $|\mathcal{A}_\varphi| = \mathsf{O}(\mathsf{lng}(\varphi))$ and $|\mathcal{S}^{\mathrm{B_{sup}},\mathrm{B_{inf}}}_{\mathrm{B},g}| = 2^{\mathsf{O}(\mathsf{lng}(\varphi)\cdot\log(\deg(\varphi)))}$. Moreover, also the alphabet $\Sigma_\varphi \times \mathrm{P}_{E_\varphi}$ of the APTS has size $2^{\mathsf{O}(\mathsf{lng}(\varphi)\cdot\log(\deg(\varphi)))}$. By Theorem 1.17, we obtain that the emptiness problem for $\langle \mathcal{A}_\varphi, \mathcal{S}^{\mathrm{B_{sup}},\mathrm{B_{inf}}}_{\mathrm{B},g} \rangle$ can be solved in time $2^{\mathsf{O}(\mathsf{lng}(\varphi)^2\cdot(\log(\mathsf{lng}(\varphi))+\log(\deg(\varphi))))} \leq 2^{\mathsf{O}(\mathsf{siz}(\varphi)^3)}$. Moreover, by recalling that GCTL subsumes CTL, the following result follows.

Theorem 1.20 (GCTL **Satisfiability Complexity**) *The satisfiability problem for* GCTL *with binary coding of degrees is* EXPTIME-COMPLETE.

1.9 Discussion

Graded modalities refine classical existential and universal quantifiers by specifying the number of elements for which the existential requirement should hold/universal requirement may not hold. Earlier work studied the extension of the μCALCULUS by graded modality on successors and shown that the complexity of the related satisfiability problem stays EXPTIME-COMPLETE. In this work, we have introduced GCTL as an extension of CTL with graded modalities on paths, in order to count the number of equivalence classes of paths satisfying a given formula. We have proposed a general framework that allows to define different kinds of "graded extensions" of GCTL, depending on the specific equivalence relation one chooses among paths. Moreover, we have described reasonable properties that such an equivalence should satisfy and, as a concrete application of our general framework, we have studied a graded logic with path prefix equivalence based on the suitable concepts of minimality and conservativeness. This choice is aimed on counting a minimal way a Kripke structure has to satisfy a given formula in such a way we can ensure its satisfiability no matter how a minimal part is extended.

One of the main features of GCTL is the capability to express properties that are weaker than those definable with the universal quantifications $A\psi$ and stronger than those definable with the existential quantifications $E\psi$. In "*planning in nondeterministic domain*" (Cimatti et al. 1998, 2003), for example, the use of strong planning (i.e., all the goals have to be satisfied by all the computations) and weak planning (i.e., all the goals have to be satisfied by some computation) are two extreme ways to achieve a given purpose. With our logic, we are able to express "graded path specification" that can be considered as a compromise between strong and weak planning.

We have studied several properties of GCTL under the path prefix equivalence and all of them hold in the general case of graded numbers coded in binary. Among the others, we have proved that this logic can be reduced to the GμCALCULUS, but that it is at least exponentially more succinct. Also, we have studied the satisfiability problem and, by using a sharp automata-theoretic approach via a binary-tree encoding of models and a refinement of the technique involving satellite automata, we have shown that this problem is EXPTIME-COMPLETE, thus no harder than the one for CTL. This result, along with the fact that GCTL is exponentially more succinct of GμCALCULUS and much more "friendly" to use, make GCTL a very useful and powerful logic to be used in practice in formal system verification. It is important to note that, all the results we have achieved for GCTL with path prefix equivalence are based on the properties we have studied for a general path equivalence. Hence, all the technical constructions can be easily lifted to any other graded extension of CTL that respects those properties.

As we have reported before, our satisfiability algorithm for GCTL uses an automata-theoretic approach on the binary-tree encoding of the models of the formula. While the automata approach results in a natural and classical one, it may be also substituted by other techniques, such as the systems of infinite tableaux (Friedman et al. 2010), turning in an algorithm with the same overall complexity we achieve. On the contrary, the binary-tree encoding seems to be unavoidable even in the case of the tableaux approach. Indeed, by using the regular models, we need to label each node with a tuple of degree functions, used for the splitting, which are not of fixed size 3 anymore, but rather linear in the degree of the formula and so exponential in its size. Then, by applying either the automata or the tableaux approach it turns in an overall double-exponential algorithm.

As future work, there are several directions that could be investigated along with graded path modalities. In particular, we left open the solution of the satisfiability problem of GCTL*. However, by a simple variation of the technique developed in this work, one can easily obtain a 3EXPTIME upper bound, while a 2EXPTIME-HARD lower bound easily derives from the satisfiability of CTL*. In this case, is also worth investigating the use of the tableaux technique to try to match the known lower bound. By exploiting a similar idea of that used for GμCALCULUS, one could also investigate whether GCTL* is equivalent to CTL* augmented with graded world modalities (Counting-CTL* (Moller and Rabinovich 2003)). However, we conjecture that GCTL* is exponentially more succinct than Counting-CTL* (for GCTL and Counting-CTL, this result holds by simply applying the same idea used for the translation from GCTL to the GμCALCULUS). This result is important for its own, as it was shown in Moller and Rabinovich 2003) that Counting-CTL* is equivalent to *monadic path logic*, which is MSOL with set quantifications restricted to paths.

Chapter 2
Minimal Model Quantifiers

Abstract Temporal logics are a well investigated formalism for the specification and verification of reactive systems. Using formal verification techniques, one can ensure the correctness of a system with respect to a desired behavior, i.e., the specification, by verifying whether a model of the former satisfies a temporal logic formula expressing the latter. In this setting, a very crucial aspect is to reasoning about substructures of the entire model. Indeed, for several fundamental problems, the formal verification approach requires to select a portion of the model of interest on which to verify a specific property. In this paper, we introduce a new logic framework that allows to select automatically desired parts of the system to be successively verified. Specifically, we extend the classical branching-time temporal logic CTL* by means of *minimal model operators* (MCTL*, for short). These operators allow to extract, from a model, minimal submodels on which we can check a specification, which is also given by an MCTL* formula. We interpret the logic under three different semantics, called *minimal* (*m*), *minimal-unwinding* (*mu*), and *unwinding-minimal* (*um*), which differ one from another on the way a substructure is extracted and then checked in the verification process. We show that both MCTL^*_m and MCTL^*_{mu} are strictly more expressive than CTL*, since these logics are sensible to unwinding and not invariant under bisimulation. Conversely, MCTL^*_{um} preserves both these properties. As far as the satisfiability concerns, we prove that MCTL^*_m and MCTL^*_{mu} are highly undecidable. We further investigate some syntactic fragments of MCTL*, such as MCTL, for which we obtain interesting results.

2.1 Introduction

Temporal logics, a special kind of *modal logics* geared towards the description of the temporal ordering of events [Pnueli (1977)], have been adopted as a powerful tool for specifying and verifying correctness of concurrent systems [Pnueli (1981)], as they allow to express the temporal ongoing behavior of a system in a well-structured

F. Mogavero, *Logics in Computer Science*, Atlantis Studies in Computing 3, DOI: 10.2991/978-94-91216-95-4_2,

way. In this field, a well-established method is *model checking*, which allows to automatically check for global system correctness [Clarke et al. (2002)], where a formal model of the system, such as a *Kripke structure*, is checked to be correct with respect to a temporal logic formula representing its desired behavior, i.e., the *specification*.

Two possible views regarding the nature of time induce two different types of temporal logics: *linear* and *branching-time* [Lamport (1980)]. In linear-time temporal logics, such as LTL [Pnueli (1977)], time is treated as if each moment in time has a unique possible future. Thus, linear temporal logic formulas are interpreted over linear sequences. In branching-time temporal logics, such as CTL [Clarke and Emerson (1981)] CTL^{+} and CTL* (Emerson and Halpern 1985), each moment in time may split into various possible futures. Accordingly, the structures over which branching temporal logic formulas are interpreted are trees. Many important parallel computer programs exhibit ongoing behavior that is characterized naturally in terms of infinite execution traces, possibly organized into tree-like structures that reflect the high degree of nondeterminism inherent in parallel computation.

From a practical point of view, a very challenging issue in formal specification and verification is to come up with automatic techniques that allow to select small critical parts of the system in order to restrict the verification process just to them. There are several factors that may induce this necessity. For example, in a concurrent setting, as the system under consideration is typically a parallel composition of many modules, we may want to verify just some of them. Also, in a huge system, we may want to restrict our attention only to a portion of its model, perhaps the one that is known to be more critical. To deal with this issue a number of attempts have been carried out in literature and some of them have required the extension of classical temporal logics by considering new semantics, i.e., by changing the interpretation of their syntactic operators, or by replacing or introducing new operators. On the semantic side, we recall CTL *with fairness constraints* [Emerson and Lei (1986)] and *module checking* over branching-time specifications [Kupferman et al. (2001)], which corresponds to model checking in the context of open system analysis (see also [Aminof et al. (2013)] for a recent work on the subject). The semantics of CTL with fairness is restricted in such a way that path quantifiers are interpreted only over fair paths. In module checking, the system is open in the sense that it maintains a continuous interaction with an external behavior. By modifying the classical modeling relations, we can require that each submodel derived by a possible interaction between the system and the environment meets the specification. On the syntactic extensions side, instead, there are the logic of public announcement [Gerbrandy and Groeneveld (1997); Plaza (2007)] sabotage logic [van Benthem (2005)] and the logics for the analysis of strategic ability, in the setting of multi-agent games, such as Alternating-Time Temporal Logic [Alur et al. (2002)] and Strategy Logic [Mogavero et al. (2010a, 2011, 2012a)].

In this paper, we address the same problem and propose a new logic framework, in which, as far as we know for the first time, it is possible to explicitly predicate on substructures of the model itself. Specifically, we introduce and study a new branching-time temporal logic, named MCTL*. This logic is an extension of the classical branching-time temporal logic CTL* by means of two special *minimal model quantifiers*, the existential $\mathbb{E}$ and the universal $\mathbb{A}$. These quantifiers allow to extract, given a model, *minimal* and *conservative* submodels of it on which we successively check a certain temporal specification, for suitable and well-founded concepts of minimality and conservativeness for Kripke structures. The goal is to check local properties of system components in order to deduce the global behavior of the entire one. In particular, the introduced logic exploits the novel idea of checking a particular module of a whole composition system while its single modules are not known in advance. By using the new quantifiers, we can write state formulas such as $\varphi_1 \mathbb{E} \varphi_2$ and $\varphi_1 \mathbb{A} \varphi_2$, which can be read, respectively, as "*there exists a minimal and conservative model of φ_2 that is model of φ_1*" and "*all minimal and conservative models of φ_2 are models of φ_1*". In accordance with this point of view, we call φ_2 the *submodel extractor*, φ_1 the *submodel verifier*, and our modular verification method an *extract-verify* paradigm. Our choice of considering only minimal and conservative submodels is justified by the fact that, in this way, we precisely select the parts of the system or of its execution that are actually responsible for the particular behavior of interest. In particular, we investigate MCTL* and its syntactic fragments MCTL$^+$, MCTL and MPML (where the M indicates the extension of the respective logics with minimal model quantifiers), from a theoretical point of view, under three different possible semantics: *minimal* (m), *minimal-unwinding* (mu), and *unwinding-minimal* (um). They differ one from the other in the use of the operation of unwinding embedded into the definition of the new kind of quantifiers.

As far as the expressiveness regards, we show that all the considered logics are strictly more expressive than the corresponding classical ones, under the m and mu semantics. We also show that MCTL under the um semantics is much more expressive than CTL, since it embeds LTL.

Unfortunately, the gained power comes at a price. Indeed, under the m and mu semantics, we prove that the satisfiability problem for MCTL is highly undecidable. Moreover, differently from CTL, we have that MCTL neither has the tree model property nor is bisimulation invariant, while it is sensible to unwinding. We also investigate the succinctness of MCTL, showing that it is at least as succinct as CTL$^+$, differently from the classical result describing that CTL$^+$ is exponentially more succinct than CTL.

Related works It is worth recalling that logics having the ability of modifying the model under evaluation, and then check the specification on the resulting part, have been also considered in other contexts. For example, we recall the arbitrary public announcement logic [Gerbrandy and Groeneveld (1997); Plaza (2007); French and van Ditmarsch (2008)] and the sabotage modal logic [van Benthem (2005); Löding and Rohde (2003)]. However, the former allows to extract, according to a sub-model extractor formula, submodels that do not necessarily satisfy the formula itself, while the latter does not extract submodels using a formula at all. Moreover, none of them is based on the concept of minimality.

Other logics that have a flavor of restricting a model under evaluation are those used for strategic reasoning, in the setting of multi-agent games. These logics, by means of the concept of strategy, allow to select a specific part of the system on which to verify a specific goal property. However, the way the subsystem is identified is explicitly described inside the model, so it is opposed to our approach.

Outline In Section 2.3, we recall the basic notions regarding the substructure ordering. Then, we have Section 2.4, in which we introduce MCTL* and define its syntax and semantics, followed by Section 2.5, in which the expressiveness and succinctness relationships of the introduced logics are studied. Finally, in Sections 2.6 and 2.7, we study the model-checking and satisfiability problems, respectively. Note that in the accompanying Appendix we recall the classical mathematical notation and some basic definitions that are used along the whole work.

2.2 Preliminaries

Substructure ordering Let $\mathcal{K}, \mathcal{K}' \in \mathrm{KS(AP)}$ be two KSs. We say that $\mathcal{K}$ is a *superstructure* of $\mathcal{K}'$ and $\mathcal{K}'$ is a *substructure* of $\mathcal{K}$, in symbols $\mathcal{K}' \preceq \mathcal{K}$, if *(i)* $\mathrm{W}_{\mathcal{K}'} \subseteq \mathrm{W}_{\mathcal{K}}$, *(ii)* $R_{\mathcal{K}'} \subseteq R_{\mathcal{K}} \cap (\mathrm{W}_{\mathcal{K}'} \times \mathrm{W}_{\mathcal{K}'})$, *(iii)* $\mathsf{L}_{\mathcal{K}'} = (\mathsf{L}_{\mathcal{K}})\restriction \mathrm{W}_{\mathcal{K}'}$, and *(iv)* $\mathsf{w}_{0\mathcal{K}'} = \mathsf{w}_{0\mathcal{K}}$. Moreover, $\mathcal{K}$ and $\mathcal{K}'$ are *comparable* if *(i)* $\mathcal{K} \preceq \mathcal{K}'$ or *(ii)* $\mathcal{K}' \preceq \mathcal{K}$ holds, otherwise they are *incomparable*. Observe that $\preceq$ represents a *partial order* on KSs, whose *strict version*, denoted by $\prec$, is such that $\mathcal{K}' \prec \mathcal{K}$ if $\mathcal{K}' \preceq \mathcal{K}$ and $\mathcal{K}' \neq \mathcal{K}$. For a given set of KSs $\aleph \subseteq \mathrm{KS(AP)}$ and a KS $\mathcal{K} \in \aleph$, we say that $\mathcal{K}$ is *minimal* in $\aleph$, in symbols $\mathcal{K} \in \min \aleph$, if there is no KS $\mathcal{K}' \in \aleph$ such that $\mathcal{K}' \prec \mathcal{K}$.

2.3 Computation Tree Logics with Minimal Model Quantifiers

In this section, we introduce a family of extensions of the classical branching-time temporal logic CTL* (Emerson and Halpern 1986) with minimal model quantifiers, which allow to extract minimal submodels on which we successively check a given property.

2.3.1 Syntax

The *full computation tree logic with minimal model quantifiers* (MCTL*, for short) extends CTL* by further using two special quantifiers, the existential $\mathbb{E}$ and the universal $\mathbb{A}$. Informally, a structure satisfies a state formula $\varphi_1 \mathbb{E} \varphi_2$ iff there is a minimal and conservative substructure satisfying φ_2 (φ_2 is the *submodel extractor*) such that it also satisfies φ_1 (φ_1 is the *submodel verifier*). By duality, a structure satisfies a state formula $\varphi_1 \mathbb{A} \varphi_2$ iff all minimal and conservative substructures satisfying φ_2 satisfy φ_1 too. As for CTL*, in MCTL* the two path quantifiers A and E can prefix a linear time formula composed by an arbitrary combination and nesting of the linear temporal operators X (“*next*”), U (“*until*”), and R (“*release*”) together with their weak version $\tilde{\mathsf{X}}$, $\tilde{\mathsf{U}}$, and $\tilde{\mathsf{R}}$. The formal syntax of MCTL* follows.

Definition 2.1 (MCTL* **Syntax**) MCTL* state *(φ) and* path *(ψ)* formulas *are built inductively from the sets of atomic propositions* AP *in the following way, where* $p \in \mathrm{AP}$*:*

(1) $\varphi ::= p \mid \neg\varphi \mid \varphi \wedge \varphi \mid \varphi \vee \varphi \mid \mathsf{E}\psi \mid \mathsf{A}\psi \mid \varphi \mathbb{E} \varphi \mid \varphi \mathbb{A} \varphi$*;*
(2) $\psi ::= \varphi \mid \neg\psi \mid \psi \wedge \psi \mid \psi \vee \psi \mid \mathsf{X}\psi \mid \psi\, \mathsf{U} \psi \mid \psi\, \mathsf{R} \psi \mid \tilde{\mathsf{X}}\psi \mid \psi\, \tilde{\mathsf{U}} \psi \mid \psi\, \tilde{\mathsf{R}} \psi$.

The class of MCTL* *formulas is the set of state formulas generated by the above grammar. In addition, the simpler classes of MCTL$^+$, MCTL, and MPML formulas are obtained, respectively, by avoiding nesting of temporal operators, by forcing each temporal operator occurring into a formula to be coupled with a path quantifier, and by excluding from MCTL path formulas the until and release operators, as in the classical case of* CTL$^+$, CTL, *and* PML.

We now introduce some auxiliary syntactical notation for MCTL*. For a formula φ, we define the *length* $\mathsf{lng}(\varphi)$ of φ as for CTL*. Formally, *(i)* $\mathsf{lng}(p) \triangleq 1$, for $p \in \mathrm{AP}$, *(ii)* $\mathsf{lng}(\mathsf{Op}\ \psi) \triangleq 1 + \mathsf{lng}(\psi)$, for all $\mathsf{Op} \in \{\neg, \mathsf{X}, \tilde{\mathsf{X}}\}$, *(iii)* $\mathsf{lng}(\psi_1 \mathsf{Op}\ \psi_2) \triangleq 1 + \mathsf{lng}(\psi_1) + \mathsf{lng}(\psi_2)$, for all $\mathsf{Op} \in \{\wedge, \vee, \mathsf{U}, \mathsf{R}, \tilde{\mathsf{U}}, \tilde{\mathsf{R}}\}$, *(iv)* $\mathsf{lng}(\mathsf{Qn}\ \psi) \triangleq 1 + \mathsf{lng}(\psi)$, for all $\mathsf{Qn} \in \{\mathsf{E}, \mathsf{A}\}$, and *(v)* $\mathsf{lng}(\varphi_1 \mathsf{Qn}\ \varphi_2) \triangleq 1 + \mathsf{lng}(\varphi_1) + \mathsf{lng}(\varphi_2)$, for all $\mathsf{Qn} \in \{\mathbb{E}, \mathbb{A}\}$. We also use $\mathsf{cl}(\psi)$ to denote the classical Fischer-Ladner *closure* [Fischer and Ladner (1979)] of ψ defined recursively in the following way: $\mathsf{cl}(\varphi) \triangleq \{\varphi\} \cup \mathsf{cl}'(\varphi)$, for all state formulas φ and $\mathsf{cl}(\psi) \triangleq \mathsf{cl}'(\psi)$, for all path formulas ψ, where *(i)* $\mathsf{cl}'(p) \triangleq \emptyset$, for $p \in \mathrm{AP}$, *(ii)* $\mathsf{cl}'(\mathsf{Op}\ \psi) \triangleq \mathsf{cl}(\psi)$, for all $\mathsf{Op} \in \{\neg, \mathsf{X}, \tilde{\mathsf{X}}\}$, *(iii)* $\mathsf{cl}'(\psi_1 \mathsf{Op}\ \psi_2) \triangleq \mathsf{cl}(\psi_1) \cup \mathsf{cl}(\psi_2)$, for all $\mathsf{Op} \in \{\wedge, \vee, \mathsf{U}, \mathsf{R}, \tilde{\mathsf{U}}, \tilde{\mathsf{R}}\}$, *(iv)* $\mathsf{cl}'(\mathsf{Qn}\ \psi) \triangleq \mathsf{cl}(\psi)$, for all $\mathsf{Qn} \in \{\mathsf{E}, \mathsf{A}\}$, and *(v)* $\mathsf{cl}'(\varphi_1 \mathsf{Qn}\ \varphi_2) \triangleq \mathsf{cl}(\varphi_1) \cup \mathsf{cl}(\varphi_2)$, for all $\mathsf{Qn} \in \{\mathbb{E}, \mathbb{A}\}$. Intuitively, $\mathsf{cl}(\varphi)$ is the set of all state formulas that are subformulas of φ.

2.3.2 Semantics

We now define the semantics of MCTL* w.r.t. a KS $\mathcal{K} = \langle \mathrm{AP}, \mathrm{W}, R, \mathsf{L}, w_0 \rangle$. In general, we write $\mathcal{K} \models \varphi$ to indicate that a state formula φ holds on $\mathcal{K}$ at its initial world w_0. However, we introduce the three different semantics for the introduced model quantifiers: *minimal*, *minimal-unwinding*, and *unwinding-minimal*. Thus, to distinguish between them we use the following modeling relations: $\models_m$, $\models_{mu}$, and $\models_{um}$. The semantics of MCTL* state and path formulas involving atomic propositions, Boolean connectives, temporal operators, and classical path quantifiers is simply defined as for CTL*. Here, we only give the semantics of the remaining minimal model quantifiers.

Definition 2.2 (MCTL* **Semantics**) *Given a* KS $\mathcal{K} = \langle \mathrm{AP}, \mathrm{W}, R, \mathsf{L}, w_0 \rangle$ *and two* GCTL* *state formulas* φ_1 *and* φ_2*, it holds that:*

(1) (a) $\mathcal{K} \models_m \varphi_1 \mathsf{E}\varphi_2$ *iff there is a* KS $\mathcal{K}' \in \min(\mathfrak{K}_m(\mathcal{K}, \varphi_2))$ *such that* $\mathcal{K}' \models_m \varphi_1$;
(b) $\mathcal{K} \models_{mu} \varphi_1 \mathsf{E}\varphi_2$ *iff there is a* KS $\mathcal{K}' \in \min(\mathfrak{K}_{mu}(\mathcal{K}, \varphi_2))$ *such that* $\mathcal{K}'_U \models_{mu} \varphi_1$;
(c) $\mathcal{K} \models_{um} \varphi_1 \mathsf{E}\varphi_2$ *iff there is a* KS $\mathcal{K}' \in \min(\mathfrak{K}_{um}(\mathcal{K}_U, \varphi_2))$ *such that* $\mathcal{K}' \models_{um} \varphi_1$;

(2) (a) $\mathcal{K} \models_m \varphi_1 \mathsf{A}\varphi_2$ *iff for all* KSs $\mathcal{K}' \in \min(\mathfrak{K}_m(\mathcal{K}, \varphi_2))$ *it holds that* $\mathcal{K}' \models_m \varphi_1$;
(b) $\mathcal{K} \models_{mu} \varphi_1 \mathsf{A}\varphi_2$ *iff for all* KSs $\mathcal{K}' \in \min(\mathfrak{K}_{mu}(\mathcal{K}, \varphi_2))$ *it holds that* $\mathcal{K}'_U \models_{mu} \varphi_1$;
(c) $\mathcal{K} \models_{um} \varphi_1 \mathsf{A}\varphi_2$ *iff for all* KSs $\mathcal{K}' \in \min(\mathfrak{K}_{um}(\mathcal{K}_U, \varphi_2))$ *it holds that* $\mathcal{K}' \models_{um} \varphi_1$;

where $\mathfrak{K}_s(\mathcal{K}, \varphi) \triangleq \{\mathcal{K}' \preceq \mathcal{K} : \forall \mathcal{K}'' \preceq \mathcal{K}. \mathcal{K}' \preceq \mathcal{K}'' \Rightarrow \mathcal{K}'' \models_s \varphi\}$, *with* $s \in \{m, mu, um\}$, *is the set of all the substructure of* $\mathcal{K}$ *that are* conservative *w.r.t.* φ.

Intuitively, by using the existential minimal model quantifier $\varphi_1 \mathsf{E}\varphi_2$, we can prove the existence of a representative substructure w.r.t. φ_2 that satisfies φ_1. The universal quantifier $\varphi_1 \mathsf{A}\varphi_2$ is simply the dual of $\varphi_1 \mathsf{E}\varphi_2$ and it allows to ensure that all representative substructures w.r.t. φ_2 satisfy φ_1. It is clear that, MCTL* (resp., MPML, MCTL, and MCTL$^+$) formulas without minimal model quantifiers are CTL* (resp., PML, CTL, and CTL$^+$) formulas.

As one can easily observe, the three semantics m, mu, and um differ one from the other only in the particular way the substructure is extracted and then used for the verification. In particular, a fundamental role is played by the operation of unwinding, which in mu is applied to the minimal substructure after its extraction, while in um it is applied to the original structure.

Let $\mathcal{K}$ be a KS, φ be an MCTL* formula, and $s \in \{m, mu, um\}$ be a symbol indicating which semantics we are interested in. Then, $\mathcal{K}$ is an *s-model* for φ iff $\mathcal{K} \models_s \varphi$. A formula φ is said *s-satisfiable* iff there exists an s-model for it. Moreover, it is an *s-invariant* for the two KSs $\mathcal{K}_1$ and $\mathcal{K}_2$ iff either $\mathcal{K}_1 \models_s \varphi$ and $\mathcal{K}_2 \models_s \varphi$ or $\mathcal{K}_1 \not\models_s \varphi$ and $\mathcal{K}_2 \not\models_s \varphi$. For all state formulas φ_1 and φ_2, we say that φ_1 *s-implies* φ_2, in symbols $\varphi_1 \overset{s}{\Rightarrow} \varphi_2$, iff, for all KS $\mathcal{K}$, it holds that if $\mathcal{K} \models_s \varphi$ then $\mathcal{K} \models_s \varphi$. Consequently, we say that φ_1 is *s-equivalent* to φ_2, in symbols $\varphi_1 \overset{s}{\equiv} \varphi_2$, iff $\varphi_1 \overset{s}{\Rightarrow} \varphi_2$ and $\varphi_2 \overset{s}{\Rightarrow} \varphi_1$. In the following, when the particular semantics represented by s is unimportant or clear from the context, we omit the relative symbol.

A substructure $\mathcal{K}'$ of $\mathcal{K}$ is *s-conservative* w.r.t. a formula φ iff, for all models $\mathcal{K}''$ extending $\mathcal{K}'$ in $\mathcal{K}$, i.e., with $\mathcal{K}' \preceq \mathcal{K}'' \preceq \mathcal{K}$, it holds that $\mathcal{K}'' \models_s \varphi$. Note that this concept of conservativeness is automatically embedded in the definition of $\mathfrak{K}_s(\mathcal{K}, \varphi)$, since we consider only models that, if extended, continue to satisfy the formula φ. To better understand the meaning and the importance of the conservativeness, consider the KS $\mathcal{K}$ built by a chain of three worlds $w_0 \to w_1 \to w_2$, in which w_2 is the only one labeled by the atomic proposition p. Moreover, consider the two submodels $\mathcal{K}'$ and $\mathcal{K}''$ built, respectively, by w_0 and $w_0 \to w_1$. Clearly, $\mathcal{K}' \preceq \mathcal{K}'' \preceq \mathcal{K}$. Moreover, for $\varphi = \mathsf{E}\tilde{\mathsf{X}}\mathsf{F}p$, we have that $\mathcal{K}' \models \varphi$, $\mathcal{K}'' \not\models \varphi$, and $\mathscr{K} \models \varphi$. Hence, we have that $\mathscr{K}'$ satisfies φ, but it is not conservative, since $\mathscr{K}''$, which extends $\mathscr{K}'$, does not satisfy φ. Intuitively, $\mathscr{K}'$ does not contain enough information about the general model $\mathscr{K}$ to be considered as one of its representative submodels w.r.t. φ.

In the rest of the paper, we mainly consider formulas in *positive normal form* (*pnf*, for short), i.e., the negation is applied only to atomic propositions, and in *existential normal form* (*enf*, for short), i.e., only existential (path and minimal model) quantifiers occur. In fact, it is to this aim that we have considered in the syntax of MCTL* both the Boolean connectives $\wedge$ and $\vee$, the path quantifiers A and E, the minimal model quantifiers $\mathbb{E}$ and $\mathbb{A}$, and temporal operators X, U, and R together with their weak version $\tilde{\mathsf{X}}$, $\tilde{\mathsf{U}}$, and $\tilde{\mathsf{R}}$. Indeed, all formulas can be linearly translated in *pnf* or *enf* by using De Morgan's laws and the following equivalences, which directly follow from the semantics of the logic: $\neg(\varphi_1 \mathbb{E} \varphi_2) \equiv (\neg\varphi_1)\mathbb{A}\varphi_2$, $\neg\mathsf{E}\psi \equiv \mathsf{A}\neg\psi$; $\neg\mathsf{X}\psi \equiv \tilde{\mathsf{X}}\neg\psi$; $\neg(\psi_1 \mathsf{U} \psi_2) \equiv (\neg\psi_1)\tilde{\mathsf{R}}(\neg\psi_2)$; $\neg(\psi_1 \mathsf{R} \psi_2) \equiv (\neg\psi_1)\tilde{\mathsf{U}}(\neg\psi_2)$. Finally, as abbreviations we also use the Boolean values $\mathfrak{t}$ ("*true*") and $\mathfrak{f}$ ("*false*").

As an example of application of the logics we have introduced, consider an arbiter system used to control a two-users access to a shared memory location (see Fig. 2.1a for a model $\mathscr{K}$ of the system), where only the request (`r`) and the acknowledge (`a`) signals are known. Suppose now that we want to verify that the idle state i and the common request state (r_1, r_2) are unique w.r.t. the order of user requests and arbiter acknowledges, respectively. We can perform this check by applying the MCTL*$_m$ or MCTL*$_{mu}$ model checking at the state i by using a formula $\varphi = \varphi_1 \wedge \varphi_2$, where $\varphi_1 = \mathsf{AG}(r_1 \wedge r_2 \rightarrow \mathsf{X}\mathfrak{t})\mathbb{A}(\mathsf{EF}(r_1 \wedge \mathsf{XF}(r_1 \wedge r_2 \wedge \mathsf{X}\mathfrak{t})) \wedge \mathsf{EX}(r_2 \wedge \mathsf{XF}r_1 \wedge r_2))$ checks whether the common request state reached by the "request subsystem" is unique and $\varphi_2 = \mathsf{AG}(i \rightarrow \mathsf{X}\mathfrak{t})\mathbb{A}\mathsf{E}(\mathsf{F}(a_1 \wedge \mathsf{XF}i) \wedge \mathsf{F}(a_2 \wedge \mathsf{XF}i))$ checks whether the "acknowledge subsystem" reaches the same idle state after two different acknowledges.

For two minimal and conservative submodels of $\mathscr{K}$ satisfying φ_1 and φ_2, respectively, see $\mathscr{K}_1$ and $\mathscr{K}_2$ in Fig. 2.1b. Observe that also their "mirror images" are models of φ_1 and φ_2. Now, one may note that the above check can not be achieved by using neither a classical logic such as CTL* nor the introduced logic MCTL*$_{um}$. Indeed, we may have a bisimilar model of $\mathscr{K}$ with more idle or common request states, for which no CTL* or MCTL*$_{um}$ formula can check that these states are not unique.

At this point, we report some basic equivalences regarding the new kind of quantifiers that are directly derived by the definition of the semantics of the logics.

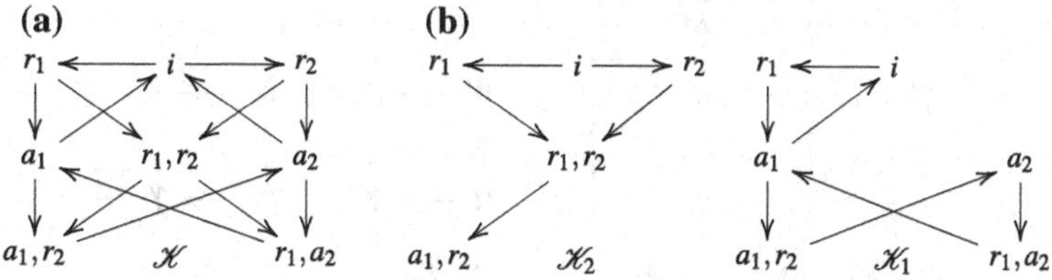

Fig. 2.1 An arbiter system for shared memory locations. **a** The model, **b** two submodels of the arbiter system

Proposition 2.1 **(Basic Equivalences)** *Let* φ, φ_1, *and* φ_2 *be state formulas and* $\mathsf{Qn}, \mathsf{Qn}' \in \{\mathsf{E\!E}, \mathsf{A\!A}\}$. *Then, the following equivalences hold:* (i) $\varphi_1 \mathsf{A\!A} \varphi_2 \equiv \neg((\neg\varphi_1)\mathsf{E\!E}\varphi_2)$; (ii) $\mathsf{t}\,\mathsf{E\!E}\varphi \equiv \varphi\mathsf{E\!E}\varphi \equiv \varphi$; (iii) $\mathsf{f}\,\mathsf{E\!E}\varphi \equiv \neg\varphi\mathsf{E\!E}\varphi \equiv \mathsf{f}$; (iv) $\mathsf{t}\mathsf{A\!A}\varphi \equiv \varphi\mathsf{A\!A}\varphi \equiv \mathsf{t}$; (v) $\mathsf{f}\mathsf{A\!A}\varphi \equiv \neg\varphi\mathsf{A\!A}\varphi \equiv \neg\varphi$; (vi) $(\varphi_1 \vee \varphi_2)\mathsf{E\!E}\varphi \equiv (\varphi_1\mathsf{E\!E}\varphi) \vee (\varphi_2\mathsf{E\!E}\varphi)$; (vii) $(\varphi_1 \wedge \varphi_2)\mathsf{A\!A}\varphi \equiv (\varphi_1\mathsf{A\!A}\varphi) \wedge (\varphi_2\mathsf{A\!A}\varphi)$; (vi) $\varphi\mathsf{E\!E}(\varphi_1 \vee \varphi_2) \equiv (\varphi\mathsf{E\!E}\varphi_1) \vee (\varphi\mathsf{E\!E}\varphi_2)$; (vii) $\varphi\mathsf{A\!A}(\varphi_1 \vee \varphi_2) \equiv (\varphi\mathsf{A\!A}\varphi_1) \wedge (\varphi\mathsf{A\!A}\varphi_2)$; (viii) $(\varphi_1 \wedge \varphi_2)\mathsf{Qn}\,\varphi_2 \equiv \varphi_1\mathsf{Qn}\ \varphi_2$; (ix) $(\varphi_1 \vee \varphi_2)\mathsf{Qn}\,\varphi_2 \equiv \varphi_2\mathsf{Qn}\ \varphi_2$; (x) $(\varphi\mathsf{Qn}\,(\varphi_1 \wedge \varphi_2)) \wedge (\varphi_1\mathsf{A\!A}\varphi_2) \Rightarrow \varphi\mathsf{Qn}\,\varphi_2$; (xi) $(\varphi\mathsf{Qn}\,\varphi_2) \wedge (\varphi_1\mathsf{A\!A}\varphi_2) \Rightarrow \varphi\mathsf{Qn}\,(\varphi_1 \wedge \varphi_2)$; (xii) $\varphi\mathsf{Qn}\,(\varphi_1\mathsf{Qn}'\varphi_2) \equiv (\varphi \wedge \varphi_1)\mathsf{Qn}\,\varphi_2$.

The following theorem summarizes the principal negative properties of MCTL* under the *m* and *mu* semantics.

Theorem 2.1 **(Negative Properties)** *For M*PML, *M*CTL, *M*CTL$^+$, *and* MCTL* *under both the m and mu semantics, it holds that:*

(1) they do not have the tree model property;
(2) they are not invariant under unwinding;
(3) they are not invariant under bisimulation.

Proof *[Item 1]* To prove the statement, we consider a formula with an existential minimal model quantifier such that it requires to extract a graph submodel that, in order to be satisfied, cannot be a tree. Consider the MPML formula $\varphi_S \triangleq \varphi_1\mathsf{E\!E}\varphi_2$, where $\varphi_1 \triangleq \mathsf{EX}(\beta \wedge \mathsf{EXEX}\gamma)$, $\varphi_2 \triangleq \alpha \wedge \mathsf{EX}(\beta \wedge \mathsf{EX}\delta) \wedge \mathsf{EX}(\gamma \wedge \mathsf{EX}(\delta \wedge \mathsf{EX}\gamma))$, $\alpha \triangleq a \wedge b$, $\beta \triangleq \neg a \wedge b$, $\gamma \triangleq a \wedge \neg b$, and $\delta \triangleq \neg a \wedge \neg b$. This formula is satisfiable. In Fig. 2.2, we show the KSs $\mathcal{K}_1$, $\mathcal{K}_2$, $\mathcal{K}_3$, and $\mathcal{K}_4$ as the only minimal models of φ_2, where only $\mathcal{K}_1$ is a tree and $\mathcal{K}_3$ and $\mathcal{K}_4$ are the only models of φ. Indeed, $\mathcal{K}_3$ and $\mathcal{K}_4$ satisfy φ_1, but $\mathcal{K}_1$ and $\mathcal{K}_2$ do not. Since any model of φ has to include $\mathcal{K}_3$ or $\mathcal{K}_4$ as submodel, it follows that no tree model can satisfy φ. Since MPML is a sublogic of MCTL, MCTL$^+$, and MCTL*, the thesis easily follows.

[Item 2] By the previous item, there exists a satisfiable MPML formula φ that does not have a tree model. Now, let $\mathcal{K}$ be its model and $\mathcal{K}_U$ the related unwinding. Then, we have that $\mathcal{K} \models \varphi$ and $\mathcal{K}_U \not\models \varphi$. Hence, MPML cannot be invariant under unwinding.

[Item 3] Since an unwinding is a particular case of a bisimilarity relation, we have also that MPML is not invariant under bisimulation, i.e., it is possible to

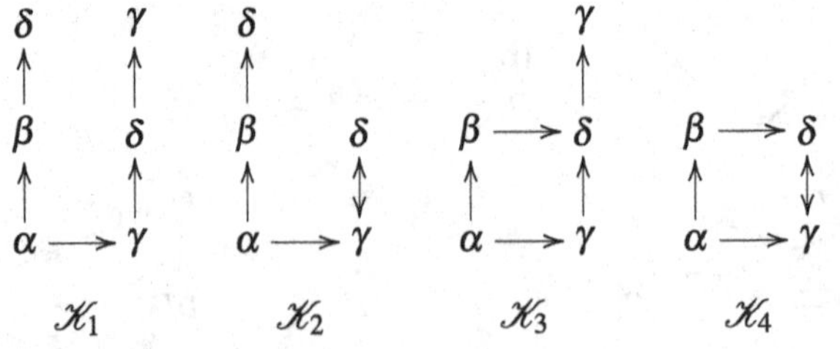

Fig. 2.2 The four minimal models of φ_S

express an MPML property satisfied on a model $\mathscr{K}$, but not on one of its bisimilar models $\mathscr{K}'$. □

We now move to the positive results about MCTL* under the *um* semantics.

Theorem 2.2 (Positive Properties) *For M*PML, *M*CTL, *M*CTL$^+$, *and* MCTL* *under the um semantics, it holds that:*

(1) they are invariant under bisimulation;
(2) they are invariant under unwinding;
(3) they have the (unbounded) tree model property.

Proof *[Item 1]* The proof proceeds by induction on the structure of the formula. In particular, here we show only the most important inductive case of $\varphi = \varphi_1 \mathbb{E} \varphi_2$. The statement that we have to prove is the following: $\mathscr{K}_1 \models \varphi$ iff $\mathscr{K}_2 \models \varphi$, for all pairs of bisimilar KSs $\mathscr{K}_1 = \langle \mathrm{AP}, \mathrm{W}_1, R_1, \mathsf{L}_1, s_{0_1} \rangle$ and $\mathscr{K}_2 = \langle \mathrm{AP}, \mathrm{W}_2, R_2, \mathsf{L}_2, s_{0_2} \rangle$. As first thing, due to the definition of the logics under the *um* semantics, one can easily note that, if $\mathscr{K} \models \varphi$, so does its unwinding $\mathscr{K}_U$, i.e., $\mathscr{K}_U \models \varphi$, for any KS $\mathscr{K}$, since $\mathscr{K}_U$ and $(\mathscr{K}_U)_U$ are isomorphic structures. So, w.l.o.g., we assume that both $\mathscr{K}_1$ and $\mathscr{K}_2$ are KTs. Now, let $\sim \subseteq \mathrm{W}_1 \times \mathrm{W}_2$ be a bisimulation relation between the two KTs and $\overline{\sim} \subseteq \mathrm{W}_1 \times \mathrm{W}_2$ be the restriction of $\sim$ to nodes of the trees that are at the same level, i.e., at the same distance from the root. Formally, $t_1 \overline{\sim} t_2$ iff $t_1 \sim t_2$ and $|t_1| = |t_2|$ (recall that a node of a tree is a finite word on a given set of directions). Moreover, associate to each subtree $\mathscr{T} \triangleq \langle \mathrm{AP}, \mathrm{T}, R, \mathsf{L}, \varepsilon \rangle \preceq \mathscr{K}_i$, with $i \in \{1, 2\}$, the maximal subtree $\widehat{\mathscr{T}_i} \triangleq \langle \mathrm{AP}, \widehat{\mathrm{T}}, \widehat{R}, \widehat{\mathsf{L}}, \varepsilon \rangle \preceq \mathscr{K}_{3-i}$ with the set of states $\widehat{\mathrm{T}} \triangleq \{t' \in \mathrm{W}_{3-1} : \exists t \in \mathrm{T}.\, t \overline{\sim} t'\}$. Note that, since both $\mathscr{T}$ and $\widehat{\mathscr{T}}$ are trees, they are bisimilar (the condition on the tree shape of the original structure is fundamental for this derivation). Thus, by the inductive hypothesis, we obtain that $\mathscr{T} \models \varphi_2$ iff $\widehat{\mathscr{T}} \models \varphi_2$. At this point, to prove the statement, it is enough to show that, for each tree $\mathscr{T} \in \min(\mathfrak{K}_{um}(\mathscr{K}_i, \varphi_2))$, there is a bisimilar tree $\mathscr{T}' \in \min(\mathfrak{K}_{um}(\mathscr{K}_{3-i}, \varphi_2))$, for all $i \in \{1, 2\}$. Indeed, by the inductive hypothesis, we have that $\mathscr{T} \models \varphi_1$ iff $\mathscr{T}' \models \varphi_1$. To do this, we prove that $\mathfrak{K}_{um}(\widehat{\mathscr{T}}, \varphi_2) \subseteq \mathfrak{K}_{um}(\mathscr{K}_{3-i}, \varphi_2)$ and so, $\min(\mathfrak{K}_{um}(\widehat{\mathscr{T}}, \varphi_2)) \subseteq \min(\mathfrak{K}_{um}(\mathscr{K}_{3-i}, \varphi_2))$, since every $\overline{\mathscr{T}} \in \mathfrak{K}_{um}(\widehat{\mathscr{T}}, \varphi_2)$ is bisimilar to $\mathscr{T}$. Indeed, suppose by contradiction that there is a tree in $\mathfrak{K}_{um}(\widehat{\mathscr{T}}, \varphi_2)$ that is not in $\mathfrak{K}_{um}(\mathscr{K}_{3-i}, \varphi_2)$. Then, due to the definition of the set $\mathfrak{K}_{um}(\cdot, \cdot)$ of conservative substructures w.r.t. a given formula, it holds that $\widehat{\mathscr{T}} \notin \mathfrak{K}_{um}(\mathscr{K}_{3-i}, \varphi_2)$, which means that there is a structure $\overline{\mathscr{T}}$ with $\widehat{\mathscr{T}} \preceq \overline{\mathscr{T}} \preceq \mathscr{K}_{3-i}$ such that $\overline{\mathscr{T}} \not\models \varphi_2$. Now, consider the related bisimilar structure $\widehat{\overline{\mathscr{T}}}$. It is evident that $\mathscr{T} \preceq \widehat{\overline{\mathscr{T}}} \preceq \mathscr{K}_i$. Moreover, $\widehat{\overline{\mathscr{T}}} \not\models \varphi_2$. This implies that $\mathscr{T} \notin \mathfrak{K}_{um}(\mathscr{K}_i, \varphi_2)$, which contradicts our assumption. Hence, the thesis holds.

[Item 2] It is known that every KS $\mathscr{K}$ is bisimilar to its unwinding $\mathscr{K}_U$. Now, by the previous item, we have that every MCTL* formula φ is an invariant for $\mathscr{K}$ and $\mathscr{K}_U$. Hence, the thesis holds.

[Item 3] Consider an MCTL* formula φ and suppose that it is satisfiable. Then, there is a KS $\mathscr{K}$ such that $\mathscr{K} \models \varphi$. By the previous item, φ is satisfied at the root of the unwinding $\mathscr{K}_U$ of $\mathscr{K}$. Thus, since $\mathscr{K}_U$ is a KT, we immediately have that φ is satisfied on a tree model. □

Finally, we show that MPML under the m semantics has the strong finite model property. To this aim, we first introduce some extra notations. For a Kripke structure $\mathcal{K}$, by $\mathsf{dep}(\mathcal{K})$ we denote the *depth* of $\mathcal{K}$, i.e., the maximal length of a path in the unwinding $\mathcal{K}[U]$, for all worlds $w \in \mathrm{W}_{\mathcal{K}}$. Moreover, given a MPML formula φ, we denote by $\mathsf{dep}(\varphi)$ the *depth* of φ, i.e., the maximal number of nested occurrences of path quantifiers in φ, but those appearing in its submodel verifiers. Formally, the depth function is inductively defined as follows: $\mathsf{dep}(p) = 0$, for $p \in \mathrm{AP}$; $\mathsf{dep}(\neg\varphi) = \mathsf{dep}(\varphi)$; $\mathsf{dep}(\varphi_1 \wedge \varphi_2) = \mathsf{dep}(\varphi_1 \vee \varphi_2) = \max\{\mathsf{dep}(\varphi_1), \mathsf{dep}(\varphi_2)\}$; $\mathsf{dep}(\varphi_1 \mathbb{E}\varphi_2) = \mathsf{dep}(\varphi_1 \mathbb{A}\varphi_2) = \mathsf{dep}(\varphi_2)$; $\mathsf{dep}(\mathsf{A}\psi) = \mathsf{dep}(\mathsf{E}\psi) = 1 + \mathsf{dep}(\varphi)$, where $\psi = \mathsf{X}\varphi$ or $\psi = \tilde{\mathsf{X}}\varphi$. It is easy to see that $\mathsf{dep}(\varphi) = \mathsf{O}(|\varphi|)$.

Theorem 2.3 *MPML under the m semantics has the strong finite model property, i.e., each MPML satisfiable formula φ has a finite model $\mathcal{K}$ with size $|\mathcal{K}| \leq g(|\varphi|)$, where g is a recursive function, and depth $\mathsf{dep}(\mathcal{K}) \leq \mathsf{dep}(\varphi)$.*

Proof We show that if there is a model $\mathcal{K}$ for φ then there exists a model $\mathcal{K}' \preceq \mathcal{K}$, with $|\mathcal{K}'| \leq g(|\varphi|) = 2^{f(|\varphi|)}$, where f is recursive and monotone, and $\mathsf{dep}(\mathcal{K}) \leq \mathsf{dep}(\varphi)$, such that for all models $\mathcal{K}''$, with $\mathcal{K}' \preceq \mathcal{K}'' \preceq \mathcal{K}$, it holds that $\mathcal{K}'' \models \varphi$, i.e., $\mathcal{K}'$ is conservative w.r.t. the model $\mathcal{K}$ and the formula φ. We proceed by mutual induction on the number of nested occurrences of minimal model quantifiers in φ (external induction) and on the structure of the formula itself (internal induction). W.l.o.g., we assume that φ is in positive normal form.

The base step for the external induction follows directly by applying to φ (since it is a PML formula) the well-known selection procedure used to prove the finite model property for PML (Blackburn et al. 2004).

We now proceed with the base step for the internal induction in the external inductive case, where φ is of the form $\varphi_1 \mathbb{E}\varphi_2$ (resp., $\varphi_1 \mathbb{A}\varphi_2$). Since φ is satisfiable, there is a model $\mathcal{K}$ for which there exists a minimal model $\mathcal{K}' \in \min \mathfrak{K}_m(\mathcal{K}, \varphi_2)$ such that $\mathcal{K}' \models \varphi_1$ (resp., for all minimal models $\mathcal{K}' \in \min \mathfrak{K}_m(\mathcal{K}, \varphi_2)$ it holds that $\mathcal{K}' \models \varphi_1$). Observe that $\mathcal{K}' \models \varphi$ too. Now, by the inductive hypothesis, since $\mathcal{K}' \models \varphi_2$ there exists a conservative model $\mathcal{K}'' \preceq \mathcal{K}'$, such that the three properties $|\mathcal{K}''| \leq 2^{f(|\varphi_2|)} \leq 2^{f(|\varphi|)}$, $\mathsf{dep}(\mathcal{K}'') \leq \mathsf{dep}(\varphi_2) = \mathsf{dep}(\varphi)$, and $\mathcal{K}'' \models \varphi_2$ hold. However, $\mathcal{K}'$ is minimal w.r.t. φ_2. Therefore, $\mathcal{K}'' = \mathcal{K}'$. Hence, the thesis follows for this base case.

Consider now the inductive steps for the internal induction in the external inductive case.

If $\varphi = \varphi_1 \vee \varphi_2$, there exists a $\mathcal{K}$ such that $\mathcal{K} \models \varphi_1$ or $\mathcal{K} \models \varphi_2$. Now, by the inductive hypothesis, there exists a conservative model $\mathcal{K}_1 \preceq \mathcal{K}$ of φ_1, with $|\mathcal{K}_1| \leq 2^{f(|\varphi_1|)}$ and $\mathsf{dep}(\mathcal{K}_1) \leq \mathsf{dep}(\varphi_1)$, or a conservative model $\mathcal{K}_2 \preceq \mathcal{K}$ of φ_2, with $|\mathcal{K}_2| \leq 2^{f(|\varphi_2|)}$ and $\mathsf{dep}(\mathcal{K}_2) \leq \mathsf{dep}(\varphi_2)$. Consider now the model $\mathcal{K}' = \mathcal{K}_i$, for some i, if $\mathcal{K}_i$ exists. It is obvious that, $\mathcal{K}' \models \varphi$. Moreover, $|\mathcal{K}'| = |\mathcal{K}_i| \leq 2^{f(|\varphi_i|)} \leq 2^{f(|\varphi|)}$ and $\mathsf{dep}(\mathcal{K}') = \mathsf{dep}(\mathcal{K}[i]) \leq \mathsf{dep}(\varphi_i) \leq \mathsf{dep}(\varphi)$, so the thesis follows.

If $\varphi = \varphi_1 \wedge \varphi_2$, there exists a $\mathcal{K}$ such that $\mathcal{K} \models \varphi_1$ and $\mathcal{K} \models \varphi_2$. Now, by inductive hypothesis, that there exist a conservative model $\mathcal{K}_1 \preceq \mathcal{K}$ of φ_1, with

$|\mathscr{K}_1| \leq 2^{f(|\varphi_1|)}$ and $\mathsf{dep}(\mathscr{K}_1) \leq \mathsf{dep}(\varphi_1)$, and a conservative model $\mathscr{K}_2 \preceq \mathscr{K}$ of φ_2, with $|\mathscr{K}_2| \leq 2^{f(|\varphi_2|)}$ and $\mathsf{dep}(\mathscr{K}_2) \leq \mathsf{dep}(\varphi_2)$. Consider now the model $\mathscr{K}' \preceq \mathscr{K}$ formed by the union[1] of $\mathscr{K}_1$ and $\mathscr{K}_2$. By conservativeness of $\mathscr{K}_1$ and $\mathscr{K}_2$, for all models $\mathscr{K}''$, with $\mathscr{K}' \preceq \mathscr{K}'' \preceq \mathscr{K}$, it holds that $\mathscr{K}'' \models \varphi_1$ and $\mathscr{K}'' \models \varphi_2$ and so $\mathscr{K}'' \models \varphi$. Moreover, $|\mathscr{K}'| \leq |\mathscr{K}_1| + |\mathscr{K}_2| \leq 2^{f(|\varphi_1|)} + 2^{f(|\varphi_2|)} \leq 2^{f(|\varphi|)}$ and $\mathsf{dep}(\mathscr{K}') = \max\{\mathsf{dep}(\mathscr{K}_1), \mathsf{dep}(\mathscr{K}_2)\} \leq \max\{\mathsf{dep}(\varphi_1), \mathsf{dep}(\varphi_2)\} = \mathsf{dep}(\varphi)$, so the thesis follows.[2]

If $\varphi = \mathsf{EX}\varphi'$, there are a model $\mathscr{K}$ and a successor $v \in \mathrm{W}_{\mathscr{K}}$ of the initial world $w\mathscr{K}_0$ such that $\mathscr{K}[v] \models \varphi$. Now, by inductive hypothesis, there exists a conservative model $\mathscr{K}' \preceq \mathscr{K}[v]$ of φ', with $|\mathscr{K}'| \leq 2^{f(|\varphi'|)}$ and $\mathsf{dep}(\mathscr{K}') \leq \mathsf{dep}(\varphi')$. If there exists a world $u \in \mathrm{W}'_{\mathscr{K}}$, with $(u, v) \in R'_{\mathscr{K}}$, then it is obvious that $\mathscr{K}'_u \models \varphi$. Moreover, $|\mathscr{K}'| \leq 2^{f(|\varphi'|)} \leq 2^{f(|\varphi|)}$ and $\mathsf{dep}(\mathscr{K}'_u) \leq \mathsf{dep}(\mathscr{K}') + 1 \leq \mathsf{dep}(\varphi') + 1 = \mathsf{dep}(\varphi)$. Hence, the thesis easily follows. If such a world u does not exist, consider the model $\mathscr{K}''$ constructed by adding to $\mathscr{K}'$ a fresh world w in such a way that $(w, v) \in R_{\mathscr{K}}$. Then, it is evident that $\mathscr{K}''$ can be used in place of $\mathscr{K}'_u$ in the previous reasoning. Therefore, the thesis holds in this case too.

For the cases $\varphi = \mathsf{AX}\varphi'$, $\varphi = \tilde{\mathsf{EX}}\varphi'$, and $\varphi = \tilde{\mathsf{AX}}\varphi'$ the proof proceeds in a similar way, so we omit the details. □

2.4 Expressiveness and Succinctness

In this section, we describe the expressiveness and succinctness relationships between the introduced logics and the classic ones.

As first immediate results, we have that all the logics under the *m* and *mu* semantics are more expressive than the classic ones.

Theorem 2.4 (*m* **and** *mu* **Expressiveness)** *M*PML, *M*CTL, *M*CTL$^+$, *and* MCTL*, *under both the m and mu semantics, are more expressive than* PML, CTL, CTL$^+$, *and* CTL*, *respectively.*

Proof The statement follows from the fact that PML, CTL, CTL$^+$, and CTL* are all invariant under bisimulation, while their extensions with minimal model quantifiers under the *m* and *mu* semantics are not. Therefore, the extended logics can characterize more models than the classical ones. Hence, they are more expressive. □

In the next theorem, we prove how the introduction of the minimal model quantifiers allows to translate in an efficient way CTL$^+$ in MCTL.

---

[1] Union of models is defined in the classical way: union of sets of worlds, union of relations, union of sets of atomic propositions, etc.

[2] Observe that this step deeply makes use of the conservativeness, indeed if $\mathscr{K}_1$ or $\mathscr{K}_2$ are not conservative we can not use the fact that their union still satisfies the formula φ. As an example, consider the formula $\varphi = \mathsf{EX}_a \wedge (\mathsf{EX}_a \rightarrow \mathsf{EX}_b)$. It is not hard to show a model with two nodes for EX_a and a non conservative model with one world for $\mathsf{EX}_a \rightarrow \mathsf{EX}_b$ whose union does not satisfy φ.

Theorem 2.5 (*m* **Reducibility of** CTL$^+$) *CTL$^+$ is polynomially reducible by satisfiability to MCTL, under the m semantics.*

Proof Given a CTL$^+$ formula φ we show that there exists an equisatisfiable MCTL formula with $\mathsf{lng}(\varphi') = \mathsf{O}(\mathsf{lng}(\varphi)^3)$. W.l.o.g., we assume that φ is in existential normal form (we recall that any CTL$^+$ formula can be linearly translated in this form). Moreover, by maintaining the satisfiability (not the equivalence) using the classical formula equivalences [Emerson and Halpern (1985)], we can transform it into another CTL$^+$ formula $\widehat{\varphi}$ that is a Boolean combination of existential quantifiers $\overline{\varphi} = \mathsf{E}\psi$, where each ψ is in turn a Boolean combination of subformulas, of the form $p_i \mathsf{U} q_i$, $\mathsf{G}r$, $\mathsf{X}s$, and $\tilde{\mathsf{X}}\mathfrak{f}$, where each p_i, q_i, r and s are atomic propositions. Note that after this reduction, $\widehat{\varphi}$ does not contain nested quantifiers, since they are replaced by apposite fresh atomic propositions, In practice, the reduction from φ to φ' turns out to use, as base case of the translation idea, the following equivalences:

- $\mathsf{E}(\bigwedge_{i=1}^{n} p_i \mathsf{U} q_i \wedge \tilde{\mathsf{X}}\mathfrak{f}) \overset{m}{\equiv} \bigwedge_{i=1}^{n} q_i \wedge \mathsf{E}\tilde{\mathsf{X}}\mathfrak{f}$;
- $\mathsf{E}(\mathsf{G}r \wedge \mathsf{X}s) \overset{m}{\equiv} r \wedge \mathsf{EX}(s \wedge \mathsf{EG}r)$;
- $\mathsf{E}(\bigwedge_{i=1}^{n} p_i \mathsf{U} q_i) \overset{m}{\equiv} \bigvee_{i=1}^{n} (\varphi_i^{ver} \mathbb{E} \varphi_i^{ext})$, where $\varphi_i^{ver} \triangleq \bigwedge_{1 \leq h<k \leq n}^{h,k \neq i} (\mathsf{EF}(q_h \wedge \mathsf{EF}q_k) \vee \mathsf{EF}(q_k \wedge \mathsf{EF}q_h))$ and $\varphi_i^{ext} \triangleq \bigwedge_{1 \leq j \leq n}^{j \neq i} \mathsf{E}((p_i \wedge p_j) \mathsf{U} (q_j \wedge \mathsf{E}(p_i \mathsf{U} q_i)))$;
- $\mathsf{E}(\bigwedge_{i=1}^{n} p_i \mathsf{U} q_i \wedge \mathsf{G}r) \overset{m}{\equiv} \bigvee_{i=1}^{n} (\varphi_i^{ver} \mathbb{E} \varphi_i^{ext})$, where φ_i^{ver} is defined as above and $\varphi_i^{ext} \triangleq \bigwedge_{1 \leq j \leq n}^{j \neq i} \mathsf{E}((r \wedge p_i \wedge p_j) \mathsf{U} (q_j \wedge \mathsf{E}((r \wedge p_i) \mathsf{U} (q_i \wedge \mathsf{EG}r))))$;
- $\mathsf{E}(\bigwedge_{i=1}^{n} p_i \mathsf{U} q_i \wedge \mathsf{X}s) \overset{m}{\equiv} (\bigwedge_{i=1}^{n} q_i \wedge \mathsf{EX}s) \vee \bigvee_{i=1}^{n} (\varphi_i^{ver} \mathbb{E} \varphi_i^{ext})$, where φ_i^{ver} is defined as above and $\varphi_i^{ext} \triangleq \bigwedge_{1 \leq j \leq n}^{j \neq i} (p_i \wedge q_j \wedge \mathsf{EX}(s \wedge \mathsf{E}(p_i \mathsf{U} q_i))) \vee (p_i \wedge p_j \wedge \mathsf{EX}(s \wedge \mathsf{E}((p_i \wedge p_j) \mathsf{U} (q_j \wedge \mathsf{E}(p_i \mathsf{U} q_i)))))$;
- $\mathsf{E}(\bigwedge_{i=1}^{n} p_i \mathsf{U} q_i \wedge \mathsf{G}r \wedge \mathsf{X}s) \overset{m}{\equiv} \bigwedge_{i=1}^{n} q_i \wedge r \wedge \mathsf{EX}(s \wedge \mathsf{EG}r) \vee \bigvee_{i=1}^{n} (\varphi_i^{ver} \mathbb{E} \varphi_i^{ext})$, where φ_i^{ver} is defined as above and $\varphi_i^{ext} \triangleq \bigwedge_{1 \leq j \leq n}^{j \neq i} (r \wedge p_i \wedge q_j \wedge \mathsf{EX}(s \wedge \mathsf{E}((r \wedge p_i) \mathsf{U} (q_i \wedge \mathsf{EG}r)))) \vee (r \wedge p_i \wedge p_j \wedge \mathsf{EX}(s \wedge \mathsf{E}((r \wedge p_i \wedge p_j) \mathsf{U} (q_j \wedge \mathsf{E}((r \wedge p_i) \mathsf{U} (q_i \wedge \mathsf{EG}r))))))$.

The first two equivalences, which do not contain the minimal model quantifier $\mathbb{E}$, are derivable by simply applying classical transformations. The proof of the last two, instead, can be obtained by simply showing that each model satisfying the first member of an equivalence must satisfy also the second one and vice versa. Here, we omit the technical details, while we give the basic intuition behind the third equivalence, which shows, as in the remaining three, how to avoid the exponential blow-up incurred by the classical translation in CTL for the corresponding case.

The key step in the translation is the selection of the right submodel of the extractor formula φ_i^{ext}, through the verifier formula φ_i^{ver}, which must satisfy $\overline{\varphi} = \mathsf{E}(\bigwedge_{i=1}^{n} p_i \mathsf{U} q_i)$.

If a Ks $\mathcal{K} = \langle \mathrm{AP}, \mathrm{W}, R, \mathsf{L}, w_0 \rangle$ satisfies the original formula $\overline{\varphi}$, we have that $\min(\mathcal{K}_m(\mathcal{K}, \overline{\varphi})) \subseteq \min(\mathcal{K}_m(\mathcal{K}, \varphi_i^{ext}))$, for a given index $i \in [1, n]$. Now, let $\mathcal{K}' \in \min(\mathcal{K}_m(\mathcal{K}, \overline{\varphi}))$. Then, for all paths $\pi \in \mathrm{Pth}(\mathcal{K}', w_0)$ such that $\mathcal{K}', \pi \models \bigwedge_{i=1}^{n} p_i \mathsf{U} q_i$, it holds that $\mathcal{K}', \pi \models \mathsf{F}(q_h \wedge \mathsf{F}q_k)$ or $\mathcal{K}', \pi \models \mathsf{F}(q_k \wedge \mathsf{F}q_h)$, for all indexes $h, k \in [1, n]$, with $h, k \neq i$. Hence, it holds that $\mathcal{K}' \models \varphi_i^{ver}$ and so, $\mathcal{K} \models \varphi_i^{ver} \mathbb{E} \varphi_i^{ext}$.

Vice versa, consider a KS $\mathscr{K} = \langle \text{AP}, \text{W}, R, \mathsf{L}, w_0 \rangle$ such that $\mathscr{K} \models \varphi_i^{ver} \mathbb{E} \varphi_i^{ext}$, for a given index $i \in [1, n]$. Then, there exists a minimal model $\mathscr{K}' \in \min(\mathscr{K}_m(\mathscr{K}, \varphi_i^{ext}))$ such that $\mathscr{K}' \models \varphi_i^{ver}$. Now, suppose by contradiction that $\mathscr{K}' \not\models \overline{\varphi}$. Consequently, there exist at least three different and not directly connected substructures $\mathscr{K}_1', \mathscr{K}_2', \mathscr{K}_3' \preceq \mathscr{K}'$ and three paths $\pi_1 \in \text{Pth}(\mathscr{K}_1', w_0)$, $\pi_2 \in \text{Pth}(\mathscr{K}_2', w_0)$, and $\pi_3 \in \text{Pth}(\mathscr{K}_3', w_0)$ such that each path formula $p_j \mathsf{U} q_j$, with $j \neq i$, is satisfied on just two of these paths. Then, each formula $\mathsf{E}((p_i \wedge p_j) \mathsf{U}(q_j \wedge \mathsf{E}(p_i \mathsf{U} q_i)))$ is satisfied in at least two ways in two different submodels of $\mathscr{K}$ and then there exists a submodel $\mathscr{K}'' \preceq \mathscr{K}$ with $\mathscr{K}'' \neq \mathscr{K}'$ such that $\mathscr{K}'' \models \varphi_i^{ext}$. Thus, $\mathscr{K}'$ is not minimal, but this contradicts the assumption. Hence, $\mathscr{K}' \models \overline{\varphi}$ and so, $\mathscr{K} \models \overline{\varphi}$. □

In the next theorem, we show how the introduction of the minimal model quantifiers allows to translate in an efficient way CTL* in MCTL.

Theorem 2.6 (*mu* **and** *um* **Reducibility of** CTL*) CTL* *is polynomially reducible by satisfiability to M*CTL*, under the mu and um semantics.*

Proof Given a CTL* formula φ we show that there exists an equisatisfiable MCTL formula φ', under the *mu* and *um* semantics, with $\mathsf{lng}(\varphi') = \mathsf{O}(\mathsf{lng}(\varphi))$. As in the previous theorem, we first consider the derived CTL* formula $\widehat{\varphi}$ in which each quantifier is of the form $\overline{\varphi} = \mathsf{E}\psi$, where ψ is a pure LTL formula without any nested quantifier. Then, in order to obtain φ', we substitute in $\widehat{\varphi}$ all the subformulas $\overline{\varphi}$ by using the equivalences $\overline{\varphi} \overset{mu}{\equiv} (\widetilde{\psi}^{\mathsf{E}} \mathbb{E} \phi) \mathbb{E} \phi$ and $\overline{\varphi} \overset{um}{\equiv} \widetilde{\psi}^{\mathsf{E}} \mathbb{E} \phi$, respectively, for the *mu* and *um* semantics, where the verifier formula $\widetilde{\psi}^{\mathsf{E}}$ is obtained from ψ by coupling each of its temporal operators with the path quantifier E and the extractor formula $\phi \triangleq \mathsf{E}((\mathsf{E}\tilde{\mathsf{X}}\mathsf{f}) \mathsf{R} \mathsf{t})$ is used to extract both finite and infinite paths from the original structure.

The correctness of the translation is due to the following reasoning that we explicit for the *um* semantics only, since the other case is similar.

For one direction of the equivalence $\overline{\varphi} \overset{um}{\equiv} \widetilde{\psi}^{\mathsf{E}} \mathbb{E} \phi$, suppose that a KS $\mathscr{K} = \langle \text{AP}, \text{W}, R, \mathsf{L}, w_0 \rangle$ is a model of $\overline{\varphi}$ and let $\pi \in \text{Pth}(\mathscr{K}, w_0)$ be a path for which $\mathscr{K}, \pi \models \psi$ holds. Then, we can assert that there exists a KT $\mathscr{T} \in \min(\mathscr{K}_{um}(\mathscr{K}_U, \phi))$ such that $\{\pi'\} = \text{Pth}(\mathscr{T}, \varepsilon)$, where $|\pi'| = |\pi|$ and $\pi'_i = \mathsf{unw}(\pi_i)$, for all $i \in [0, |\pi|[$. Since ψ is a pure LTL formula, it is evident that $\mathscr{T}, \pi' \models \psi$ and so, $\mathscr{T} \models \overline{\varphi}$. Consequently, $\mathscr{T} \models \widetilde{\psi}^{\mathsf{E}}$ and thus, $\mathscr{K} \models \widetilde{\psi}^{\mathsf{E}} \mathbb{E} \phi$.

The other direction is simply the converse of the previous one. The crucial point resides in the fact that, since each KT $\mathscr{T} \in \min(\mathscr{K}_{um}(\mathscr{K}_U, \phi))$ contains just one path, we can surely assert that if $\mathscr{T} \models \widetilde{\psi}^{\mathsf{E}}$ then $\mathscr{T} \models \overline{\varphi}$. □

In the case of the *mu* and *um* semantics, we can also prove that MCTL subsumes LTL, as we show in the next theorem.

Theorem 2.7 (*mu* **and** *um* **Reducibility of** LTL) LTL *is polynomially reducible by equivalence to M*CTL*, under the mu and um semantics.*

Proof Differently from the previous two theorems, we now show that, given an LTL formula ψ in the CTL* form $\mathsf{A}\psi$, there exists an equivalent and not only equisatisfiable

MCTL formula φ', with $\mathsf{lng}(\varphi') = \mathsf{O}(\mathsf{lng}(\psi))$. In particular, we show that $\mathsf{A}\psi \overset{mu}{\equiv} (\widetilde{\psi}^{\mathsf{A}} \mathbb{A} \phi) \mathbb{A} \phi$ and $\mathsf{A}\psi \overset{um}{\equiv} \widetilde{\psi}^{\mathsf{A}} \mathbb{A} \phi$, where $\widetilde{\psi}^{\mathsf{A}}$ is obtained from ψ by coupling each of its temporal operators with the path quantifier A and $\phi \triangleq \mathsf{E}((\mathsf{E}\tilde{\mathsf{X}}\mathsf{f})\,\mathsf{R}\mathsf{t})$.

Indeed, every formula $\mathsf{A}\psi$ is equivalent to $\neg\mathsf{E}\neg\psi$. Now, by applying to $\mathsf{E}\neg\psi$ the equivalences proved in Theorem 2.6, we obtain that $\mathsf{E}\neg\psi \overset{mu}{\equiv} (\widetilde{\neg\psi}^{\mathsf{E}} \mathbb{E} \phi) \mathbb{E} \phi$ and $\mathsf{E}\neg\psi \overset{um}{\equiv} \widetilde{\neg\psi}^{\mathsf{E}} \mathbb{E} \phi$. At this point, by recalling that $\neg(\varphi_1 \mathbb{E} \varphi_2) \equiv (\neg\varphi_1) \mathbb{A} \varphi_2$ and observing that $\neg(\widetilde{\neg\psi}^{\mathsf{E}}) = \widetilde{\psi}^{\mathsf{A}}$, we have that $\mathsf{A}\psi \overset{mu}{\equiv} (\widetilde{\psi}^{\mathsf{A}} \mathbb{A} \phi) \mathbb{A} \phi$ and $\mathsf{A}\psi \overset{um}{\equiv} \widetilde{\psi}^{\mathsf{A}} \mathbb{A} \phi$. □

By the previous theorem, we directly derive that MCTL, also under the *um* semantics, is more expressive than CTL and CTL$^+$.

Corollary 2.1 (*um* **Expressiveness of** *M*CTL) *M*CTL *is more expressive than* CTL *and* CTL$^+$.

Proof It is known that the LTL formulas $\psi = \mathsf{FG}p$ in the CTL* form $\mathsf{A}\psi$ does not have any equivalent formula in CTL and so, in CTL$^+$ [Clarke and Draghicescu (1988)]. However, by Theorem 2.7, $\mathsf{A}\psi \overset{um}{\equiv} (\mathsf{AFAG}p) \mathbb{A}\, \mathsf{E}((\mathsf{E}\tilde{\mathsf{X}}\mathsf{f})\,\mathsf{R}\mathsf{t})$. Thus, we can express in MCTL the property ψ. Hence, the statement follows.

Finally, by a model-theoretic reasoning, we prove that MCTL is at least exponentially more expressive than CTL.

Corollary 2.2 (**Succinctness of** *M*CTL) *M*CTL *is exponentially more succinct than* CTL.

Proof By Theorems 2.5 and 2.6, it holds that CTL$^+$ is polynomially reducible to MCTL. Such a translation, preserves the structure of the model, since from the one of the CTL$^+$ formula we construct a new model that differs from the first at most by its enriched labeling. Now, it is known that there exists a sequence of CTL$^+$ formulas φ_n, with $\mathsf{lng}(\varphi_n) = \mathsf{O}(n)$ and $n \in \mathbb{N}$, whose minimal models have size $\mathsf{O}(2^n \cdot 2^{2^n})$ (Lange 2008). Thus, also in MCTL, we can write a related sequences with the same property. However, by the small model property of CTL (Emerson and Halpern 1985), every formula of this logic has minimal models whose size is at most exponential in its length. Hence, the statement follows. □

2.5 Model Checking

In this section, we solve the model checking for the introduced logics under the *m* semantics, showing that the considered extract-verify paradigm retains the decidability of this problem.

We start with a lemma that shows how to calculate a polynomial certificate for particular MCTL and MCTL* formulas. This result will be then useful to show the corresponding upper bound results for the addressed model checking problems.

Lemma 2.1 *Let $\mathcal{K}$ be a Kripke structure and $\varphi = \varphi_1 \mathbb{E} \varphi_2$ be a* MCTL *(resp.,* MCTL*) *formula, with φ_1 and φ_2* CTL *(resp.,* CTL*) *formulas. Then, there exists a polynomial certificate $\mathcal{K}'$ of the testing $\mathcal{K} \models \varphi$, which is verifiable in* PTIME *w.r.t. both φ and $\mathcal{K}$ (resp., in* PSPACE *w.r.t. φ and in* PTIME *w.r.t. $\mathcal{K}$).*

Proof To check that the test $\mathcal{K} \models \varphi$ is in NPTIME (resp., in PSPACE), we verify that there exists a minimal and conservative submodel $\mathcal{K}'$ of $\mathcal{K}$ satisfying φ_2 (the certificate of the test) of polynomial size (since $|\mathcal{K}'| \leq |\mathcal{K}|$) such that $\mathcal{K}' \models \varphi_1$.

To this aim, we split the verification procedure into the following four phases: *(i)* testing of $\mathcal{K}' \models \varphi_2$, *(ii)* checking the minimality of $\mathcal{K}'$, *(iii)* checking the conservativeness for $\mathcal{K}'$, and *(iv)* testing of $\mathcal{K}' \models \varphi_1$.

The first and last items are easily achievable in PTIME (resp., in PSPACE) w.r.t. the formula and in NLOGSPACE w.r.t. the model, by applying a classical CTL (resp., CTL*) model checking algorithm [Kupferman et al. (2000)].

To verify that $\mathcal{K}'$ is minimal w.r.t. the formula φ_2, we check that, for all maximal and proper submodels $\mathcal{K}''$ of $\mathcal{K}'$, it holds that $\mathcal{K}'' \not\models \varphi_2$. Now, note that all such models $\mathcal{K}''$ are in number $O(|\mathcal{K}'|)$, since each of them is obtained by removing only one component from $\mathcal{K}'$ at the time. So, we deduce that also the check for minimality can be done in PTIME w.r.t. both the length of the formula and the size of the model (resp., in PSPACE w.r.t. the formula and PTIME w.r.t. the model).

Finally, it remains to verify whether $\mathcal{K}'$ is conservative, i.e., for all models $\mathcal{K}''$, with $\mathcal{K}' \preceq \mathcal{K}''$, it holds that $\mathcal{K}'' \models \varphi_2$. To do this, we can check that, for all subformula φ' of φ_2 and worlds $w \in W'_{\mathcal{K}}$, it holds that $\mathcal{K}'_w \models \varphi'$ iff $\mathcal{K}_w \models \varphi'$. Since the number of all subformulas φ' is polynomial in the size of φ_2, and thus in the size of φ, it follows that also the check for conservativeness can be done in PTIME w.r.t. both the length of the formula and the size of the model (resp., in PSPACE w.r.t. the formula and PTIME w.r.t. the model). □

Using the above result, we are now able to prove the following two theorems.

Theorem 2.8 MCTL* *has a* PSPACE *model checking problem both in the length of the formula and in the size of the model.*

Proof Let $\mathcal{K}$ be a Kripke structure and φ an MCTL* in existential normal form, we construct a recursive algorithm that checks in PSPACE whether $\mathcal{K} \models \varphi$.

First of all, we enumerate all subformulas $\overline{\varphi} = \varphi' \mathbb{E} \varphi''$ of φ and associate to each of them a fresh different atomic proposition. More formally, suppose we have a sequence $\langle \overline{\varphi_1}, \ldots, \overline{\varphi_n} \rangle$ of such subformulas, then we associate to each $\overline{\varphi_i}$ the proposition ep_i. Now, consider Θ_m as the set of all formulas $\overline{\varphi} = \varphi' \mathbb{E} \varphi''$ subformulas of φ such that φ' contains just m and φ'' contains at most m nested occurrences of the $\mathbb{E}$ quantifier, or vice versa. Also, consider Θ'_m as the set of formulas $\tilde{\varphi}$ obtained from each $\overline{\varphi} \in \Theta_m$ by replacing every occurrence of a minimal model quantifier, but the most external one, with the relative atomic proposition. It is evident that, for all m, each $\tilde{\varphi} \in \Theta'_m$ is an MCTL* formula of the type $\tilde{\varphi} = \varphi' \mathbb{E} \varphi''$, with φ' and φ'' CTL* formulas. Moreover, $|\Theta'_m| = O(|\varphi|)$.

At this point, set $\mathscr{K}^0 = \mathscr{K}$, we construct a sequence of Kripke structures $\langle \mathscr{K}^0, \ldots, \mathscr{K}^n \rangle$ such that, for all $\tilde{\varphi} \in \Theta_m$ and $w \in \mathrm{W}_{\mathscr{K}}$, it holds that $ep \in \mathsf{L}_{\mathscr{K}^m}(w)$ iff $\mathscr{K}^{m-1} \models \tilde{\varphi}$, where ep is the atomic proposition relative to $\overline{\varphi}$. The latter can be checked by applying the PSPACE procedure of Lemma 2.1.

Consequently, the result follows by recursively applying the above procedure. □

Theorem 2.9 *M*CTL *has a model checking problem that is* Δ_2^P-COMPLETE *w.r.t. the length of the formula and* Δ_2^P *in the size of the model.*

Proof First note that, since by Theorem 2.5 it holds that CTL$^+$ is reducible to MCTL and the first has a Δ_2^P-COMPLETE model checking problem (Laroussinie et al. 2001), we have that the same problem for MCTL is Δ_2^P-HARD.

It remains to show that it is solvable in Δ_2^P. To prove this, we use a variation of the deterministic algorithm of Theorem 2.8, which, instead to call a PSPACE procedure to check whether $\mathscr{K} \models \varphi_1 \mathbb{E} \varphi_2$ or not, call an NPTIME-COMPLETE oracle (in accordance with Lemma 2.1), which solves the check in a single step. Now, since all other instructions of the algorithm are based on a classical CTL model checking procedure that can be executed in PTIME, we easily obtain a Δ_2^P model checking procedure for MCTL. □

2.6 Satisfiability

In this section, we first show the undecidability of the satisfiability problem for MCTL, MCTL$^+$, and MCTL* under the *m* and *mu* semantics through a reduction of the *recurrent domino problem*. Then, we show the decidability of MPML under the *m* semantics by means of a brute force procedure via strong finite model property (Blackburn et al. 2004).

The well-known *domino problem*, proposed for the first time by Wang (1961), consists of placing a given number of tile types on an infinite grid, satisfying a predetermined set of constraints on adjacent tiles. Its standard version asks for a compatible tiling of the whole plane $\mathbb{Z} \times \mathbb{Z}$. However, as stated by Knuth (1968), a compatible tiling of the first quadrant yields compatible tilings of arbitrary large finite rectangles, which in turn yields a compatible tiling of the whole plane. Since the existence of a solution for the original problem is known to be Π_0^1-COMPLETE [Berger (1966); Robinson (1971)], we have undecidable results also for the above variants of the classical domino problem. A formal definition of the $\mathbb{N} \times \mathbb{N}$ tiling problem follows.

Definition 2.3 (Domino System) *An* $\mathbb{N} \times \mathbb{N}$ domino system $\mathscr{D} = \langle \mathrm{D}, H, V \rangle$ *consists of a finite non-empty set* D *of* domino types *and two* horizontal *and* vertical matching relations $H, V \subseteq \mathrm{D} \times \mathrm{D}$. *The domino problem asks for an* admissible tiling *of* $\mathbb{N} \times \mathbb{N}$, *which is a* solution mapping $\partial : \mathbb{N} \times \mathbb{N} \to \mathrm{D}$ *such that, for all* $x, y \in \mathbb{N}$, *it holds that* (i) $(\partial(x, y), \partial(x + 1, y)) \in H$ *and* (ii) $(\partial(x, y), \partial(x, y + 1)) \in V$.

In the literature, an extension of the above problem has been also introduced as the *recurrent domino problem*. This problem, in addition to the tiling of the semiplane $\mathbb{N} \times \mathbb{N}$, asks whether there exists a distinguished tile type that occurs infinitely often in the first row of the grid. This problem is known to be more complex of the classical one. Indeed, it turns out to be Σ_1^1-COMPLETE [Harel (1984)]. The formal definition follows.

Definition 2.4 (Recurrent Domino System). *An* $\mathbb{N} \times \mathbb{N}$ *recurrent tiling system* $\langle \mathcal{D}, t^* \rangle$ *is a structure in which* $\mathcal{D} = \langle \mathrm{D}, H, V \rangle$ *is an* $\mathbb{N} \times \mathbb{N}$ *domino system and* $t^* \in \mathrm{D}$ *is a distinguished tile type. The recurrent domino problem asks for a solution mapping* $\partial : \mathbb{N} \times \mathbb{N} \to \mathrm{D}$ *such that* (i) ∂ *is an admissible tiling for* $\mathcal{D}$ *and* (ii) $|\{x \in \mathbb{N} : \partial(x, 0) = t^*\}| = \omega$.

By showing a reduction from the recurrent domino problem, we prove, in particular, that the satisfiability problem for MCTL* is Σ_1^1-HARD, which implies that it is even not computably enumerable. We achieve this reduction by describing how a given recurrent tiling system $\langle \mathcal{D}, t^* \rangle$ with $\mathcal{D} = \langle \mathrm{D}, H, V \rangle$ can be "*embedded*" into a model of a particular sentence $\varphi^{dom} \triangleq a \wedge b \wedge r \wedge \varphi^{rch}$ over AP $\triangleq \{a, b, r\} \cup \mathrm{D}$, where $a, b, r \notin \mathrm{D}$, in such a way that φ^{dom} is satisfiable iff $\mathcal{D}$ allows an admissible tiling. For the sake of clarity, we split the reduction into four tasks where we explicit the structure of the formula φ^{rch} built on the three formulas φ^{grd}, φ^{til}, and φ^{rec}.

Grid specification It is needed to represent a "square structure" of $\mathbb{N} \times \mathbb{N}$, which consists of the four points (x, y), $(x+1, y)$, $(x, y+1)$, and $(x+1, y+1)$, in order to yield a complete covering of the semi-plane via a repeating regular grid structure. The basic idea is to use the minimal model quantifiers to force the horizontal successor of $(x, y+1)$ and the vertical successor of $(x+1, y)$ to correspond to the unique point $(x+1, y+1)$, with the aim to represent a square structure model on which to place the domino types. Formally, this can be expressed by using the formula $\varphi^{grd} \triangleq \varphi_S \wedge \varphi_{U_H} \wedge \varphi_{U_V} \wedge \varphi_A$, with $\alpha \triangleq a \wedge b$, $\beta \triangleq \neg a \wedge b$, $\gamma \triangleq a \wedge \neg b$, and $\delta \triangleq \neg a \wedge \neg b$, where φ_S, φ_{U_H}, φ_{U_V}, and φ_A are defined as follows:

- $\varphi_H(\varphi) \triangleq (\alpha \rightarrow \mathsf{EX}(\gamma \wedge \varphi)) \wedge (\beta \rightarrow \mathsf{EX}(\delta \wedge \varphi)) \wedge (\gamma \rightarrow \mathsf{EX}(\alpha \wedge \varphi)) \wedge (\delta \rightarrow \mathsf{EX}(\beta \wedge \varphi))$;
- $\varphi_V(\varphi) \triangleq (\alpha \rightarrow \mathsf{EX}(\beta \wedge \varphi)) \wedge (\beta \rightarrow \mathsf{EX}(\alpha \wedge \varphi)) \wedge (\gamma \rightarrow \mathsf{EX}(\delta \wedge \varphi)) \wedge (\delta \rightarrow \mathsf{EX}(\gamma \wedge \varphi))$;
- $\varphi_S \triangleq \varphi_V(\varphi_H(\varphi_V(\mathsf{t})))\mathbb{E}(\varphi_V(\varphi_H(\mathsf{t})) \wedge \varphi_H(\varphi_V(\varphi_V(\mathsf{t}))))$;
- $\varphi_{U_H} \triangleq \varphi_H(\varphi_H(\mathsf{t}) \wedge \varphi_V(\mathsf{t}))\mathbb{A}(\varphi_H(\varphi_H(\mathsf{t})) \wedge \varphi_H(\varphi_V(\mathsf{t})))$;
- $\varphi_{U_V} \triangleq \varphi_V(\varphi_H(\mathsf{t}) \wedge \varphi_V(\mathsf{t}))\mathbb{A}(\varphi_V(\varphi_H(\mathsf{t})) \wedge \varphi_V(\varphi_V(\mathsf{t})))$;
- $\varphi_A \triangleq ((\alpha \vee \delta) \rightarrow \mathsf{AX}(\beta \vee \gamma)) \wedge ((\beta \vee \gamma) \rightarrow \mathsf{AX}(\alpha \vee \delta))$.

Compatible tiling It is needed to express that a tiling is locally compatible, i.e., the two h*orizontal and vertical neighborhoods of a given point have admissible domino types with respect to that one. The idea here is to associate to each domino type an atomic proposition and express the horizontal and vertical matching conditions via suitable object labeling. Note that these constraints are very easy to express.

Indeed, they can be simply expressed in PML. Formally, we have $\varphi^{til} \triangleq \bigvee_{t \in \mathrm{D}}(t \wedge \bigwedge_{t' \in \mathrm{D}}^{t' \neq t} \neg t' \wedge \bigvee_{(t,t') \in H} \varphi_H(t') \wedge \bigvee_{(t,t') \in V} \varphi_V(t'))$.

Recurrent tile It is required to assert that the distinguished tile type t^* occurs infinitely often on the first row of the semi-plane. This task can be easily achieved by using the kind of recursion available in the basic logic CTL. By means of this recursion, we can impose that the relative atomic proposition is satisfied in an infinite number of worlds linearly reachable from the origin of the grid. Formally, we have $\varphi^{rec} \triangleq \varphi_V(\mathsf{AG}\neg r) \wedge (r \rightarrow \varphi_H(\mathsf{EF}(r \wedge t^*)))$.

Global Reachability Finally, we need to impose that the above three conditions hold on all points of the $\mathbb{N} \times \mathbb{N}$ grid. As for the recurrent tile condition, also this task can be achieved by the simple recursion given by CTL. Formally, we have $\varphi^{rch} \triangleq \mathsf{AG}(\varphi^{grd} \wedge \varphi^{til} \wedge \varphi^{rec})$.

Construction correctness At this point, we have all the tools to formally prove the correctness of the undecidability reduction, by showing the equivalence between finding the solution of the recurrent tiling problem and the satisfiability of the sentence φ^{dom}.

Theorem 2.10 (*M*CTL, *M*CTL$^+$, **and** MCTL* **Satisfiability for** *m* **and** *mu*) *The satisfiability problem for M*CTL, *M*CTL$^+$, *and* MCTL*, *under the m and mu semantics, is highly undecidable. In particular, it is* Σ_1^1-HARD.

Proof Assume, for the direct reduction, that there exists a solution mapping $\partial : \mathbb{N} \times \mathbb{N} \rightarrow \mathrm{D}$ for the given domino system $\mathscr{D}$. Then, we can build a KS $\mathcal{K}_\partial^\star \triangleq \langle \mathrm{AP}, \mathrm{W}, R, \mathsf{L}, w_0 \rangle$ satisfying the sentence φ^{dom} in the following way: *(i)* $\mathrm{W} \triangleq \mathbb{N} \times \mathbb{N}$; *(ii)* $R \triangleq \{((x, y), (x + 1, y)), ((x, y), (x, y + 1)) : x, y \in \mathbb{N}\}$; *(iii)* $a \in \mathsf{L}((x, y))$ iff $y \equiv 0 \pmod 2$, $b \in \mathsf{L}((x, y))$ iff $x \equiv 0 \pmod 2$, $r \in \mathsf{L}((x, y))$ iff $y = 0$ and $\partial(x, 0) = t^*$, and $\mathsf{L}((x, y)) \cap \mathrm{D} = \{\partial(x, y)\}$, for all $x, y \in \mathbb{N}$; *(iv)* $w_0 = (0, 0)$ and $r \in \mathsf{L}((0, 0))$. By a simple case analysis on the subformulas of φ^{dom}, it is possible to see that $\mathcal{K}_\partial^\star \models \varphi^{dom}$.

Conversely, let $\mathcal{K} = \langle \mathrm{AP}, \mathrm{W}, R, \mathsf{L}, w_0 \rangle$ be a model of the sentence φ^{dom}. First, we show that $\mathcal{K}$ is a *grid-like model* and then that is possible to construct a solution mapping ∂ from it. In fact, since $\mathcal{K}, w_0 \models \varphi^{dom}$, we have that for all worlds $v \in \mathrm{W}$ reachable from w_0, i.e., $(w_0, v) \in R^n$ for some $n \in \mathbb{N}$, it holds that $\mathcal{K}, v \models \varphi^{grd}$ and thus $\mathcal{K}, v \models \varphi_S$. Now, it is not difficult to see that $\mathcal{K}$ must contain a square submodel rooted in v. Indeed, there exist only four different minimal models of the extractor formula $\varphi_e \triangleq (\varphi_V(\varphi_H(\mathsf{t})) \wedge \varphi_H(\varphi_V(\varphi_V(\mathsf{t}))))$ (see Fig. 2.2 for the possible submodels rooted in a node v such that $\mathcal{K}, v \models \alpha$) among which only the two models of the verifier formula $\varphi_v \triangleq \varphi_V(\varphi_H(\varphi_V(\mathsf{t})))$ have a square shape. Moreover, $\mathcal{K}, v \models \varphi_A$, so there are only two kinds of successors for v, i.e., if $\mathcal{K}, v \models \alpha$ or $\mathcal{K}, v \models \delta$ then, for all worlds $u \in \mathrm{W}$ with $(v, u) \in R$, it holds that $\mathcal{K}, u \models \beta$ or $\mathcal{K}, u \models \gamma$ and vice versa. Finally, since $\mathcal{K}, v \models \varphi_{U_H} \wedge \varphi_{U_V}$, if $\mathcal{K}, v \models \alpha$ or $\mathcal{K}, v \models \delta$ then there exists just one world $u_1 \in \mathrm{W}$ with $(v, u_1) \in R$ such that $\mathcal{K}, u_1 \models \beta$ and just one world $u_2 \in \mathrm{W}$ with $(v, u_2) \in R$ such that $\mathcal{K}, u_2$ and vice versa. Now, it is clear that each world v reachable from w (including w itself) has only two successors u_1 and u_2, which have a common successor o. Hence, $\mathscr{K}$ is a grid-like model. At this point, the extraction of a solution mapping ∂ from $\mathscr{K}$ is a routine task and it is left to the reader. □

Finally, we report the decidability result for MPML.

Theorem 2.11 *The satisfiability problem for* MPML *under the m semantics is decidable.*

Proof By Theorem 2.3, MPML has the strong finite model property w.r.t. a precise recursive function g. So for a given MPML formula φ we can construct a non deterministic Turing machine that, once the value of the function $g(|\varphi|)$ is computed, it guesses a model $\mathscr{K}$ of size not greater to this value and then checks if it satisfies the formula by applying the decidable model checking procedure given by Theorem 2.9. Since φ is satisfiable iff it is satisfied on a model $\mathscr{K}$ of size at most $g(|\varphi|)$ and the built machine systematically examines all these kinds of models, the thesis easily follows. □

2.7 Discussion

In this paper, we have introduced a new branching-time temporal logic, namely MCTL*, as an extension of the classic branching-time temporal logic CTL*, by means of minimal model quantifiers. By predicating directly on substructures, these quantifiers allow to extract minimal submodels of a system model, even when the modularity of the system is not known in advance, on which we successively check a given property of the introduced logic.

We have deeply investigated MCTL* and some of its sublogics, under three different semantics: minimal (m), minimal-unwinding (mu), and unwinding-minimal (um). They differ on the way a substructure is extracted and then checked in the verification process. In particular, in the mu semantics, we use the unwinding of a minimal substructure only applied after its extraction, while in the um one it is applied to the original structure.

As far as the expressiveness regards, we have showed that MCTL* is strictly more expressive than CTL*. Unfortunately, this power comes at a price. Indeed, the satisfiability problem for MCTL* (under the m and mu semantics), as well as for its sublogic MCTL, has been proved to be highly undecidable. Moreover, MCTL* (under the m and mu semantics) does not have the tree model property, it is not bisimulation-invariant, and it is sensible to unwinding, differently from CTL*. On the contrary, under the um semantics, the intended logic preserves all the mentioned model-theoretic properties.

As good news, we have showed that the sublogic MPML of MCTL* retains both the finite model property and the decidability of the satisfiability problem. Moreover, we have proved that the model checking problem for MCTL* (under the m semantics) remains decidable and in PSPACE. In more details and differently from CTL*, the

PSPACE upper bound we provide is in both the size of the system and the length of the specification. Since for CTL it is only PSPACE in the size of the formula, it is left as an open question whether this extra complexity can be avoided. Anyway, although practical applications of MCTL* are not in the target of this paper, we argue that the extra blow-up for MCTL* should not have any consequence in practical applications, as it can be absorbed in classical symbolic model checking algorithms. Last but not least, we have investigated succinctness and the model checking problems for MCTL. We have shown that, differently from the classical case of CTL, MCTL is as succinct as CTL$^+$. Moreover, as for CTL$^+$, the model checking problem for MCTL (under the m semantics) is Δ_2^P, i.e., PTIME^{NPTIME}.

As future work, it would be worth investigating if the bisimulation-invariant fragment of MCTL* (under the m semantics), i.e., the set of formulas that agree on bisimilar Kripke structures, is equally expressive as CTL*. In other words, we would like to check whether there exists an MCTL* formula which does not distinguish bisimilar structures, but it is still not expressible in CTL*. Then, if such a fragment is not equivalent to CTL*, it would be also relevant to investigate the related decidability problems.

Part II
Logics for Strategies

General Preliminaries II

In this section we introduce some more preliminary definitions and further notation used in the second part of the book.

Concurrent game structures A *concurrent game structure* (CGS, for short) (Alur et al. 2002) is a tuple $\mathcal{G} \triangleq \langle \text{AP}, \text{Ag}, \text{Ac}, \text{St}, \lambda, \tau, s_0 \rangle$, where AP and Ag are finite non-empty sets of *atomic propositions* and *agents*, Ac and St are enumerable non-empty sets of *actions* and *states*, $s_0 \in \text{St}$ is a designated *initial state*, and $\lambda : \text{St} \to 2^{\text{AP}}$ is a *labeling function* that maps each state to the set of atomic propositions true in that state. Let $\text{Dc} \triangleq \text{Ac}^{\text{Ag}}$ be the set of *decisions*, i.e., functions from Ag to Ac representing the choices of an action for each agent. Then, $\tau : \text{St} \times \text{Dc} \to \text{St}$ is a *transition function* mapping a state and a decision to a state. Intuitively, CGSs provide a generalization of *labeled transition systems* and *Kripke structures*, modeling *multi-agent systems*, viewed as *multi-player games* in which players perform *concurrent actions*, chosen strategically as a function of the history of the game. Note that elements in St are not global states of the system, but states of the environment in which the agents operate. Thus, they can be viewed as states of the game, which do not include the local states of the agents. We say that a CGS $\mathcal{G}$ is *turn-based* iff there is an additional function $\eta : \text{St} \to \text{Ag}$, named *owner function*, such that if $\mathsf{d}_1(\eta(s)) = \mathsf{d}_2(\eta(s))$ then $\tau(s, \mathsf{d}_1) = \tau(s, \mathsf{d}_2)$, for all $s \in \text{St}$ and $\mathsf{d}_1, \mathsf{d}_2 \in \text{Dc}$. Intuitively, a CGS is turn-based iff it is possible to associate at each state an agent, the owner of the state, which is the only responsible for the choice of the successor of that state. It is immediate to note that the function η introduce a partitioning of the set of states into $|\mathsf{rng}(\eta)|$ components. By $\|\mathcal{G}\| \triangleq |\text{St}| \cdot |\text{Dc}|$ we denote the *size* of $\mathcal{G}$, which also corresponds to the size $|\mathsf{dom}(\tau)|$ of the transition function τ. If the set of actions is finite, i.e., $b = |\text{Ac}| < \infty$, we say that $\mathcal{G}$ is *b-bounded*, or simply *bounded*. If both the sets of actions and states are finite, we say that $\mathcal{G}$ is *finite*. It is immediate to note that $\mathcal{G}$ is finite iff it has a finite size.

Concurrent game trees A *concurrent game tree* (CGT, for short) is a CGS $\mathcal{T} = \langle \mathrm{AP}, \mathrm{Ag}, \mathrm{Ac}, \mathrm{St}, \lambda, \tau, \mathrm{s}_0 \rangle \varepsilon$, where (i) $\mathrm{St} \subseteq \Delta^*$ is a Δ-tree for a given set Δ of directions and (ii) $t \cdot d \in \mathrm{St}$ iff there is a decision $\mathsf{d} \in \mathrm{Dc}$ such that $\tau(t, \mathsf{d}) = t \cdot d$, for all $t \in \mathrm{St}$ and $d \in \Delta$.

Tracks and paths A *track* (resp., *path*) in $\mathcal{G}$ is a finite (resp., an infinite) sequence of states $\rho \in \mathrm{St}^*$ (resp., $\pi \in \mathrm{St}^\omega$) such that, for all $i \in [0, |\rho|[$ (resp., $i \in \mathbb{N}$), there exists $\mathsf{d} \in \mathrm{Dc}$ such that $(\rho)_{i+1} = \tau((\rho)_i, \mathsf{d})$ (resp., $(\pi)_{i+1} = \tau((\pi)_i, \mathsf{d})$). Intuitively, tracks and paths of a CGS $\mathcal{G}$ are legal sequences of reachable states in $\mathcal{G}$ that can be seen, respectively, as a partial and complete description of the possible *outcomes* of the game modeled by $\mathcal{G}$. A track ρ is said *non-trivial* iff $|\rho| > 0$, i.e., $\rho \neq \varepsilon$. We use $\mathrm{Trk}(\mathcal{G}) \subseteq \mathrm{St}^+$ (resp., $\mathrm{Pth}(\mathcal{G}) \subseteq \mathrm{St}^\omega$) to indicate the set of all non-trivial tracks (resp., paths) of the CGS $\mathcal{G}$. Moreover, by $\mathrm{Trk}(\mathcal{G}, \mathrm{s}) \subseteq \mathrm{Trk}(\mathcal{G})$ (resp., $\mathrm{Pth}(\mathcal{G}, \mathrm{s}) \subseteq \mathrm{Pth}(\mathcal{G})$) we denote the subsets of tracks (resp., paths) starting at the state s.

Chapter 3
Reasoning About Strategies

Abstract In open systems verification, to formally check for reliability, one needs an appropriate formalism to model the interaction between open entities and express that the system is correct no matter how the environment behaves. An important contribution in this context is given by the *modal logics for strategic ability*, in the setting of *multi-agent games*, such as ATL, ATL*, and the like. Recently, Chatterjee, Henzinger, and Piterman introduced *Strategy Logic*, which we denote here by CHP-SL, with the aim of getting a powerful framework for reasoning explicitly about strategies. CHP-SL is obtained by using first-order quantifications over strategies and it has been investigated in the specific setting of two-agents turned-based game structures where a non-elementary model-checking algorithm has been provided. While CHP-SL is a very expressive logic, we claim that it does not fully capture the strategic aspects of multi-agent systems. In this work, we introduce and study a more general strategy logic, denoted SL, for reasoning about strategies in multi-agent concurrent systems. We prove that SL strictly includes CHP-SL, while maintaining a decidable model-checking problem. Indeed, we show that it is 2EXPTIME-COMPLETE under a reasonable semantics, thus not harder than that for ATL* and a remarkable improvement of the same problem for CHP-SL. We also consider the satisfiability problem and show that it is undecidable already for the sub-logic CHP-SL under the concurrent game semantics.

3.1 Introduction

In system design, *model checking* is a well-established formal method that allows to automatically check for global system correctness (Clarke and Emerson 1981; Queille and Sifakis 1981; Clarke et al. 2002). In such a framework, in order to check whether a system satisfies a required property, we express the system in a formal model (such as a *Kripke* structure), specify the property with a formula of a temporal logic (such as LTL (Pnueli 1977), CTL (Clarke and Emerson 1981), or CTL* (Emerson

F. Mogavero, *Logics in Computer Science*, Atlantis Studies in Computing 3, DOI: 10.2991/978-94-91216-95-4_3,

and Halpern 1986)), and check formally that the model satisfies the formula. In the last decade, interest has arisen in analyzing the behavior of individual components and sets of components in systems with several entities. This interest has started in reactive systems, which are systems that interact continually with their environments. In *module checking* (Kupferman et al. 2001) the system is modeled as a module that interacts with its environment and correctness means that a desired property holds with respect to all such interactions.

Starting from the study of module checking, researchers have looked for logics focusing on strategic behavior of agents in multi-agent systems (Alur et al. 2002; Pauly 2002; Jamroga and van der Hoek 2004). One of the most important development in this field is *Alternating-Time Temporal Logic* (ATL*, for short), introduced by Alur et al. 2002. ATL* allows reasoning about strategies for agents with temporal goals. Formally, it is obtained as a generalization of CTL* in which the path quantifiers, "E" (*there exists*) and "A" (*for all*) are replaced with "*strategic modalities*" of the form $\langle\langle A\rangle\rangle$ and $[[A]]$, where A is a set of *agents* (a.k.a. *players*). Strategic modalities over agent sets are used to express cooperation and competition among agents in order to achieve certain goals. In particular, these modalities express selective quantifications over those paths that are the results of infinite games between the coalition and its complement. ATL* formulas are interpreted over *concurrent game structures* (CGS, for short), which model interacting processes. Given a CGS $\mathcal{G}$ and a set A of agents, the ATL* formula $\langle\langle A\rangle\rangle\psi$ is satisfied at a state s of $\mathcal{G}$ if there is a *strategy* for agents in A such that, no matter the strategy that is executed by agents not in A, the resulting outcome of the interaction in $\mathcal{G}$ satisfies ψ at s. Thus, ATL* can express properties related to the interaction among agents, while CTL* can only express property of the global system. As an example, consider the property "processes α and β cooperate to ensure that a system (having more than two processes) never enters a failure state". This property can be be expressed by the ATL* formula $\langle\langle\{\alpha, \beta\}\rangle\rangle \mathsf{G}\neg \mathit{fail}$, where G is the classical temporal modality "*globally*". CTL*, in contrast, cannot express this property (Alur et al. 2002). Indeed, CTL* can only say whether the set of all agents can or cannot prevent the system from entering a fail state. The price that one have to pay for the expressiveness of ATL* is increased complexity. Indeed, both model checking and satisfiability checking are 2EXPTIME-COMPLETE (Alur et al. 2002; Schewe 2008).

Despite its powerful expressiveness, ATL* suffers of the strong limitation that strategies are treated only implicitly, through modalities that refer to games between competing coalitions. To overcome this problem, Chatterjee, Henzinger, and Piterman introduced Strategy Logic (CHP-SL, for short) (Chatterjee et al. 2007), a logic that treats strategies in two-player games as explicit first-order objects. In CHP-SL, the ATL* formula $\langle\langle\alpha\rangle\rangle\psi$, for a system modeled by a CGS with agents α and β, becomes $\exists x.\forall y.\psi(x, y)$, i.e., "there exists a player-α strategy x such that for all player-β strategies y, the unique infinite path resulting from the two players following the strategies x and y satisfies the property ψ". The explicit treatment of strategies in CHP-SL allows to state many properties not expressible in ATL*. In particular, it is shown in (Chatterjee et al. 2007) that ATL* corresponds to the proper one-alternation fragment of CHP-SL. Chatterjee et al. have shown that the model-checking problem for CHP-SL is decidable, although only a non-elementary algorithm for it, both in the size of the system and the size formula, has been provided, leaving as open the question whether an algorithm with a better complexity exists or not. The question about the decidability of satisfiability checking for CHP-SL was also left open in Chatterjee et al. (2007).

While the basic idea exploited in Chatterjee et al. (2007) to quantify over strategies, and thus to commit agent explicitly to certain strategies, turns out to be very powerful, as discussed above, the logic CHP-SL introduced there has been defined and investigated only under the weak framework of two-players and turn-based games. Also, the specific syntax considered for CHP-SL allows only a weak kind of strategy commitment. For example, CHP-SL does not allow different players to share, in different contexts, the same strategy. These considerations, as well as all questions left open about CHP-SL, have led us to introduce and investigate a new *Strategy Logic*, denoted SL, as a more general framework than CHP-SL, for explicit reasoning about strategies in multi-player concurrent game structures. Syntactically, SL extends LTL by means of two *strategy quantifiers*, the existential $\langle\langle x\rangle\rangle$ and the universal $[[x]]$, and an *agent binding* (α, x), where α is an agent and x is variable. Intuitively, these elements can be respectively read as "*there exists a strategy* x", "*for all strategies* x", and "*bind agent* α *to the strategy associated with* x". For example, in a CGS with three agents α, β, γ, the previous ATL* formula $\langle\langle\{\alpha, \beta\}\rangle\rangle \mathsf{G}\neg \mathit{fail}$ can be translated in the SL formula $\langle\langle x\rangle\rangle\langle\langle y\rangle\rangle[[z]](\alpha, x)(\beta, y)(\gamma, z)(\mathsf{G}\neg \mathit{fail})$. The variables x and y are used to select two strategies for the agents α and β, respectively, and z is used to select all strategies for agent γ such that the composition of all these strategies results in a play where *fail* is never meet. Note that we can also require (by means of agent binding) that agents α and β share the same strategy, using the formula $\langle\langle x\rangle\rangle[[z]](\alpha, x)(\beta, x)(\gamma, z)(\mathsf{G}\neg \mathit{fail})$. We can also vary the structure of the game by changing the way the quantifiers alternate, for example, in the formula $\langle\langle x\rangle\rangle[[z]]\langle\langle y\rangle\rangle(\alpha, x)(\beta, y)(\gamma, z)(\mathsf{G}\neg \mathit{fail})$. In this case, x remains uniform w.r.t. z, but y becomes dependent on z. The last two examples show that SL is a proper extension of both ATL* and CHP-SL. It is worth to note that the pattern of modal quantifications over strategies and binding to agents can be extended to other logics than LTL, such as the linear μCALCULUS (Vardi 1988). In fact, the use of LTL here is only a matter of simplicity in presenting our framework, and changing the embedded temporal logic involves only few side-changes in the decision procedures.

As a main result in this paper, we show that the model-checking problem for SL, under a reasonable restrict semantics, is decidable and precisely PTIME in the size of the model and 2EXPTIME-COMPLETE in the size of the specification, thus not harder than that for ATL*. Remarkably, this result improves significantly the complexity of the model-checking problem for CHP-SL, for which only a non-elementary upper-bound was known (Chatter-jee et al. 2007). The lower bound for the addressed problem immediately follows from ATL*, which SL includes. For the upper bound, we follow an *automata-theoretic approach* (Kupferman et al. 2000), by reducing the decision problem for the logic to the emptiness problem of automata. To this aim, we use *alternating parity tree automata*, which are *alternating tree automata* (see Grädel et al. 2002, for a survey) along with a parity acceptance condition (Muller and

Schupp 1995). Due to the exponential size of the required automaton and the EXPTIME complexity required for checking its emptiness, we get the desired 2EXPTIME upper bound.

As another important issue in this paper, we address the satisfiability problem for SL. By using a reduction from the *recurrent domino problem*, we show that this problem is highly undecidable, and in fact Σ_1^1-HARD, (i.e., it is not computably enumerable). Interestingly, the reduction we propose also holds for the fragment of CHP-SL in which only the next temporal operator is used, under the concurrent game semantics. Thus, we show that in this setting also CHP-SL is highly undecidable, while it remains an open question whether it is decidable or not in the turn-based framework. A key point to prove the undecidability of SL has been to show that this logic lacks of the bounded-tree model property, which does hold for ATL* (Schewe 2008).

Since the rise of temporal and modal program logics in the mid-to-late 1970s, we have learned to expect such logics to have a decidable satisfiability problem. In the context of temporal logic, decidability results were extended from LTL to CTL* and ATL*. SL deviates from this pattern. It has a decidable model-checking problem, but an undecidable satisfiability problem. In this, it is similar to first-order logic. The decidability of model checking for first-order logic is the foundation for query evaluation in relational databases, and undecidability of satisfiability is a challenge we need to contend with. At the same time, it is clear that SL has nontrivial fragments, for example ATL*, which do have a decidable satisfiability problem. Identifying larger fragments of SL with a decidable satisfiability problem is an important research problem.

Related works Several works have focused on extensions of ATL* to incorporate more powerful strategic constructs. Among them, we recall the logics *Alternating-Time* μCALCULUS (AMC, for short) (Alur et al. 2002), *Game Logic* (GL, for short) (Alur et al. 2002), *Quantified Decision Modality* μCALCULUS (QDμ, for short) (Pinchinat 2007), *Coordination Logic* (CL, for short) (Finkbeiner and Schewe 2010), and some extensions of ATL* considered in Brihaye et al. (2009). AMC and QDμ are intrinsically different from SL (as well as CHP-SL and ATL*) as they are obtained by extending the propositional μ-calculus (Kozen 1983) with strategic modalities. CL is similar to QDμ but with LTL temporal operators instead of explicit fixpoint constructs. GL is strictly included in CHP-SL, but does not use any explicit treatment of strategies. Also the extensions of ATL* considered in Brihaye et al. (2009) do not use any explicit treatment of strategies. Rather, they consider restrictions on the memory for strategy quantifiers. Thus, all the above logics are different from SL, which aims it at being a minimal but powerful logic to reason about strategic behavior in multi-agent systems.

Outline In Sect. 3.2, we recall the basic notions regarding strategies, assignments, and plays. Then, we have Sect. 3.3, in which we introduce SL and define its syntax and semantics, followed by Sects. 3.4 and 3.5, in which we study the basic properties of the logic. In Sect. 3.6, we describe the ATA automaton model. Finally, in Sects. 3.7 and 3.8 we describe, respectively, the procedure used to solve the model-checking problem, and the undecidability proof of the satisfiability problem. Note that, in the accompanying Appendix, we recall standard mathematical notation and some basic definitions that are used in the work.

3.2 Preliminaries

Strategies Let $\mathcal{G} = \langle \text{AP}, \text{Ag}, \text{Ac}, \text{St}, \lambda, \tau, s_0 \rangle$ be a CGS. A *strategy* for $\mathcal{G}$ is a partial function $\mathsf{f} : \text{Trk}(\mathcal{G}) \rightharpoonup \text{Ac}$ whose domain is a St-tree, non associated to any particular agent, which maps each non-trivial track in its domain to an action. Intuitively, a strategy is a *plan* for an agent that contains all choices of moves as a function of the history of the current outcome. For a state s, we say that f is s-*total* iff it is defined on all non-trivial tracks starting in s, i.e., $\mathsf{dom}(\mathsf{f}) = \{\rho \in \text{Trk}(\mathcal{G}) : \mathsf{fst}(\rho) = s\}$. We use $\text{Str}(\mathcal{G})$ (resp., $\text{Str}(\mathcal{G}, s)$ with $s \in \text{St}$) to indicate the set of all the (resp., s-total) strategies of the CGS $\mathcal{G}$. For a track $\rho \in \mathsf{dom}(\mathsf{f})$, by f_ρ we denote the *translation* of f along ρ, i.e., the $\mathsf{lst}(\rho)$-total strategy such that $\mathsf{f}_\rho(\mathsf{lst}(\rho) \cdot \rho') \triangleq \mathsf{f}(\rho \cdot \rho')$, for all $\mathsf{lst}(\rho) \cdot \rho' \in \mathsf{dom}(\mathsf{f}_\rho)$.

Assignments Let $\text{Var} = \{x, x_0, x_1, \ldots, y, \ldots\}$ be a fixed set of *variables*. An *assignment* for $\mathcal{G}$ is a partial function $\chi : \text{Ag} \cup \text{Var} \rightharpoonup \text{Str}(\mathcal{G})$ mapping every agent and variable, a.k.a. *placeholders*, to a strategy. An assignment χ is *complete* iff $\text{Ag} \subseteq \mathsf{dom}(\chi)$. For a state s, we say that χ is s-*total* iff all strategies $\chi(l)$ are s-total too, for $l \in \mathsf{dom}(\chi)$. We use $\text{Asg}(\mathcal{G})$ (resp., $\text{Asg}(\mathcal{G}, s)$ with $s \in \text{St}$) to indicate the set of all (resp., s-total) assignments of the CGS $\mathcal{G}$. Moreover, by $\text{Asg}(\mathcal{G}, \text{V})$ (resp., $\text{Asg}(\mathcal{G}, \text{V}, s)$ with $s \in \text{St}$) we indicate the subsets of (resp., s-total) assignments defined on $\text{V} \subseteq \text{Ag} \cup \text{Var}$. Let ρ be a track and χ be an $\mathsf{fst}(\rho)$-total assignment. By χ_ρ we denote the *translation* of χ along ρ, i.e., the $\mathsf{lst}(\rho)$-total assignment with $\mathsf{dom}(\chi_\rho) \triangleq \mathsf{dom}(\chi)$, such that $\chi_\rho(l) \triangleq \chi(l)_\rho$, for all $l \in \mathsf{dom}(\chi)$. Intuitively, the translation χ_ρ is the update of all strategies contained into the assignment χ, after that the history of the game becomes ρ. Let χ be an assignment, a be an agent, x be a variable, and f be a strategy. Then, by $\chi_{a \mapsto \mathsf{f}}$ and $\chi_{x \mapsto \mathsf{f}}$ we denote, respectively, the new assignments defined on $\mathsf{dom}(\chi) \cup \{a\}$ and $\mathsf{dom}(\chi) \cup \{x\}$ that return f on a and x and are equal to χ on the remaining part of their domain. Note that, if χ and f are s-total, $\chi_{a \mapsto \mathsf{f}}$ and $\chi_{x \mapsto \mathsf{f}}$ are s-total, too.

Plays Finally, a path π starting in a state s is a *play* w.r.t. a complete s-total assignment χ ((χ, s)-*play*, for short) iff, for all $i \in \mathbb{N}$, it holds that $\pi_{i+1} = \tau(\pi_i, \mathsf{d})$, where $\mathsf{d}(a) = \chi(a)(\pi_{\leq i})$, for all $a \in \text{Ag}$. Note that there is a unique (χ, s)-play. Intuitively, a play is the outcome of the game determined by all the agent strategies participating to the game.

In the sequel, we use the Greek letters "α, β, γ" possibly with indexes to indicate specific agents of a CGS, while we use the Latin letter "a" as a meta-variable on the agents themselves.

3.3 Strategy Logic

In this section, we formally introduce an extension of the classical linear-time temporal logic LTL (Pnueli 1977) with the concepts of strategy quantification and binding and discuss its main properties. In particular, we show that it has a kind of tree model property, different to that proved to hold for ATL*, but not the relative bounded version, which is usually required in order to obtain a decidable satisfiability problem. Differently from CHP-SL, to formally define the extended logic, we do not use the CTL* formulas framework but the LTL one.

3.3.1 Syntax

Strategy logic (SL, for short) syntactically extends LTL by means of two *strategy quantifiers*, the existential $\langle\langle x\rangle\rangle$ and the universal $[[x]]$, and an *agent binding* (a, x), where a is an agent and x is a variable. Intuitively, these new elements can be read, respectively, as "*there exists a strategy* x", "*for all strategies* x", and "*bind agent* a *to the strategy associated with variable* x". The formal syntax of SL follows.

Definition 3.1 (SL **Syntax**) SL formulas *are built inductively from the sets of atomic propositions* AP, *variables* Var, *and agents* Ag, *in the following way, where* $p \in \mathrm{AP}$, $x \in \mathrm{Var}$, *and* $a \in \mathrm{Ag}$*:*

$$\varphi ::= p \mid \neg\varphi \mid \varphi \wedge \varphi \mid \varphi \vee \varphi \mid \mathsf{X}\varphi \mid \varphi\,\mathsf{U}\varphi \mid \varphi\,\mathsf{R}\varphi \mid \langle\langle x\rangle\rangle\varphi \mid [[x]]\varphi \mid (a, x)\varphi.$$

We now introduce some auxiliary syntactical notation. For a formula φ, we define the *length* $\mathsf{lng}(\varphi)$ of φ as for LTL. Formally, *(i)* $\mathsf{lng}(p) \triangleq 1$, for $p \in \mathrm{AP}$, *(ii)* $\mathsf{lng}(\mathsf{Op}\ \psi) \triangleq 1+\mathsf{lng}(\psi)$, for all $\mathsf{Op} \in \{\neg, \mathsf{X}\}$, *(iii)* $\mathsf{lng}(\psi_1 \mathsf{Op}\ \psi_2) \triangleq 1+\mathsf{lng}(\psi_1)+\mathsf{lng}(\psi_2)$, for all $\mathsf{Op} \in \{\wedge, \vee, \mathsf{U}, \mathsf{R}\}$, and *(iv)* $\mathsf{lng}(\mathsf{Qn}\ \psi) \triangleq 1 + \mathsf{lng}(\psi)$, for all $\mathsf{Qn} \in \{\langle\langle x\rangle\rangle, [[x]], (a, x)\}$. We also use $\mathsf{free}(\varphi)$ we denote the set of *free agents/variables*, a.k.a. *free placeholders*, of φ defined as the subset of $\mathrm{Ag} \cup \mathrm{Var}$ containing *(i)* all the agents for which there is no variable application after the occurrence of a temporal operator and *(ii)* all the variables for which there is an application but no quantifications. For example, let $\varphi = \langle\langle x\rangle\rangle(\alpha, x)(\beta, y)(\mathsf{F}p)$ be a formula on the agents $\mathrm{Ag} = \{\alpha, \beta, \gamma\}$. Then, we have $\mathsf{free}(\varphi) = \{\gamma, y\}$, since γ is an agent without any application after $\mathsf{F}p$ and y has no quantification at all. Formally, *(i)* $\mathsf{free}(p) = \emptyset$, for $p \in \mathrm{AP}$; *(ii)* $\mathsf{free}(\neg\varphi) = \mathsf{free}(\varphi)$; *(iii)* $\mathsf{free}(\varphi_1 \mathsf{Op}\ \varphi_2) = \mathsf{free}(\varphi_1) \cup \mathsf{free}(\varphi_2)$, where $\mathsf{Op} \in \{\wedge, \vee\}$; *(iv)* $\mathsf{free}(\mathsf{X}\varphi) = \mathrm{Ag} \cup \mathsf{free}(\varphi)$; *(v)* $\mathsf{free}(\varphi_1 \mathsf{Op}\ \varphi_2) = \mathrm{Ag} \cup \mathsf{free}(\varphi_1) \cup \mathsf{free}(\varphi_2)$, where $\mathsf{Op} \in \{\mathsf{U}, \mathsf{R}\}$; *(vi)* $\mathsf{free}(\mathsf{Qn}\ \varphi) = \mathsf{free}(\varphi) \setminus \{x\}$, where $\mathsf{Qn} \in \{\langle\langle x\rangle\rangle, [[x]]\}$; and *(vii)* if $a \in \mathsf{free}(\varphi)$ then $\mathsf{free}((a, x)\varphi) = (\mathsf{free}(\varphi) \setminus \{a\}) \cup \{x\}$ else $\mathsf{free}((a, x)\varphi) = \mathsf{free}(\varphi)$. A formula φ without free agents (resp., variables), i.e., with $\mathsf{free}(\varphi) \cap \mathrm{Ag} = \emptyset$ (resp., $\mathsf{free}(\varphi) \cap \mathrm{Var} = \emptyset$), is named *agent-closed* (resp., *variable-closed*). If φ is both agent- and variable-closed, it is referred to as a *sentence*.

3.3.2 Semantics

As for ATL* and differently from CHP-SL, we define the semantics of SL w.r.t. concurrent game structures. For a CGS $\mathcal{G}$, a state s, and an s-total assignment χ with $\mathsf{free}(\varphi) \subseteq \mathsf{dom}(\chi)$, we write $\mathcal{G}, \chi, s \models \varphi$ to indicate that the formula φ holds at s under the assignment χ. Similarly, if χ is a complete assignment, for the (χ, s)-play π and a natural number k, we write $\mathcal{G}, \chi, \pi, k \models \varphi$ to indicate that φ holds at the position k of π. The semantics of the SL formulas involving atomic propositions, the Boolean connectives $\neg$, $\wedge$, and $\vee$, as well as that for the temporal operators X, U, and R, is defined as usual in LTL. The novel part resides in the semantics of strategy quantifications and agent binding.

Definition 3.2 (SL **Semantics**) *Given a* CGS $\mathcal{G} = \langle \mathrm{AP}, \mathrm{Ag}, \mathrm{Ac}, \mathrm{St}, \lambda, \tau, s_0 \rangle$, *for all* SL *formulas* φ, *states* $s \in \mathrm{St}$, *and s-total assignments* $\chi \in \mathrm{Asg}(\mathcal{G}, s)$ *with* $\mathsf{free}(\varphi) \subseteq \mathsf{dom}(\chi)$, *the relation* $\mathcal{G}, \chi, s \models \varphi$ *is inductively defined as follows.*

(1) $\mathcal{G}, \chi, s \models p$ *iff* $p \in \lambda(s)$, *with* $p \in \mathrm{AP}$.
(2) For all formulas φ, φ_1, *and* φ_2, *it holds that:*

(a) $\mathcal{G}, \chi, s \models \neg\varphi$ *iff not* $\mathcal{G}, \chi, s \models \varphi$, *that is* $\mathcal{G}, \chi, s \not\models \varphi$;
(b) $\mathcal{G}, \chi, s \models \varphi_1 \wedge \varphi_2$ *iff* $\mathcal{G}, \chi, s \models \varphi_1$ *and* $\mathcal{G}, \chi, s \models \varphi_2$;
(c) $\mathcal{G}, \chi, s \models \varphi_1 \vee \varphi_2$ *iff* $\mathcal{G}, \chi, s \models \varphi_1$ *or* $\mathcal{G}, \chi, s \models \varphi_2$.

(3) For an agent $a \in \mathrm{Ag}$, *a variable* $x \in \mathrm{Var}$, *and a formula* φ, *it holds that:*

(a) $\mathcal{G}, \chi, s \models \langle\langle x \rangle\rangle\varphi$ *iff there exists a strategy* $\mathsf{f} \in \mathrm{Str}(\mathcal{G}, s)$ *such that* $\mathcal{G}, \chi_{x \mapsto \mathsf{f}}, s \models \varphi$;
(b) $\mathcal{G}, \chi, s \models [[x]]\varphi$ *iff for all strategies* $\mathsf{f} \in \mathrm{Str}(\mathcal{G}, s)$ *it holds that* $\mathcal{G}, \chi_{x \mapsto \mathsf{f}}, s \models \varphi$;
(c) $\mathcal{G}, \chi, s \models (a, x)\varphi$ *iff* $\mathcal{G}, \chi_{a \mapsto \chi(x)}, s \models \varphi$.

(4) Finally, if χ *is also complete, for all formulas* φ, φ_1, *and* φ_2, *where* π *is the* (χ, s)*-play and* $k \in \mathbb{N}$, *it holds that:*

(a) $\mathcal{G}, \chi, s \models \varphi$ *iff* $\mathcal{G}, \chi, \pi, 0 \models \varphi$;
(b) $\mathcal{G}, \chi, \pi, k \models \mathsf{X}\varphi$ *iff* $\mathcal{G}, \chi, \pi, k+1 \models \varphi$;
(c) $\mathcal{G}, \chi, \pi, k \models \varphi_1 \mathsf{U} \varphi_2$ *iff there is an index* $i \in \mathbb{N}$ *with* $k \leq i$ *such that* $\mathcal{G}, \chi, \pi, i \models \varphi_2$ *and, for all indexes* $j \in \mathbb{N}$ *with* $k \leq j < i$, *it holds that* $\mathcal{G}, \chi, \pi, j \models \varphi_1$;
(d) $\mathcal{G}, \chi, \pi, k \models \varphi_1 \mathsf{R} \varphi_2$ *iff, for all indexes* $i \in \mathbb{N}$ *with* $k \leq i$, *it holds that* $\mathcal{G}, \chi, \pi, i \models \varphi_2$ *or there is an index* $j \in \mathbb{N}$ *with* $k \leq j < i$ *such that* $\mathcal{G}, \chi, \pi, j \models \varphi_1$;
(e) $\mathcal{G}, \chi, \pi, k \models \varphi$ *iff* $\mathcal{G}, \chi_{\pi_{\leq k}}, \pi_k \models \varphi$.

Intuitively, at Items 3a and 3b, respectively, we evaluate existential and universal quantifiers over strategies. At Item 3c, by means of an agent binding (a, x), we commit the agent a to a strategy contained in the variable x. Finally, Items 4a and 4e can be easily understood by looking at their analogous path and state formulas in ATL*.

In fact, Item 4a can be viewed as the rule that allows to move the evaluation process from states to plays and, vice versa, Item 4e from plays to states.

We say that a CGS $\mathcal{G}$ is a *model* of an SL sentence φ, in symbols $\mathcal{G} \models \varphi$, iff $\mathcal{G}, \varnothing, s_0 \models \varphi$, where $\varnothing$ is the empty assignment. In this case, we also say that $\mathcal{G}$ is a model for φ on s_0. A sentence φ is said *satisfiable* iff there is a model for it. Moreover, it is an *invariant* for the two CGSs $\mathcal{G}_1$ and $\mathcal{G}_2$ iff either $\mathcal{G}_1 \models \varphi$ and $\mathcal{G}_2 \models \varphi$ or $\mathcal{G}_1 \not\models \varphi$ and $\mathcal{G}_2 \not\models \varphi$. For two SL formulas φ_1 and φ_2 we say that φ_1 *implies* φ_2, formally $\varphi_1 \Rightarrow \varphi_2$, iff, for all CGSs $\mathcal{G}$, states s, and s-defined assignments $\chi \in \mathrm{Asg}(\mathcal{G}, s)$ with $\mathsf{free}(\varphi_1) \cup \mathsf{free}(\varphi_2) \subseteq \mathsf{dom}(\chi)$, it holds that if $\mathcal{G}, \chi, s \models \varphi_1$ then $\mathcal{G}, \chi, s \models \varphi_2$. Consequently, we say that φ_1 is *equivalent* to φ_2, in symbols $\varphi_1 \equiv \varphi_2$, iff $\varphi_1 \Rightarrow \varphi_2$ and $\varphi_2 \Rightarrow \varphi_1$.

To get attitude to the introduced logic framework, let us consider the simple sentence $\varphi = \langle\langle x\rangle\rangle[[y]]\langle\langle z\rangle\rangle(\alpha, x)(\beta, y)(\mathsf{X}p) \wedge (\alpha, y)(\beta, z)(\mathsf{X}q)$ to see how to evaluate it. First, note that α and β both use the strategy associated with y to achieve the goals $\mathsf{X}q$ and $\mathsf{X}p$, respectively. A model for φ is $\mathcal{G} \triangleq \langle\{p, q\}, \{\alpha, \beta\}, \{0, 1\}, \{s_0, s_1, s_2, s_3\}, \lambda, \tau, s_0\rangle$, where $\lambda(s_0) \triangleq \emptyset$, $\lambda(s_1) \triangleq \{p\}$, $\lambda(s_2) \triangleq \{p, q\}$, $\lambda(s_3) \triangleq \{q\}$, $\tau(s_0, (0, 0)) \triangleq s_1$, $\tau(s_0, (0, 1)) \triangleq s_2$, $\tau(s_0, (1, 0)) \triangleq s_3$, and all the remaining transitions (with any action) go to s_0 (see Fig. 3.1). Clearly, $\mathcal{G}, s_0 \models \varphi$ by letting, on s_0, the variables x to chose action 0 (the goal $\mathsf{X}p$ is satisfied for any choice of y, since we can move from s_0 to s_1 or s_2, both labeled with p) and z to choose action 1 when y has action 0 and, vice versa, 0 when y has 1 (in both the cases, the goal $\mathsf{X}q$ is satisfied, since one can move from s_0 to s_2 or s_3, both labeled with q).

An important property that is possible to express in SL, but neither in ATL* nor in CHP-SL, is the existence of *deterministic multi-player Nash equilibria*. For example, consider n agents $\alpha_1, \ldots, \alpha_n$ each of them having the LTL goals $\psi_1, \ldots, \psi_n$. Then, we can express the existence of a *strategy profile* $(x_1, \ldots, x_n)$ that is a Nash equilibrium for $\alpha_1, \ldots, \alpha_n$ w.r.t. $\psi_1, \ldots, \psi_n$ by using the sentence $\langle\langle x_1\rangle\rangle \cdots \langle\langle x_n\rangle\rangle(\alpha_1, x_1) \cdots (\alpha_n, x_n)(\bigwedge_{i=1}^{n}(\langle\langle y\rangle\rangle(\alpha_i, y)\psi_i) \rightarrow \psi_i)$. Informally, this sentence asserts that every agent α_i has the "best" strategy w.r.t. the goal ψ_i once all the other strategies of the remaining agents have been fixed. Note that here we have only considered equilibria under deterministic strategies.

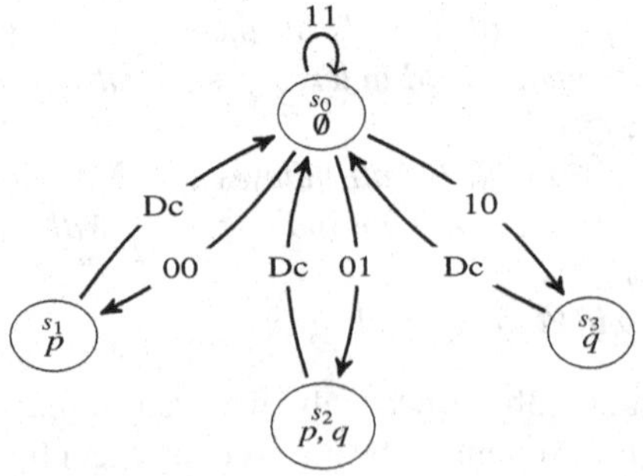

Fig. 3.1 The CGS $\mathcal{G}$ model of φ

In the following, we also consider the case in which SL has its semantics defined on turn-based CGS only. In such an eventuality, we call the logic *Turn-based strategy logic* (TB-SL, for short).

3.4 Basic Properties

We now investigate some basic properties of SL that turn out to be important for their own and useful to prove the decidability of the model checking and the undecidability of the satisfiability. In particular, for the introduced logics we investigate the concepts of bisimulation, local-isomorphism, and unwinding as well as the tree and finite model properties.

3.4.1 Basic Definitions

As principal definition, we formally state the concept of bisimilarity between CGSs. Intuitively, two CGSs $\mathcal{G}_1$ and $\mathcal{G}_2$ are bisimilar iff we can build an association of each state of the first structure with a state of the second one, and vice versa, in a way that each play in $\mathcal{G}_1$ has an equivalent play in $\mathcal{G}_2$ and vice versa. As we show later, such a concept results to be or not enough strong to characterize equivalent structures in dependence of the logic we want to consider.

Definition 3.3 (Bisimulation) *Let* $\mathcal{G}_1 = \langle \mathrm{AP}, \mathrm{Ag}, \mathrm{Ac}_1, \mathrm{St}_1, \lambda_1, \tau_1, s_{0_1} \rangle$ *and* $\mathcal{G}_2 = \langle \mathrm{AP}, \mathrm{Ag}, \mathrm{Ac}_2, \mathrm{St}_2, \lambda_2, \tau_2, s_{0_2} \rangle$ *be two* CGSs. *Then,* $\mathcal{G}_1$ *and* $\mathcal{G}_2$ *are* bisimilar *iff there are a relation* $\sim \subseteq \mathrm{St}_1 \times \mathrm{St}_2$ *between states, called* bisimulation relation, *and a function* $\mathsf{g} : \sim \rightarrow 2^{\mathrm{Ac}_1 \times \mathrm{Ac}_2}$ *mapping pairs of states in* $\sim$ *to relations between actions, called* bisimulation function, *such that the following holds:*

(1) $s_{0_1} \sim s_{0_2}$;
(2) for all $s_1 \in \mathrm{St}_1$ *and* $s_2 \in \mathrm{St}_2$, *if* $s_1 \sim s_2$ *then*
   (a) $\lambda_1(s_1) = \lambda_2(s_2)$;
   (b) for all $c_1 \in \mathrm{Ac}_1$, *there is* $c_2 \in \mathrm{Ac}_2$ *such that* $(c_1, c_2) \in \mathsf{g}(s_1, s_2)$;
   (c) for all $c_2 \in \mathrm{Ac}_2$, *there is* $c_1 \in \mathrm{Ac}_1$ *such that* $(c_1, c_2) \in \mathsf{g}(s_1, s_2)$;
   (d) for all $(\mathsf{d}_1, \mathsf{d}_2) \in \widehat{\mathsf{g}}(s_1, s_2)$, *it holds that* $\tau_1(s_1, \mathsf{d}_1) \sim \tau_2(s_2, \mathsf{d}_2)$, *where* $\widehat{\mathsf{g}} : \sim \rightarrow 2^{\mathrm{Dc}_1 \times \mathrm{Dc}_2}$ *is the lifting of* g *to decisions, i.e., it is the function mapping pairs of states in* $\sim$ *to relations between decisions such that* $(\mathsf{d}_1, \mathsf{d}_2) \in \widehat{\mathsf{g}}(s_1, s_2)$ *iff, for all* $a \in \mathrm{Ag}$, *it holds that* $(\mathsf{d}_1(a), \mathsf{d}_2(a)) \in \mathsf{g}(s_1, s_2)$.

The bisimulation relation extends the classical concepts of bisimilarity defined for *Kripke structures* by replacing the *forth and back conditions* considered there by means of Items 2b–2d defined above, which intuitively state the following: Item 2b, the forth clause, (resp., Item 2c, the back clause) says that for each action in $\mathcal{G}_1$

(resp., $\mathcal{G}_2$), there exists a bisimilar action in $\mathcal{G}_2$ (resp., in $\mathcal{G}_1$), while Item 2d asserts that bisimilar states are mapped to bisimilar successors through bisimilar decisions.

It is easy to see that the bisimulation of two structures implies the existence of a bisimulation between their decisions, as stated in the following proposition. However, note that the existence of a bisimulation between decisions, on the converse, does not imply the existence of a bisimulation function for the actions on which these decisions are built.

Proposition 3.1 (Decision Bisimulation) *Let* $\mathcal{G}_1 = \langle \mathrm{AP}, \mathrm{Ag}, \mathrm{Ac}_1, \mathrm{St}_1, \lambda_1, \tau_1, s_{0_1} \rangle$ *and* $\mathcal{G}_2 = \langle \mathrm{AP}, \mathrm{Ag}, \mathrm{Ac}_2, \mathrm{St}_2, \lambda_2, \tau_2, s_{0_2} \rangle$ *be two bisimilar* CGSs*. Then, for all* $s_1 \in \mathrm{St}_1$ *and* $s_2 \in \mathrm{St}_2$ *with* $s_1 \sim s_2$*, the following holds:*

(1) for all $\mathsf{d}_1 \in \mathrm{Dc}_1$*, there is* $\mathsf{d}_2 \in \mathrm{Dc}_2$ *such that* $(\mathsf{d}_1, \mathsf{d}_2) \in \widehat{\mathsf{g}}(s_1, s_2)$*;*
(2) for all $\mathsf{d}_2 \in \mathrm{Dc}_2$*, there is* $\mathsf{d}_1 \in \mathrm{Dc}_1$ *such that* $(\mathsf{d}_1, \mathsf{d}_2) \in \widehat{\mathsf{g}}(s_1, s_2)$*.*

We now introduce a strengthening of the bisimulation concept that allows us to characterize the models that are invariant w.r.t. SL sentences.

Definition 3.4 (Local-Isomorphism) *Let* $\mathcal{G}_1 = \langle \mathrm{AP}, \mathrm{Ag}, \mathrm{Ac}_1, \mathrm{St}_1, \lambda_1, \tau_1, s_{0_1} \rangle$ *and* $\mathcal{G}_2 = \langle \mathrm{AP}, \mathrm{Ag}, \mathrm{Ac}_2, \mathrm{St}_2, \lambda_2, \tau_2, s_{0_2} \rangle$ *be two* CGSs*. Then,* $\mathcal{G}_1$ *and* $\mathcal{G}_2$ *are* locally-isomorphic *iff there is a bisimulation relation* $\sim \subseteq \mathrm{St}_1 \times \mathrm{St}_2$ *satisfying all the requirements of Definition 3.3 such that* $\sim \cap (\{\tau_1(s_1, \mathsf{d}) : \mathsf{d} \in \mathrm{Dc}_1\} \times \{\tau_2(s_2, \mathsf{d}) : \mathsf{d} \in \mathrm{Dc}_2\})$ *is a bijective function between the successors of* s_1 *and those of* s_2*, for all* $s_1 \in \mathrm{St}_1$ *and* $s_2 \in \mathrm{St}_2$ *with* $s_1 \sim s_2$*.*

The local-isomorphism restricts the previous definition of the concept of bisimilarity, by asserting that bisimilar states have the same number of successors, in order to ensure that strategies over bisimilar structures maintain the same information. In this way, the branching degree of the subtrees of Trk that are domains of the strategies does not change when we pass from a strategy to a bisimilar one.

At this point, we define two generalizations for CGS of the classical concept of unwinding of labeled transition systems, which allows us to show that SL has the (unbounded) tree model property (Fig. 3.2).

Definition 3.5 (State-Unwinding) *Let* $\mathcal{G} = \langle \mathrm{AP}, \mathrm{Ag}, \mathrm{Ac}, \mathrm{St}, \lambda, \tau, s_0 \rangle$ *be a* CGS*. Then, the* state-unwinding *of* $\mathcal{G}$ *is the* CGT $\mathcal{G}_{SU} \triangleq \langle \mathrm{AP}, \mathrm{Ag}, \mathrm{Ac}, \mathrm{St}', \lambda', \tau', \varepsilon \rangle$*, where*

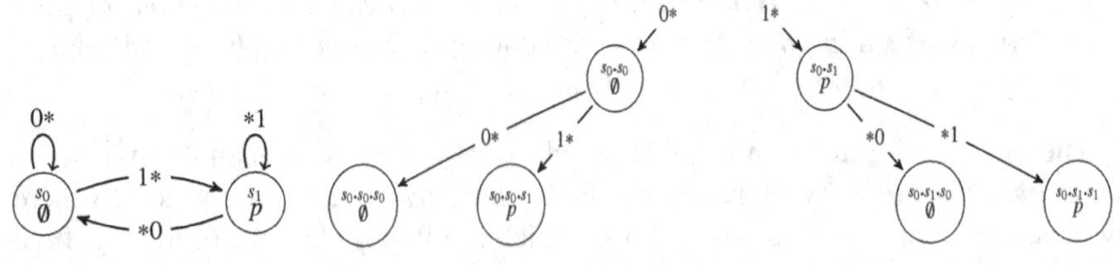

Fig. 3.2 A CGS and its state-unwinding. **a** CGS $\mathcal{G}$, **b** part of the CGT $\mathcal{G}_{SU}$

(i) St *is the set of directions,* (ii) *the states in* $\mathrm{St}' = \{\rho_{\geq 1} \in \mathrm{St}^* : \rho \in \mathrm{Trk}(\mathcal{G}, s_0)\}$ *are the suffixes of the tracks starting in* s_0, (iii) $\tau'(t, \mathsf{d}) = t \cdot \tau(\mathsf{lst}(s_0 \cdot t), \mathsf{d})$, *and* (iv) *there is a surjective function* $\mathsf{unw} : \mathrm{St}' \to \mathrm{St}$ *such that* (iv.i) $\mathsf{unw}(t) = \mathsf{lst}(s_0 \cdot t)$, *and* (iv.ii) $\lambda'(t) = \lambda(\mathsf{unw}(t))$, *for all* $t \in \mathrm{St}'$ *and* $\mathsf{d} \in \mathrm{Dc}$.

Definition 3.6 (Decision-Unwinding) *Let* $\mathcal{G} = \langle \mathrm{AP}, \mathrm{Ag}, \mathrm{Ac}, \mathrm{St}, \lambda, \tau, s_0 \rangle$ *be a* CGS. *Then, the* decision-unwinding *of* $\mathcal{G}$ *is the* CGT $\mathcal{G}_{DU} \triangleq \langle \mathrm{AP}, \mathrm{Ag}, \mathrm{Ac}, \mathrm{St}', \lambda', \tau', \varepsilon \rangle$, *where* (i) Dc *is the set of directions,* (ii) *the states in* $\mathrm{St}' = \mathrm{Dc}^*$ *are words over decisions,* (iii) $\tau'(t, \mathsf{d}) = t \cdot \mathsf{d}$, *and* (iv) *there is a surjective function* $\mathsf{unw} : \mathrm{St}' \to \mathrm{St}$ *such that* (iv.i) $\mathsf{unw}(\varepsilon) = s_0$, (iv.ii) $\mathsf{unw}(\tau'(t, \mathsf{d})) = \tau(\mathsf{unw}(t), \mathsf{d})$, *and* (iv.iii) $\lambda'(t) = \lambda(\mathsf{unw}(t))$, *for all* $t \in \mathrm{St}'$ *and* $\mathsf{d} \in \mathrm{Dc}$.

Note that each CGS $\mathcal{G}$ has unique associated state- and decision-unwindings $\mathcal{G}_{SU}$ and $\mathcal{G}_{DU}$. Moreover, it is important to observe that the state-unwinding preserves the turn-based property, i.e., if $\mathcal{G}$ is turn-based then $\mathcal{G}_{SU}$ is turn-based, too, while every decision-unwinding $\mathcal{G}_{DU}$ cannot be turn-based.

Before to continue, we have to show the main properties of the unwinding operations we have just defined. These properties are simply a translation in the CGS framework of what we have in the case of Kripke structures.

Theorem 3.1 (Unwinding Properties) *For every* CGS $\mathcal{G}$, *it holds that* $\mathcal{G}$ *and* $\mathcal{G}_{SU}$ *are local-isomorphic and* $\mathcal{G}$ *and* $\mathcal{G}_{DU}$ *are bisimilar. Moreover, there is a* CGS $\mathcal{G}$ *such that* $\mathcal{G}$ *and* $\mathcal{G}_{DU}$ *are not local-isomorphic.*

Proof To see that $\mathcal{G} = \langle \mathrm{AP}, \mathrm{Ag}, \mathrm{Ac}, \mathrm{St}, \lambda, \tau, s_0 \rangle$ and $\mathcal{G}_{SU} = \langle \mathrm{AP}, \mathrm{Ag}, \mathrm{Ac}, \mathrm{St}', \lambda', \tau', \varepsilon \rangle$ are local-isomorphic, consider the unwinding function unw between them. Now, let $\sim \triangleq \{(\mathsf{unw}(t), t) : t \in \mathrm{St}'\}$ and $\mathsf{g}(\mathsf{unw}(t), t) \triangleq \{(c, c) : c \in \mathrm{Ac}\}$, for each $t \in \mathrm{St}'$. Then, it is not hard to see that, due to the Definition 3.5 of state-unwinding, $\sim$ and g satisfy, respectively, all constraints on the bisimulation relation and bisimulation function of Definitions 3.3 and 3.4. Hence, the thesis holds. By doing the same reasoning for $\mathcal{G}$ and $\mathcal{G}_{DU}$, we obtain that they are bisimilar. Finally, to show that, in general, the construction of the decision-unwinding does not preserve enough information about the original structure, consider a CGS $\mathcal{G} = \langle \mathrm{AP}, \mathrm{Ag}, \mathrm{Ac}, \mathrm{St}, \lambda, \tau, s_0 \rangle$, having at least two states $s_0, s_1 \in \mathrm{St}$ and two actions, such that $\tau(s_0, \mathsf{d}) = s_1$, for all $\mathsf{d} \in \mathrm{Dc}$ (see Fig. 3.3a). Moreover, consider its decision-unwinding $\mathcal{G}_{DU}$ (see Fig. 3.3b). It is evident that every possible bisimulation relation $\sim$ between $\mathcal{G}$ and $\mathcal{G}_{DU}$ cannot satisfy the constraint of Definition 3.4, since the initial states of the two structures have necessarily different numbers of successors. □

3.4.2 Positive Properties

We now are able to prove the unbounded tree model property for SL by showing a more general property of this logic, i.e., that it is invariant under local-isomorphism and, consequently, under state-unwinding.

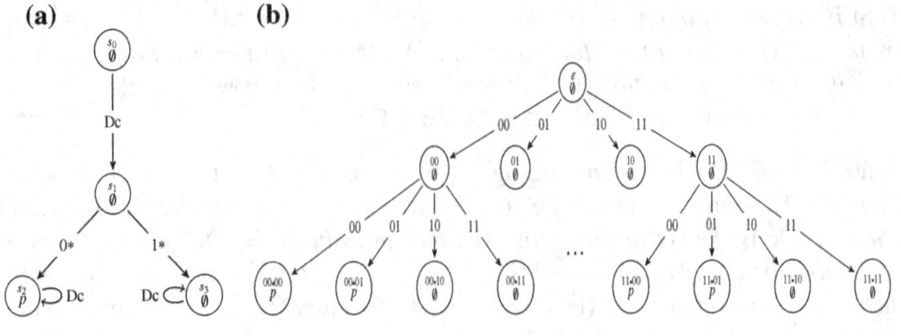

Fig. 3.3 A CGS and its decision-unwinding. **a** CGS $\mathcal{G}$, **b** part of the CGT $\mathcal{G}_{DU}$

Theorem 3.2 (SL **Positive Properties**) *For* SL, *it holds that:*

(1) it is invariant under local-isomorphism;
(2) it is invariant under state-unwinding;
(3) it has the (unbounded) tree model property.

Proof [*Item 1*]. The statement asserts that, every sentences φ is invariant w.r.t. all pairs of local-isomorphic CGSs $\mathcal{G}_1$ and $\mathcal{G}_2$, i.e., that $\mathcal{G}_1 \models \varphi$ iff $\mathcal{G}_2 \models \varphi$. Actually, we prove a stronger result, which asserts that such an invariance property holds not only at the initial state of the structures under empty assignment, but for any possible assignment and state. To this aim, we first extend the concept of local-isomorphism to tracks, paths, strategies, and assignments. Then, we use the new concepts to prove the statement, by induction on the structure of φ.

Two tracks $\rho_1 \in \mathrm{Trk}(\mathcal{G}_1)$ and $\rho_2 \in \mathrm{Trk}(\mathcal{G}_2)$ (resp., paths $\pi_1 \in \mathrm{Pth}(\mathcal{G}_1)$ and $\pi_2 \in \mathrm{Pth}(\mathcal{G}_2)$) are local-isomorphic, in symbols $\rho_1 \sim \rho_2$ (resp., $\pi_1 \sim \pi_2$), iff *(i)* $|\rho_1| = |\rho_2|$ and *(ii)* for all $0 \leq i < |\rho_1|$ (resp., $i \in \mathbb{N}$), it holds that $(\rho_1)_i \sim (\rho_2)_i$ (resp., $(\pi_1)_i \sim (\pi_2)_i$). Two strategies $\mathsf{f}_1 \in \mathrm{Str}(\mathcal{G}_1)$ and $\mathsf{f}_2 \in \mathrm{Str}(\mathcal{G}_2)$ are local-isomorphic, in symbols $\mathsf{f}_1 \sim \mathsf{f}_2$, iff, for all $k \in \{1, 2\}$ and $\rho_k \in \mathsf{dom}(\mathsf{f}_k)$ there is $\rho_{3-k} \in \mathsf{dom}(\mathsf{f}_{3-k})$ with $\rho_1 \sim \rho_2$ such that $(\mathsf{f}_1(\rho_1), \mathsf{f}_2(\rho_2)) \in \mathsf{g}(\mathsf{lst}(\rho_1), \mathsf{lst}(\rho_2))$. Finally, two assignments $\chi_1 \in \mathrm{Asg}(\mathcal{G}_1)$ and $\chi_2 \in \mathrm{Asg}(\mathcal{G}_2)$ are local-isomorphic, in symbols $\chi_1 \sim \chi_2$, iff *(i)* $\mathsf{dom}(\chi_1) = \mathsf{dom}(\chi_2)$ and *(ii)* $\chi_1(l) \sim \chi_2(l)$, for all $l \in \mathsf{dom}(\chi_1)$. Observe that, if $\chi_1 \sim \chi_2$ and $\mathsf{f}_1 \sim \mathsf{f}_2$, then $\chi_1[l \rightarrow \mathsf{f}_1] \sim \chi_2[l \rightarrow \mathsf{f}_2]$. Moreover, if χ_1 and χ_2 are also complete, χ_1 is s_1-total, and χ_2 is s_2-total, with $s_1 \sim s_2$, we have that $\pi_1 \sim \pi_2$ and $(\chi_1)_{(\pi_1)_{\leq k}} \sim (\chi_2)_{(\pi_2)_{\leq k}}$, for all $k \in \mathbb{N}$, where π_1 and π_2 are the (χ_1, s_1)-play and (χ_2, s_2)-play, respectively.

Now, the statement we prove is the following: for all formulas φ in *existential normal form*[1] and local-isomorphic CGSs $\mathcal{G}_1$, $\mathcal{G}_2$, states $s_1 \in \mathrm{St}_1$, $s_2 \in \mathrm{St}_2$, and assignments $\chi_1 \in \mathrm{Asg}(\mathcal{G}_1, s_1)$, $\chi_2 \in \mathrm{Asg}(\mathcal{G}_2, s_2)$, where $\mathsf{free}(\varphi) \subseteq \mathsf{dom}(\chi_1) = \mathsf{dom}(\chi_2)$, it holds that $\mathcal{G}_1, \chi_1, s_1 \models \varphi$ iff $\mathcal{G}_2, \chi_2, s_2 \models \varphi$. The base case of atomic

[1] An SL formula is in existential normal form iff it has only existential quantifiers and no release temporal operators. Using classical reasoning, it is not hard to see that every SL can be translated into this specific form.

propositions directly follows from Item 2a of the Definition 3.3 of bisimulation, while the cases of Boolean connectives are immediate from the inductive hypothesis. There are left to prove the cases of existential quantification, agent binding, and of the two temporal operator next and until.

- $\varphi = \langle\langle x\rangle\rangle\varphi'$. *[Only if]*. By Item 3a of Definition 3.2 of semantics, if $\mathcal{G}_1, \chi_1, s_1 \models \varphi$ then there is a strategy $\mathsf{f}_1 \in \mathrm{Str}(\mathcal{G}_1, s_1)$ such that $\mathcal{G}_1, \chi_1[x \to \mathsf{f}_1], s_1 \models \varphi'$. By Item 2b of Definition 3.3, Definition 3.4, and the concept of local-isomorphism for strategies, there is a strategy $\mathsf{f}_2 \in \mathrm{Str}(\mathcal{G}_2, s_2)$ such that $\mathsf{f}_1 \sim \mathsf{f}_2$. By the inductive hypothesis, $\mathcal{G}_1, \chi_1[x \to \mathsf{f}_1], s_1 \models \varphi'$ iff $\mathcal{G}_2, \chi_2[x \to \mathsf{f}_2], s_2 \models \varphi'$. Hence, $\mathcal{G}_1, \chi_1, s_1 \models \varphi$ implies that there is a strategy $\mathsf{f}_2 \in \mathrm{Str}(\mathcal{G}_2, s_2)$ such that $\mathcal{G}_2, \chi_2[x \to \mathsf{f}_2], s_2 \models \varphi'$, i.e., $\mathcal{G}_2, \chi_2, s_2 \models \varphi$. *[If]*. The converse direction easily follows by switching indexes 1 and 2 and using Item 2c of Definition 3.3 instead of Item 2b.
- $\varphi = (a, x)\varphi'$. By Item 3c of Definition 3.2 of semantics, it holds that $\mathcal{G}_1, \chi_1, s_1 \models \varphi$ iff $\mathcal{G}_1, \chi_1[a \to \chi_1(x)], s_1 \models \varphi'$. By the inductive hypothesis, $\mathcal{G}_1, \chi_1[a \to \chi_1(x)], s_1 \models \varphi'$ iff $\mathcal{G}_2, \chi_2[a \to \chi_2(x)], s_2 \models \varphi'$, since $\chi_1[a \to \chi_1(x)] \sim \chi_2[a \to \chi_2(x)]$. Hence $\mathcal{G}_1, \chi_1, s_1 \models \varphi$ iff $\mathcal{G}_2, \chi_2, s_2 \models \varphi$.
- $\varphi = \mathsf{X}\varphi'$. By Item 4a of Definition 3.2 of semantics, it holds that $\mathcal{G}_1, \chi_1, s_1 \models \varphi$ iff $\mathcal{G}_1, \chi_1, \pi_1, 0 \models \varphi$, where π_1 is the (χ_1, s_1)-play. Now, by Items 4b and 4e of Definition 3.2, we have that $\mathcal{G}_1, \chi_1, \pi_1, 0 \models \varphi$ iff $\mathcal{G}_1, (\chi_1)_{(\pi_1)_{\leq 1}}, (\pi_1)_1 \models \varphi'$. Consider now the (χ_2, s_2)-play π_2. By Item 2d of Definition 3.3, it follows that the $(\pi_1)_1$ is local-isomorphic to $(\pi_2)_1$. Hence, by the inductive hypothesis, it holds that $\mathcal{G}_1, (\chi_1)_{(\pi_1)_{\leq 1}}, (\pi_1)_1 \models \varphi'$ iff $\mathcal{G}_2, (\chi_2)_{(\pi_2)_{\leq 1}}, (\pi_2)_1 \models \varphi'$. Now, again by Items 4e, 4b, and 4a of Definition 3.2, it follows that $\mathcal{G}_2, (\chi_2)_{(\pi_2)_{\leq 1}}, (\pi_2)_1 \models \varphi'$ iff $\mathcal{G}_2, \chi_2, s_2 \models \varphi$. Hence, we obtain that $\mathcal{G}_1, \chi_1, s_1 \models \varphi$ iff $\mathcal{G}_2, \chi_2, s_2 \models \varphi$.
- $\varphi = \varphi_1 \mathsf{U} \varphi_2$. The proof is similar to the previous one. The only difference is in the result of the inductive hypothesis: $\mathcal{G}_1, (\chi_1)_{(\pi_1)_{\leq k}}, (\pi_1)_k \models \varphi_i$ iff $\mathcal{G}_2, (\chi_2)_{(\pi_2)_{\leq k}}, (\pi_2)_k \models \varphi_i$, for all $k \in \mathbb{N}$ and $i \in \{1, 2\}$.

[Item 2]. By Theorem 3.1, we know that $\mathcal{G}$ and $\mathcal{G}_{SU}$ are local-isomorphic, for every CGS $\mathcal{G}$. Now, by the previous item, we have that every sentence φ is an invariant for $\mathcal{G}$ and $\mathcal{G}_{SU}$. Hence, the thesis holds.

[Item 3]. Consider a sentence φ and suppose that it is satisfiable. Then, there is a CGS $\mathcal{G}$ such that $\mathcal{G} \models \varphi$. By the previous item, φ is satisfied at the root of the state-unwinding $\mathcal{G}_{SU}$ of $\mathcal{G}$. Thus, since $\mathcal{G}_{SU}$ is a CGT, we immediately have that φ is satisfied on a tree model. □

3.4.3 Negative Properties

We now move to the negative results about SL and their sublogics. In particular, we first show that TB-SL, and so SL, is not invariant under bisimulation.

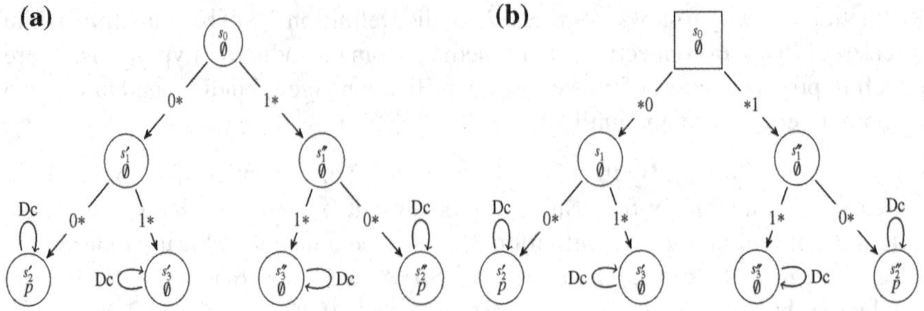

Fig. 3.4 Two bisimilar but not local-isomorphic turn-based CGSs. **a** CGS $\mathcal{G}_1$, **b** CGS $\mathcal{G}_2$

Theorem 3.3 (TB-SL **Negative Properties**) TB-SL *it is not invariant under bisimulation.*

Proof Consider the two CGSs $\mathcal{G}_1 = \langle \text{AP}, \text{Ag}, \text{Ac}, \text{St}, \lambda, \tau_1, s_0 \rangle$ and $\mathcal{G}_2 = \langle \text{AP}, \text{Ag}, \text{Ac}, \text{St}, \lambda, \tau_2, s_0 \rangle$, with $\text{AP} = \{p\}$, $\text{Ag} = \{\alpha, \beta\}$, $\text{Ac} = \{0, 1\}$, and $\text{St} = \{s_0, s'_1, s''_1, s'_2, s''_2, s'_3, s''_3\}$, of Fig. 3.4. It is immediate to see that they are bisimilar, by simply assuming $\sim \triangleq \{(s, s) : s \in \text{St}\}$ and $\text{g}(s, s) \triangleq \{(c, c) : c \in \text{Ac}\}$, for each $s \in \text{St}$, since they satisfy all constraints on the bisimulation relation and bisimulation function of Definitions 3.3. Moreover, they are turn-based, too. Indeed, in $\mathcal{G}_1$ all states are owned by agent α, i.e., $\eta_1(s) = \alpha$, for all $s \in \text{St}$, while in $\mathcal{G}_2$ the initial state s_0 is the only one owned by player β, i.e., $\eta_2(s_0) = \beta$ and $\eta_2(s) = \alpha$, for all $s \in \text{St} \setminus \{s_0\}$.

Now, consider the formula $\varphi = \langle\!\langle x \rangle\!\rangle(\alpha, x)(\langle\!\langle y \rangle\!\rangle(\beta, y)(\mathsf{X}\mathsf{X}p)) \wedge (\langle\!\langle y \rangle\!\rangle(\beta, y)(\mathsf{X}\mathsf{X}\neg p))$. It is easy to see that $\mathcal{G}_1 \not\models \varphi$ while $\mathcal{G}_2 \models \varphi$, so TB-SL cannot be invariant under bisimulation. Indeed, each strategy $\text{f} \in \text{Str}(\mathcal{G}_1, s_0)$ of the agent α in $\mathcal{G}_1$ forces to reach only one state at a time among s'_2, s''_2, s'_3, and s''_3. Thus, it is impossible to satisfy both the goals $\mathsf{X}\mathsf{X}p$ and $\mathsf{X}\mathsf{X}\neg p$ with the same strategy of α. On the contrary, since s_0 in $\mathcal{G}_2$ is owned by the agent β, we can reach both s'_1 and s''_1 with the same strategy $\text{f} \in \text{Str}(\mathcal{G}_2, s_0)$ of α. Thus, if $\text{f}(s_0 \cdot s'_1) \neq \text{f}(s_0 \cdot s''_1)$, we can reach, at the same time, either the pair of states s'_2 and s''_3 or s'_3 and s''_2. Hence, we can satisfy both the goals $\mathsf{X}\mathsf{X}p$ and $\mathsf{X}\mathsf{X}\neg p$ with the same strategy of α. □

It is interesting to note that the two structure of Fig. 3.4 are not local-isomorphic, although they are bisimilar and have the same number of successors for each state. Indeed, as shown in the previous theorem, there is a formula that is not invariant between them.

We now show that SL does have neither the bounded-tree nor the finite model property. We recall that a modal logic has the bounded-tree model property (resp., finite model property) if whenever a formula is satisfiable, it is so on a model with a finite number of actions having a tree shape (resp., finite states). Clearly, if a modal logic invariant under unwinding has the finite model property, it has the bounded-tree model property as well. The other direction may not hold, instead.

To prove the results, we introduce, in the following definition, the formula φ^{ord} to be used as a counterexample.

Definition 3.7 (Ordering Sentence) *Let* $x_1 < x_2 \triangleq \langle\langle y \rangle\rangle\, \varphi(x_1, x_2, y)$ *be an agent-closed formula, named* partial order, *on the sets* $\mathrm{AP} = \{p\}$ *and* $\mathrm{Ag} = \{\alpha, \beta\}$, *where* $\varphi(x_1, x_2, y) \triangleq (\beta, y)((\alpha, x_1)(\mathsf{X}p) \wedge (\alpha, x_2)(\mathsf{X}\neg p))$. *Then, the* order sentence $\varphi^{ord} \triangleq \varphi^{unb} \wedge \varphi^{trn}$ *is the conjunction of the following two sentences, called* unboundedness *and* transitivity *strategy requirements:*

(1) $\varphi^{unb} \triangleq [[x_1]]\langle\langle x_2 \rangle\rangle\, x_1 < x_2$;
(2) $\varphi^{trn} \triangleq [[x_1]][[x_2]][[x_3]]\, (x_1 < x_2 \wedge x_2 < x_3) \rightarrow x_1 < x_3$.

Intuitively, φ^{unb} asserts that, for each strategy x_1, there is a different strategy x_2 in relation of $<$ w.r.t. the first one, i.e., $<$ has no upper bound. Moreover, φ^{trn} expresses the fact that the relation $<$ is transitive. Note also that, by definition, $<$ is not reflexive.

Obviously, the formula φ^{ord} needs to be satisfiable, as reported in the following lemma.

Lemma 3.1 (Ordering Satisfiability) *The* SL *sentence* φ^{ord} *is satisfiable.*

Proof To prove that φ^{ord} is satisfiable, consider the unbounded CGS $\mathcal{G}^\star \triangleq \langle \mathrm{AP}, \mathrm{Ag}, \mathrm{Ac}, \mathrm{St}, \lambda, \tau, s_0 \rangle$, where *(i)* $\mathrm{Ac} \triangleq \mathbb{N}$, *(ii)* $\mathrm{St} \triangleq \{s_0, s_1, s_2\}$, *(iii)* λ is such that $\lambda(s_0) = \lambda(s_2) \triangleq \emptyset$ and $\lambda(s_1) \triangleq \{p\}$, and *(iv)* τ is such that if $\mathsf{d} \in \mathrm{Dc}' \triangleq \{\mathsf{d} \in \mathrm{Dc} : \mathsf{d}(\alpha) \leq \mathsf{d}(\beta)\}$ then $\tau(s_0, \mathsf{d}) = s_1$ else $\tau(s_0, \mathsf{d}) = s_2$, and $\tau(s_i, \mathsf{d}) = s_i$, for all $\mathsf{d} \in \mathrm{Dc}$ and $i \in \{1, 2\}$ (see Fig. 3.5). Now, it is easy to see that $\mathcal{G}^\star, \varnothing, s_0 \models \varphi^{unb}$, since for every strategy $\mathsf{f}_{x_1} \in \mathrm{Str}(\mathcal{G}^\star, s_0)$ for x_1, consisting of picking a natural number $n = \mathsf{f}_{x_1}(s_0)$ as an action at the initial state, we can reply with the strategy $\mathsf{f}_{x_2} \in \mathrm{Str}(\mathcal{G}^\star, s_0)$ for x_2 having $\mathsf{f}_{x_2}(s_0) > n$ and the strategy $\mathsf{f}_y \in \mathrm{Str}(\mathcal{G}^\star, s_0)$ for y having $\mathsf{f}_y(s_0) = n$. Formally, we have that $\mathcal{G}^\star, \chi, s_0 \models \varphi(x_1, x_2, y)$, where $\chi(x_2)(s_0) > \chi(x_1)(s_0)$ and $\chi(y)(s_0) = \chi(x_1)(s_0)$, for all assignments $\chi \in \mathrm{Asg}(\mathcal{G}^\star, \{x_1, x_2, y\}, s_0)$. By a similar reasoning, we can see that $\mathcal{G}^\star, \varnothing, s_0 \models \varphi^{trn}$. Indeed, consider three strategies $\mathsf{f}_{x_1}, \mathsf{f}_{x_2}, \mathsf{f}_{x_3} \in \mathrm{Str}(\mathcal{G}^\star, s_0)$ for the variables x_1, x_2, and x_3, which respectively correspond to picking three natural numbers $n_1 = \mathsf{f}_{x_1}(s_0)$, $n_2 = \mathsf{f}_{x_2}(s_0)$, and $n_3 = \mathsf{f}_{x_3}(s_0)$. Now, if $\mathcal{G}^\star, \chi, s_0 \models x_1 < x_2$ and $\mathcal{G}^\star, \chi, s_0 \models x_2 < x_3$, where $\chi(x_1) = \mathsf{f}_{x_1}$, $\chi(x_2) = \mathsf{f}_{x_2}$, and $\chi(x_3) = \mathsf{f}_{x_3}$, we have that $n_1 < n_2$ and $n_2 < n_3$, and then $n_1 < n_3$, for all assignments $\chi \in \mathrm{Asg}(\mathcal{G}^\star, \{x_1, x_2, x_3\}, s_0)$. Hence, using a strategy

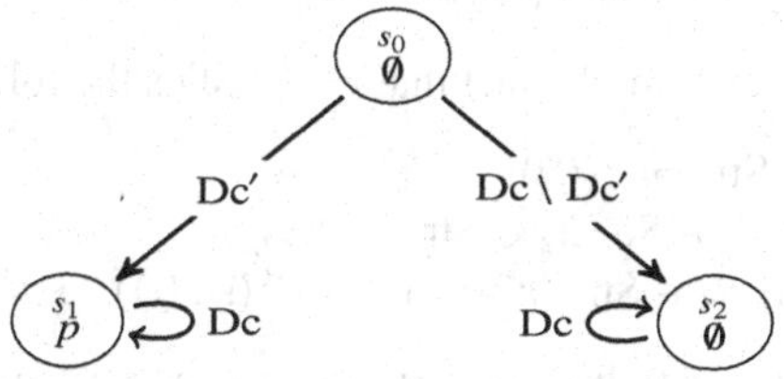

Fig. 3.5 The CGS $\mathcal{G}^\star$ model of φ^{ord}

$\mathsf{f}_y \in \mathrm{Str}(\mathcal{G}^\star, s_0)$ for y with $\mathsf{f}_y(s_0) = \mathsf{f}_{x_1}(s_0)$ we have $\mathcal{G}^\star, \chi_{y \to \mathsf{f}_y}, s_0 \models \varphi(x_1, x_3, y)$ and thus $\mathcal{G}^\star, \chi, s_0 \models x_1 < x_3$. □

However, it is also important to observe that φ^{ord} cannot have turn-based models.

Lemma 3.2 (Ordering Turn-Based Unsatisfiability) *The* SL *sentence* φ^{ord} *is unsatisfiable over turn-based* CGSs.

Proof Let $\mathcal{G} = \langle \mathrm{AP}, \mathrm{Ag}, \mathrm{Ac}, \mathrm{St}, \lambda, \tau, s_0 \rangle$ be a model of φ^{ord}. Then, we have $\mathcal{G} \models \varphi^{unb}$ and so, $\mathcal{G} \models [[x_1]]\langle\langle x_2 \rangle\rangle\langle\langle y \rangle\rangle\varphi(x_1, x_2, y)$. Directly from the satisfiability concept, we derive the existence of an assignment $\chi \in \mathrm{Asg}(\mathcal{G}, \{x_1, x_2, y\}, s_0)$ such that $\mathcal{G}, \chi, s_0 \models \varphi(x_1, x_2, y)$. Now, consider the two decisions $\mathsf{d}_1, \mathsf{d}_2 \in \mathrm{Dc}$ given by the following settings: $\mathsf{d}_1(\alpha) \triangleq \chi(x_1)(s_0)$, $\mathsf{d}_2(\alpha) \triangleq \chi(x_2)(s_0)$, and $\mathsf{d}_1(\beta) = \mathsf{d}_2(\beta) \triangleq \chi(y)(s_0)$, It is easy to observe that $\lambda(\tau(s_0, \mathsf{d}_1)) = \{p\}$ and $\lambda(\tau(s_0, \mathsf{d}_2)) = \emptyset$. So, $\tau(s_0, \mathsf{d}_1) \neq \tau(s_0, \mathsf{d}_2)$. Then, since $\mathsf{d}_1(\beta) = \mathsf{d}_2(\beta)$, by definition of owner function $\eta : \mathrm{St} \to \mathrm{Ag}$, it is evident that β cannot be the owner of state s_0, i.e., $\eta(s_0) \neq \beta$. At this point, again from the satisfiability concept, we derive the existence of another assignment $\chi' \in \mathrm{Asg}(\mathcal{G}, \{x_1, x_2, y\}, s_0)$ such that $\mathcal{G}, \chi', s_0 \models \varphi(x_1, x_2, y)$, with $\chi'(x_1) = \chi(x_2)$. Now, consider the decision $\mathsf{d}_3 \in \mathrm{Dc}$ given by the following settings: $\mathsf{d}_1(\alpha) \triangleq \chi'(x_1)(s_0)$ and $\mathsf{d}_1(\beta) \triangleq \chi'(y)(s_0)$. Also in this case, it is easy to observe that $\lambda(\tau(s_0, \mathsf{d}_3)) = \{p\}$. So, $\tau(s_0, \mathsf{d}_2) \neq \tau(s_0, \mathsf{d}_3)$. Then, since $\mathsf{d}_2(\alpha) = \mathsf{d}_3(\alpha)$, again by definition of owner function, it is evident that also α cannot be the owner of state s_0, i.e., $\eta(s_0) \neq \alpha$. Consequently, it's impossible to find the function η with required conditions, which implies that $\mathcal{G}$ cannot be turn-based. □

Next two lemmas report two important properties of the formula φ^{ord}, for the negative statements we want to show. Namely, they state that, in order to be satisfied, φ^{ord} must require the existence of strict partial order relations on strategies and actions that do not admit any maximal element. From this, as stated in Theorem 3.4, we directly derive that φ^{ord} needs an infinite chain of actions to be satisfied (i.e., it cannot have a bounded model).

Lemma 3.3 (Strategy Order) *Let* $\mathcal{G}$ *be a model of* φ^{ord} *and* $r^< \subseteq \mathrm{Str}(\mathcal{G}, s_0) \times \mathrm{Str}(\mathcal{G}, s_0)$ *be a relation between* s_0*-total strategies such that* $r^<(\mathsf{f}_1, \mathsf{f}_2)$ *holds iff* $\mathcal{G}, \chi, s_0 \models x_1 < x_2$, *where* $\chi(x_1) = \mathsf{f}_1$ *and* $\chi(x_2) = \mathsf{f}_2$, *for all strategies* $\mathsf{f}_1, \mathsf{f}_2 \in \mathrm{Str}(\mathcal{G}, s_0)$ *and assignments* $\chi \in \mathrm{Asg}(\mathcal{G}, \{x_1, x_2\}, s_0)$, *with* s_0 *as the initial state of* $\mathcal{G}$. *Then,* $r^<$ *is a* strict partial order without maximal element.

Proof The proof derives from the fact that $r^<$ satisfies the following properties:

(1) *Irreflexivity*: $\forall \mathsf{f} \in \mathrm{Str}.\ \neg r^<(\mathsf{f}, \mathsf{f})$;
(2) *Unboundedness*: $\forall \mathsf{f}_1 \in \mathrm{Str}\ \exists \mathsf{f}_2 \in \mathrm{Str}.\ r^<(\mathsf{f}_1, \mathsf{f}_2)$;
(3) *Transitivity*: $\forall \mathsf{f}_1, \mathsf{f}_2, \mathsf{f}_3 \in \mathrm{Str}.\ (r^<(\mathsf{f}_1, \mathsf{f}_2) \wedge r^<(\mathsf{f}_2, \mathsf{f}_3)) \to r^<(\mathsf{f}_1, \mathsf{f}_3)$.

Indeed, Items *(ii)* and *(iii)* are directly derived from the strategy unboundedness and strategy transitivity requirements. The proof of Item *(i)* derives from the following reasoning. By contradiction, suppose that $r^<$ is not a strict order, i.e., there is a strategy $\mathsf{f} \in \mathrm{Str}(\mathcal{G}^\star, s_0)$ for which $r^<(\mathsf{f}, \mathsf{f})$ holds. This means that, at the initial state s_0 in $\mathcal{G}$, there exists an assignment $\chi \in \mathrm{Asg}(\mathcal{G}^\star, \{x_1, x_2, y\}, s_0)$ for which $\mathcal{G}, \chi, s_0 \models \varphi(x_1, x_2, y)$, where $\chi(x_1) = \chi(x_2) = \mathsf{f}$. This implies the existence of a successor of s_0 in which both p and $\neg p$ hold, which is clearly impossible. □

Lemma 3.4 (Action Order) *Let $\mathcal{G}$ be a model of φ^{ord} and $s^< \subseteq \mathrm{Ac} \times \mathrm{Ac}$ be a relation between actions such that $s^<(c_1, c_2)$ holds iff $r^<(\mathsf{f}_1, \mathsf{f}_2)$ holds, where $c_1 = \mathsf{f}_1(s_0)$ and $c_2 = \mathsf{f}_2(s_0)$, for all actions $c_1, c_2 \in \mathrm{Ac}$ and strategies $\mathsf{f}_1, \mathsf{f}_2 \in \mathrm{Str}(\mathcal{G}, s_0)$, with s_0 as the initial state of $\mathcal{G}$. Then, $s^<$ is a* strict partial order without maximal element.

Proof The irreflexivity and transitivity of $s^<$ are directly derived from the fact that, by Lemma 3.3, $r^<$ is irreflexive and transitive too. The proof of the unboundedness property derives, instead, from the following reasoning. As first thing, observe that, since the formula $x_1 < x_2$ relies on $\mathsf{X}p$ and $\mathsf{X}\neg p$ as the only temporal operators, it holds that $r^<(\mathsf{f}_1, \mathsf{f}_2)$ implies $r^<(\mathsf{f}'_1, \mathsf{f}'_2)$, for all strategies $\mathsf{f}_1, \mathsf{f}_2, \mathsf{f}'_1, \mathsf{f}'_2 \in \mathrm{Str}(\mathcal{G}, s_0)$ such that $\mathsf{f}_1(s_0) = \mathsf{f}'_1(s_0)$ and $\mathsf{f}_2(s_0) = \mathsf{f}'_2(s_0)$. Now, suppose by contradiction that $s^<$ does not satisfy the unboundedness property, i.e., there is an action $c \in \mathrm{Ac}$ such that, for all actions $c' \in \mathrm{Ac}$, $s^<(c, c')$ does not hold. Then, by the definition of $s^<$ and the previous observation, we derive the existence of a strategy $\mathsf{f} \in \mathrm{Str}(\mathcal{G}, s_0)$ with $\mathsf{f}(s_0) = c$ such that $r^<(\mathsf{f}, \mathsf{f}')$ does not hold, for all strategies $\mathsf{f}' \in \mathrm{Str}(\mathcal{G}, s_0)$, which is clearly impossible. □

Now, we have all tools to prove also that SL lacks of the finite and bounded-tree model properties, which hold in several commonly used multi-agent logics, such as ATL*.

Theorem 3.4 (SL **Negative Properties)** *For* SL, *it holds that:*

(1) it is not invariant under decision-unwinding;
(2) it is not invariant under bisimulation;
(3) it does not have the bounded-tree model property;
(4) it does not have the finite-model property.

Proof *[Item 1].* Assume by contraddiction that the logic is invariant under decision-unwinding and consider the two structures $\mathcal{G}_1$ and $\mathcal{G}_2$ (see Fig. 3.4) used in the proof of Theorem 3.3. Also, observe that $\mathcal{G}_1$ and $\mathcal{G}_2$ have the same decision unwinding, i.e., $\mathcal{G}_{1DU} = \mathcal{G}_{2DU}$ (see Fig. 3.3b). Then, it is evident that $\mathcal{G}_1 \models \varphi$ iff $\mathcal{G}_2 \models \varphi$, in particular for the sentence φ of the proof of Theorem 3.3, but this is in contraddiction with what we have yet proved there.

[Item 2]. The thesis directly follows from the fact that yet the turn-based fragment is not invariant under bisimulation, as shown in Theorem 3.3.

[Item 3]. To prove the statement, we show that φ^{ord} cannot be satisfied on a bounded CGS. Consider a CGS $\mathcal{G} = \langle \mathrm{AP}, \mathrm{Ag}, \mathrm{Ac}, \mathrm{St}, \lambda, \tau, s_0 \rangle$ such that $\mathcal{G}, \varnothing, s_0 \models \varphi$. The existence of such a model is ensured by Lemma 3.1. Now, consider the strict partial order without maximal element between actions $s^<$ described in Lemma 3.4.

By a classical result on first order logic model theory (Ebbinghaus and Flum 1995), the relation $s^{<}$ cannot be defined on a finite set. Hence, $|\mathrm{Ac}| = \infty$.

[Item 4]. Consider again the formula φ^{ord}. We have already proved in Item *(i)* that each CGS $\mathcal{G}$ model of φ^{ord} must have an infinite number of actions. Hence, the number of its decisions $|\mathrm{Dc}|$ is infinite, and so $|\mathcal{G}| = \infty$. □

3.5 Strategy Quantification

In this section, we introduce the concepts of quantification prefix and spectrum and show how any strategy quantification of an SL formula can be represented by an adequate choice of a quantification spectrum. The main idea here is inspired by what Skolem proposed for the first order logic in order to eliminate each existential quantification over variables, by substituting them with second order quantifications over functions, whose choice is uniform w.r.t. the universal variables.

Definition 3.8 (Quantification Prefixes) *A* quantification prefix *over a set of n placeholders* $\mathrm{P} \subseteq \mathrm{Ag} \cup \mathrm{Var}$ *is a finite word* $\wp \in \{\langle\langle x\rangle\rangle, [[x]] : x \in \mathrm{P}\}^{n}$ *of length n such that each placeholder* $x \in \mathrm{P}$ *occurs once and only once in* $\wp$*, i.e., there are no indexes* $i, j \in [0, n]$ *with* $i \neq j$ *such that* $\wp_i, \wp_j \in \{\langle\langle x\rangle\rangle, [[x]]\}$.

Let $x \in \mathrm{P}$. Recall that with $\langle\langle x\rangle\rangle$ and $[[x]]$ we represent the *existential* and *universal* quantification of x, respectively. By $\Xi(\wp) \triangleq \{x \in \mathrm{Var} : \exists i \in [0, n].\wp_i = \langle\langle x\rangle\rangle\}$ and $\Lambda(\wp) \triangleq \mathrm{Var} \setminus \Xi(\wp)$ we denote, respectively, the sets of existential and universal placeholders in $\wp$. For two placeholders x and y, we say that x *precedes* y in $\wp$, in symbols $x <_{\wp} y$, iff there are two indexes $i, j \in [0, n]$ such that $i < j$, $\wp_i \in \{\langle\langle x\rangle\rangle, [[x]]\}$, and $\wp_j \in \{\langle\langle y\rangle\rangle, [[y]]\}$. Moreover, we say that y is *functional dependent* on x iff y is existentially quantified after that x is universally quantified, so there may be a dependence between the value chosen by x and that chosen by y. Formally, this definition induces the relation $\Upsilon(\wp) \triangleq \{(x, y) \in \mathrm{Var} \times \mathrm{Var} : x <_{\wp} y \wedge x \in \Lambda(\wp) \wedge y \in \Xi(\wp)\}$. In the following, we also use $\Upsilon(\wp, y) \triangleq \{x \in \mathrm{Var} : (x, y) \in \Upsilon(\wp)\}$ to denote the sets of placeholders from which y depends.

As an example, let $\wp = [[x]]\langle\langle y\rangle\rangle\langle\langle z\rangle\rangle[[w]]\langle\langle v\rangle\rangle$. Then, we have $\Xi(\wp) = \{y, z, v\}$, $\Lambda(\wp) = \{x, w\}$, and $\Upsilon(\wp) = \{(x, y), (x, z), (x, v), (w, v)\}$.

We now give the semantics of the quantification prefixes by means of the following definition.

Definition 3.9 (Quantification Spectra) *Let* $\wp$ *be a quantification prefix over a set of placeholders* P*, and* D *be a set. Then, a* quantification spectrum *for* $\wp$ *over* D *is a function* $\theta : \mathrm{D}^{\Lambda(\wp)} \to \mathrm{D}^{\mathrm{P}}$ *such that the following properties hold:*

(1) $\theta(\mathsf{d})_{\restriction \Lambda(\wp)} = \mathsf{d}$*, for all* $\mathsf{d} \in \mathrm{D}^{\Lambda(\wp)}$*, i.e.,* θ *takes the same values of its argument w.r.t. the universal placeholders in* $\wp$*;*

(2) $\theta(\mathsf{d}_1)(x) = \theta(\mathsf{d}_2)(x)$, *for all* $\mathsf{d}_1, \mathsf{d}_2 \in \mathrm{D}^{\Lambda(\wp)}$ *and* $x \in \Xi(\wp)$ *such that* $\mathsf{d}_{1\restriction \Upsilon(\wp,x)} = \mathsf{d}_{2\restriction \Upsilon(\wp,x)}$, *i.e., the value of* θ *w.r.t. an existential placeholder* x *in* $\wp$ *does not depend on placeholders not in* $\Upsilon(\wp, x)$.

By $\Theta_{\mathrm{D}}(\wp)$ *we denote the set of all quantification spectra* θ *for* $\wp$ *over* D.

Intuitively, a quantification spectrum θ for $\wp$ can be considered as a set of *Skolem functions* that, given a value for each placeholder in P that is universally quantified in $\wp$, returns a possible value for all the existential placeholders in $\wp$ in a way that is coherent w.r.t. the order of quantification. Observe that, for all $\theta \in \Theta_{\mathrm{D}}(\wp)$, we have $|\mathsf{rng}(\theta)| = |\mathrm{D}|^{|\Lambda(\wp)|}$. Moreover, $|\Theta_{\mathrm{D}}(\wp)| = \prod_{x\in\Xi(\wp)} |\mathrm{D}|^{|\mathrm{D}|^{|\Upsilon(\wp,x)|}}$.

As an example, let $\mathrm{D} = \{0, 1\}$ and $\wp = [[x]]\langle\langle y\rangle\rangle[[z]]$ be a quantification prefix over $\mathrm{P} = \{x, y, z\}$. Then, we have $|\Theta_{\mathrm{D}}(\wp)| = 4$. Moreover, the quantification spectra $\theta_i \in \Theta_{\mathrm{D}}(\wp)$ with $i \in [1, 4]$ (in a particular order) are such that $\theta_0(\mathsf{d})(y) = 0$, $\theta_1(\mathsf{d})(y) = \mathsf{d}(x)$, $\theta_2(\mathsf{d})(y) = 1 - \mathsf{d}(x)$, and $\theta_3(\mathsf{d})(y) = 1$, for all $\mathsf{d} \in \mathrm{D}^{\{x,z\}}$.

We now prove how to eliminate a strategy quantification of a formula by substituting it with a choice of a quantification spectrum. This procedure can be seen as the equivalent of the *Skolemization* in first order logic.

Theorem 3.5 (Strategy Quantification) *Let* $\mathcal{G}$ *be a* CGS *with initial state* s_0 *and* $\varphi = \wp \cdot \psi$ *be a formula being* $\wp$ *a quantification prefix over a set of placeholders* $\mathrm{P} \subseteq \mathsf{free}(\psi) \cap \mathrm{Var}$. *Then, for all assignments* $\chi \in \mathrm{Asg}(\mathcal{G}, \mathsf{free}(\varphi), s_0)$, *the following holds:* $\mathcal{G}, \chi, s_0 \models \varphi$ *iff there exists a quantification spectrum* $\theta \in \Theta_{\mathrm{Str}(\mathcal{G},s_0)}(\wp)$ *such that* $\mathcal{G}, \chi \uplus \theta(\chi'), s_0 \models \psi$, *for all* $\chi' \in \mathrm{Asg}(\mathcal{G}, \Lambda(\wp), s_0)$.

Proof The proof proceeds by induction on the length of the quantification prefix $\wp$. For the base case $|\wp| = 0$, the thesis immediately follows, since $\Lambda(\wp) = \emptyset$ and, consequently, both $\Theta_{\mathrm{Str}}(\wp)$ and $\mathrm{Asg}(\mathcal{G}, \Lambda(\wp), s_0)$ contain only the empty function (we are assuming $\varnothing(\varnothing) \triangleq \varnothing$).

We now prove, separately, the two directions of the inductive case.

[Only if]. Suppose that $\mathcal{G}, \chi, s_0 \models \varphi$, where $\wp = \mathsf{Qn} \cdot \wp'$. Then, we have two possible cases: either $\mathsf{Qn} = \langle\langle x\rangle\rangle$ or $\mathsf{Qn} = [[x]]$. On one hand, if $\mathsf{Qn} = \langle\langle x\rangle\rangle$, by Item 3a of Definition 3.2 of semantics, there is a strategy $\mathsf{f} \in \mathrm{Str}(\mathcal{G}, s_0)$ such that $\mathcal{G}, \chi_{x\mapsto\mathsf{f}}, s_0 \models \wp' \cdot \psi$. Note that $\Lambda(\wp) = \Lambda(\wp')$. By the inductive hypothesis, we have that there exists a quantification spectrum $\theta \in \Theta_{\mathrm{Str}(\mathcal{G},s_0)}(\wp')$ such that $\mathcal{G}, \chi_{x\mapsto\mathsf{f}} \uplus \theta(\chi'), s_0 \models \psi$, for all $\chi' \in \mathrm{Asg}(\mathcal{G}, \Lambda(\wp'), s_0)$. Now, consider the function $\widehat{\theta} : \mathrm{Asg}(\mathcal{G}, \Lambda(\wp), s_0) \to \mathrm{Asg}(\mathcal{G}, \mathrm{P}, s_0)$ defined by $\widehat{\theta}(\chi') \triangleq \theta(\chi')[x \mapsto \mathsf{f}]$, for all $\chi' \in \mathrm{Asg}(\mathcal{G}, \Lambda(\wp), s_0)$. It is easy to check that $\widehat{\theta}$ is a quantification spectrum for $\wp$ over $\mathrm{Str}(\mathcal{G}, s_0)$, i.e., $\widehat{\theta} \in \Theta_{\mathrm{Str}(\mathcal{G},s_0)}(\wp)$. Moreover, $\chi_{x\mapsto\mathsf{f}} \uplus \theta(\chi') = \chi \uplus \theta(\chi')[x \mapsto \mathsf{f}] = \chi \uplus \widehat{\theta}(\chi')$, for $\chi' \in \mathrm{Asg}(\mathcal{G}, \Lambda(\wp), s_0)$. Hence, $\mathcal{G}, \chi \uplus \widehat{\theta}(\chi'), s_0 \models \psi$, for all $\chi' \in \mathrm{Asg}(\mathcal{G}, \Lambda(\wp), s_0)$. On the other hand, if $\mathsf{Qn} = [[x]]$, by Item 3b of Definition 3.2, we have that, for all strategies $\mathsf{f} \in \mathrm{Str}(\mathcal{G}, s_0)$, it holds that $\mathcal{G}, \chi_{x\mapsto\mathsf{f}}, s_0 \models \wp' \cdot \psi$. Note that $\Lambda(\wp) = \Lambda(\wp') \cup \{x\}$. By the inductive hypothesis, we derive that, for each $\mathsf{f} \in \mathrm{Str}(\mathcal{G}, s_0)$, there exists a quantification spectrum $\theta_{\mathsf{f}} \in \Theta_{\mathrm{Str}(\mathcal{G},s_0)}(\wp')$ such that $\mathcal{G}, \chi_{x\mapsto\mathsf{f}} \uplus \theta_{\mathsf{f}}(\chi'), s_0 \models \psi$, for all $\chi' \in \mathrm{Asg}(\mathcal{G}, \Lambda(\wp'), s_0)$. Now, consider the function $\widehat{\theta} : \mathrm{Asg}(\mathcal{G}, \Lambda(\wp), s_0) \to \mathrm{Asg}(\mathcal{G}, \mathrm{P}, s_0)$ defined by

$\widehat{\theta}(\chi') \triangleq \theta_{\chi'(x)}(\chi'_{\upharpoonright\Lambda(\wp')})[x \mapsto \chi'(x)]$, for all $\chi' \in \text{Asg}(\mathcal{G}, \Lambda(\wp), s_0)$. It is evident that $\widehat{\theta}$ is a quantification spectrum for $\wp$ over $\text{Str}(\mathcal{G}, s_0)$, i.e., $\widehat{\theta} \in \Theta_{\text{Str}(\mathcal{G},s_0)}(\wp)$. Moreover, $\chi_{x\mapsto f} \uplus \theta_f(\chi') = \chi \uplus \theta_f(\chi')[x \mapsto f] = \chi \uplus \widehat{\theta}(\chi'[x \mapsto f])$, for $f \in \text{Str}(\mathcal{G}, s_0)$ and $\chi' \in \text{Asg}(\mathcal{G}, \Lambda(\wp'), s_0)$. Hence, $\mathcal{G}, \chi \uplus \widehat{\theta}(\chi'), s_0 \models \psi$, for all $\chi' \in \text{Asg}(\mathcal{G}, \Lambda(\wp), s_0)$.

[If]. Suppose that there exists a quantification spectrum $\theta \in \Theta_{\text{Str}(\mathcal{G},s_0)}(\wp)$ such that $\mathcal{G}, \chi \uplus \theta(\chi'), s_0 \models \psi$, for all $\chi' \in \text{Asg}(\mathcal{G}, \Lambda(\wp), s_0)$, where $\wp = \mathsf{Qn} \cdot \wp'$. Then, we have two possible cases: either $\mathsf{Qn} = \langle\langle x\rangle\rangle$ or $\mathsf{Qn} = [[x]]$. On one hand, if $\mathsf{Qn} = \langle\langle x\rangle\rangle$, there is $f \in \text{Str}(\mathcal{G}, s_0)$ such that $f = \theta(\chi')(x)$, for all $\chi' \in \text{Asg}(\mathcal{G}, \Lambda(\wp), s_0)$. Note that $\Lambda(\wp) = \Lambda(\wp')$ and consider the function $\widehat{\theta} : \text{Asg}(\mathcal{G}, \Lambda(\wp'), s_0) \to \text{Asg}(\mathcal{G}, \text{P} \setminus \{x\}, s_0)$ defined by $\widehat{\theta}(\chi') \triangleq \theta(\chi')_{\upharpoonright(\text{P}\setminus\{x\})}$, for all $\chi' \in \text{Asg}(\mathcal{G}, \Lambda(\wp'), s_0)$. It is easy to check that $\widehat{\theta}$ is a quantification spectrum for $\wp'$ over $\text{Str}(\mathcal{G}, s_0)$, i.e., $\widehat{\theta} \in \Theta_{\text{Str}(\mathcal{G},s_0)}(\wp')$. Moreover, $\chi \uplus \theta(\chi') = \chi \uplus \widehat{\theta}(\chi')[x \mapsto f] = \chi_{x\mapsto f} \uplus \widehat{\theta}(\chi')$, for $\chi' \in \text{Asg}(\mathcal{G}, \Lambda(\wp'), s_0)$. Then, it is evident that $\mathcal{G}, \chi_{x\to f} \uplus \widehat{\theta}(\chi'), s_0 \models \psi$, for all $\chi' \in \text{Asg}(\mathcal{G}, \Lambda(\wp'), s_0)$. By the inductive hypothesis, we derive that $\mathcal{G}, \chi_{x\mapsto f}, s_0 \models \wp' \cdot \psi$, which, by Item 3a of Definition 3.2 of semantics, means that $\mathcal{G}, \chi, s_0 \models \varphi$. On the other hand, if $\mathsf{Qn} = [[x]]$, note that $\Lambda(\wp) = \Lambda(\wp') \cup \{x\}$ and consider the functions $\widehat{\theta_f}$: $\text{Asg}(\mathcal{G}, \Lambda(\wp'), s_0) \to \text{Asg}(\mathcal{G}, \text{P}\setminus\{x\}, s_0)$ defined by $\widehat{\theta_f}(\chi') \triangleq \theta(\chi'[x \mapsto f])_{\upharpoonright(\text{P}\setminus\{x\})}$, for each $f \in \text{Str}(\mathcal{G}, s_0)$ and $\chi' \in \text{Asg}(\mathcal{G}, \Lambda(\wp'), s_0)$. It is evident that every $\widehat{\theta_f}$ is a quantification spectrum for $\wp'$ over $\text{Str}(\mathcal{G}, s_0)$, i.e., $\widehat{\theta_f} \in \Theta_{\text{Str}(\mathcal{G},s_0)}(\wp')$. Moreover, $\chi \uplus \theta(\chi') = \chi \uplus \widehat{\theta_{\chi'(x)}}(\chi'_{\upharpoonright(\text{P}\setminus\{x\})})[x \mapsto \chi'(x)] = \chi_{x\mapsto\chi'(x)} \uplus \widehat{\theta_{\chi'(x)}}(\chi'_{\upharpoonright(\text{P}\setminus\{x\})})$, for $\chi' \in \text{Asg}(\mathcal{G}, \Lambda(\wp), s_0)$. Then, it is evident that $\mathcal{G}, \chi_{x\to f} \uplus \widehat{\theta_f}(\chi'), s_0 \models \psi$, for all $f \in \text{Str}(\mathcal{G}, s_0)$ and $\chi' \in \text{Asg}(\mathcal{G}, \Lambda(\wp'), s_0)$. By the inductive hypothesis, we derive that $\mathcal{G}, \chi_{x\mapsto f}, s_0 \models \wp' \cdot \psi$ for all $f \in \text{Str}(\mathcal{G}, s_0)$, which, by Item 3b of Definition 3.2, means that $\mathcal{G}, \chi, s_0 \models \varphi$. □

The above theorem substantially characterizes SL semantics by means of the concept of quantification spectrum. Such a characterization enables the definition of alternative semantics, based on the choice of a subset of quantification spectra that ensures better model properties and easier decision problems. Here, we consider the set of quantification spectra that are *elementary*, which allows us to greatly simplify the reasoning about strategy quantifications by reducing them to a set of quantifications over actions, one for each track in their domains.

In the following, we give some fundamental definitions and two relative lemmas that are used to define the concept of elementariness.

Definition 3.10 (Adjoint Functions) *Let $\wp$ be a quantification prefix over a set of placeholders* P, D *and* T *be two sets, and* $\theta : (\text{T} \to \text{D})^{\Lambda(\wp)} \to (\text{T} \to \text{D})^{\text{P}}$ *and* $\widehat{\theta} : \text{T} \to (\text{D}^{\Lambda(\wp)} \to \text{D}^{\text{P}})$ *be two functions. Then, we say that $\widehat{\theta}$ is the* adjoint *of θ w.r.t.* D, T, *and $\wp$ iff $\theta(\mathsf{h})(x)(t) = \widehat{\theta}(t)(\overline{\mathsf{h}}(t))(x)$, for all $\mathsf{h} \in (\text{T} \to \text{D})^{\Lambda(\wp)}$, $x \in \text{P}$, and $t \in \text{T}$, where $\overline{\mathsf{h}} : \text{T} \to \text{D}^{\Lambda(\wp)}$ is such that $\overline{\mathsf{h}}(t)(y) = \mathsf{h}(y)(t)$, for each $y \in \text{P}$ and $t \in \text{T}$.*

The formal meaning of the elementariness of a quantification spectrum over generic functions follows.

Definition 3.11 (Elementary Quantification Spectrum) *Let $\wp$ be a quantification prefix over a set of placeholders* P *and* $\theta \in \Theta_{T \to D}(\wp)$ *be a quantification spectrum for* $\wp$ *over* T $\to$ D. *Then,* θ *is* elementary *if it admits an adjoint function.*

Next lemma formally states that each elementary quantification spectrum over a set T $\to$ D can be seen as a set of quantification spectra over D, one for each element of T and vice versa.

Lemma 3.5 (Adjoint Functions) *Let $\wp$ be a quantification prefix over a set of placeholders* P, D *and* T *be two sets, and* $\theta : (T \to D)^{\Lambda(\wp)} \to (T \to D)^{P}$ *and* $\widehat{\theta} : T \to (D^{\Lambda(\wp)} \to D^{P})$ *be two functions such that* $\widehat{\theta}$ *is the adjoint of* θ *w.r.t.* D, T, *and* $\wp$. *Then,* $\theta \in \Theta_{T \to D}(\wp)$ *iff, for all* $t \in T$, *it holds that* $\widehat{\theta}(t) \in \Theta_{D}(\wp)$.

Proof To prove the statement, it is enough to show separately that Items 1 and 2 of Definition 3.9 hold for one function if all the others satisfy the same items, and vice versa.

[Item 1, if]. Assume that $\widehat{\theta}(t)$ satisfies Item 1, for each $t \in T$, i.e., $\widehat{\theta}(t)(\mathsf{d})_{\restriction \Lambda(\wp)} = \mathsf{d}$, for all $\mathsf{d} \in D^{\Lambda(\wp)}$. Then, we have that $\widehat{\theta}(t)(\overline{\mathsf{h}}(t))(x) = \overline{\mathsf{h}}(t)(x)$, for all $\mathsf{h} \in (T \to D)^{\Lambda(\wp)}$ and $x \in \Lambda(\wp)$. By hypothesis, we have that $\theta(\mathsf{h})(x)(t) = \widehat{\theta}(t)(\overline{\mathsf{h}})(t)(x)$, thus $\theta(\mathsf{h})(x)(t) = \mathsf{h}(x)(t)$, which means that $\theta(\mathsf{h})_{\restriction \Lambda(\wp)} = \mathsf{h}$, for all $\mathsf{h} \in (T \to D)^{\Lambda(\wp)}$.

[Item 1, only if]. Assume now that θ satisfies Item 1, i.e., $\theta(\mathsf{h})_{\restriction \Lambda(\wp)} = \mathsf{h}$, for all $\mathsf{h} \in (T \to D)^{\Lambda(\wp)}$. Then, we have that $\theta(\mathsf{h})(x)(t) = \mathsf{h}(x)(t)$, for all $x \in \Lambda(\wp)$ and $t \in T$. By hypothesis, we have that $\widehat{\theta}(t)(\overline{\mathsf{h}}(t))(x) = \theta(\mathsf{h})(x)(t)$, so $\widehat{\theta}(t)(\overline{\mathsf{h}}(t))(x) = \overline{\mathsf{h}}(t)(x)$, which means that $\widehat{\theta}(t)(\overline{\mathsf{h}}(t))_{\restriction \Lambda(\wp)} = \overline{\mathsf{h}}(t)$. Now, since for each $\mathsf{d} \in D^{\Lambda(\wp)}$, there is an $\mathsf{h} \in (T \to D)^{\Lambda(\wp)}$ such that $\overline{\mathsf{h}}(t) = \mathsf{d}$, we obtain that $\widehat{\theta}(t)(\mathsf{d})_{\restriction \Lambda(\wp)} = \mathsf{d}$, for all $\mathsf{d} \in D^{\Lambda(\wp)}$ and $t \in T$.

[Item 2, if]. Assume that $\widehat{\theta}(t)$ satisfies Item 2, for each $t \in T$, i.e., $\widehat{\theta}(t)(\mathsf{d}_1)(x) = \widehat{\theta}(t)(\mathsf{d}_2)(x)$, for all $\mathsf{d}_1, \mathsf{d}_2 \in D^{\Lambda(\wp)}$ and $x \in \Xi(\wp)$ such that $\mathsf{d}_{1 \restriction \Upsilon(\wp, x)} = \mathsf{d}_{2 \restriction \Upsilon(\wp, x)}$. Then, we have that $\widehat{\theta}(t)(\overline{\mathsf{h}_1}(t))(x) = \widehat{\theta}(t)(\overline{\mathsf{h}_2}(t))(x)$, for all $\mathsf{h}_1, \mathsf{h}_2 \in (T \to D)^{\Lambda(\wp)}$ such that $\mathsf{h}_{1 \restriction \Upsilon(\wp, x)} = \mathsf{h}_{2 \restriction \Upsilon(\wp, x)}$. By hypothesis, we have that $\theta(\mathsf{h}_1)(x)(t) = \widehat{\theta}(t)(\overline{\mathsf{h}_1}(t))(x)$ and $\widehat{\theta}(t)(\overline{\mathsf{h}_2}(t))(x) = \theta(\mathsf{h}_2)(x)(t)$, thus $\theta(\mathsf{h}_1)(x)(t) = \theta(\mathsf{h}_2)(x)(t)$. Hence, $\theta(\mathsf{h}_1)(x) = \theta(\mathsf{h}_2)(x)$, for all $\mathsf{h}_1, \mathsf{h}_2 \in (T \to D)^{\Lambda(\wp)}$ and $x \in \Xi(\wp)$ such that $\mathsf{h}_{1 \restriction \Upsilon(\wp, x)} = \mathsf{h}_{2 \restriction \Upsilon(\wp, x)}$.

[Item 2, only if]. Assume that θ satisfies Item 2, i.e., $\theta(\mathsf{h}_1)(x) = \theta(\mathsf{h}_2)(x)$, for all $\mathsf{h}_1, \mathsf{h}_2 \in (T \to D)^{\Lambda(\wp)}$ and $x \in \Xi(\wp)$ such that $\mathsf{h}_{1 \restriction \Upsilon(\wp, x)} = \mathsf{h}_{2 \restriction \Upsilon(\wp, x)}$. Then, we have that $\theta(\mathsf{h}_1)(x)(t) = \theta(\mathsf{h}_2)(x)(t)$, for all $t \in T$. By hypothesis, we have that $\widehat{\theta}(t)(\overline{\mathsf{h}_1}(t))(x) = \theta(\mathsf{h}_1)(x)(t)$ and $\theta(\mathsf{h}_2)(x)(t) = \widehat{\theta}(t)(\overline{\mathsf{h}_2}(t))(x)$, hence $\widehat{\theta}(t)(\overline{\mathsf{h}_1}(t))(x) = \widehat{\theta}(t)(\overline{\mathsf{h}_2}(t))(x)$. Now, since for each $\mathsf{d}_1, \mathsf{d}_2 \in D^{\Lambda(\wp)}$ there are $\mathsf{h}_1, \mathsf{h}_2 \in (T \to D)^{\Lambda(\wp)}$ such that $\overline{\mathsf{h}_1}(t) = \mathsf{d}_1$ and $\overline{\mathsf{h}_2}(t) = \mathsf{d}_2$, we obtain that $\widehat{\theta}(t)(\mathsf{d}_1)(x) = \widehat{\theta}(t)(\mathsf{d}_2)(x)$, for all $\mathsf{d}_1, \mathsf{d}_2 \in D^{\Lambda(\wp)}$ and $x \in \Xi(\wp)$ such that $\mathsf{d}_{1 \restriction \Upsilon(\wp, x)} = \mathsf{d}_{2 \restriction \Upsilon(\wp, x)}$. □

Before to state the last result, we have to define the following concept.

Definition 3.12 (Binding Functions) *Let $\wp$ be a quantification prefix over a set of placeholders* P. *Then, a* binding function *for $\wp$ is a function* $\zeta : \mathrm{Ag} \mapsto P$ *that assigns* *to each agent a placeholder, with the proviso that if* $\zeta(a) \in \mathrm{Ag}$*, for* $a \in \mathrm{Ag}$*, then* $\zeta(a) = a$*. By* $\mathrm{Bnd}(\wp)$ *we denote the set of all binding functions of* $\wp$*.*

Now, we show how each play w.r.t. a complete assignment derived from a quantification spectrum can be characterized, in an equivalent way, in base of the adjoint function of that spectrum. This fact is used in the construction of the model checking procedure and, in particular, in the automaton to which we relay for the building of the pruning of the unwinding of the model. This pruning needs to be coherent with the quantification of the sentence that is represented as an action quantification in each node of the tree.

Lemma 3.6 (Adjoint Paths) *Let* $\mathcal{G}$ *be a* CGS*,* s *be one of its states and* $\wp$ *be a quantification prefix over a set of placeholders* P*. Moreover, let* $\theta \in \Theta_{\mathrm{Str}(\mathcal{G},s)}(\wp)$ *be a quantification spectrum for* $\wp$*,* $\chi \in \mathrm{Asg}(\mathcal{G}, \Lambda(\wp), s)$ *be a assignment on* $\Lambda(\wp)$*,* $\zeta \in \mathrm{Bnd}(\wp)$ *be a binding function, and* $\pi \in \mathrm{Pth}(\mathcal{G}, s)$ *be a path. Then, it holds that* π *is a* $(\zeta \circ \theta(\chi), s)$*-play iff* $\pi_{i+1} = \tau(\pi_i, \zeta \circ \widehat{\theta}(\pi_{\leq i})(\overline{\chi}(\pi_{\leq i})))$*, for all* $i \in \mathbb{N}$*.*

Proof By definition, a path π is a $(\zeta \circ \theta(\chi), s)$-play iff, for all $i \in \mathbb{N}$, it holds that $\pi_{i+1} = \tau(\pi_i, \mathsf{d}_i)$, where $\mathsf{d}_i(a) = (\zeta \circ \theta(\chi))(a)(\pi_{\leq i})$, for all $a \in \mathrm{Ag}$. Hence, to prove the statement, we have to show that $(\zeta \circ \theta(\chi))(a)(\pi_{\leq i}) = (\zeta \circ \widehat{\theta}(\pi_{\leq i})(\overline{\chi}(\pi_{\leq i})))(a)$ holds. Indeed, by the meaning of composition of functions, we have that $(\zeta \circ \widehat{\theta}(\pi_{\leq i})(\overline{\chi}(\pi_{\leq i})))(a) = \widehat{\theta}(\pi_{\leq i})(\overline{\chi}(\pi_{\leq i}))(\zeta(a))$. Now, by Definition 3.10 of adjoint function, it holds that $\widehat{\theta}(\pi_{\leq i})(\overline{\chi}(\pi_{\leq i}))(\zeta(a)) = \theta(\chi)(\zeta(a))(\pi_{\leq i})$. Finally, again by the meaning of composition, we obtain that $\theta(\chi)(\zeta(a)) = (\zeta \circ \theta(\chi))(a)$ and so, $\theta(\chi)(\zeta(a))(\pi_{\leq i}) = (\zeta \circ \theta(\chi))(a)(\pi_{\leq i})$. Hence, the thesis holds. □

Finally, we introduce a notion of *elementary satisfiability*, in symbols $\models_{\mathrm{E}}$, which requires the elementariness of quantification spectra over strategies.

Definition 3.13 (SL **Elementary Semantics**) *Let* $\mathcal{G}$ *be a* CGS *and* $\varphi = \wp\psi$ *an* SL *sentence, where* ψ *is agent-closed and* $\wp$ *is quantification prefix over* $\mathsf{free}(\psi)$*. Then,* $\mathcal{G} \models_{\mathrm{E}} \varphi$ *iff there exists an elementary quantification spectrum* $\theta \in \Theta_{\mathrm{Str}(\mathcal{G},s_0)}(\wp)$ *such that* $\mathcal{G}, \theta(\chi), s_0 \models \psi$*, for all* $\chi \in \mathrm{Asg}(\Lambda(\wp), s_0)$*.*

3.6 Alternating Tree Automata

Nondeterministic tree automata are a generalization to infinite trees of the classical *nondeterministic word automata* (see Thomas 1990, for an introduction). *Alternating tree automata* are a further generalization of nondeterministic tree automata (Muller and Schupp 1987). Intuitively, on visiting a node of the input tree, while the latter sends exactly one copy of itself to each of the successors of the node, an ATA can send several copies of itself to the same successor. Here we use, in particular, *alternating parity tree automata*, which are ATAs along with a *parity acceptance condition* (see Grädel et al. 2002, for a survey).

We now give the formal definition of alternating tree automata.

Definition 3.14 (Alternating Tree Automata) *An* alternating tree automaton *(*ATA, *for short) is a tuple* $\mathcal{A} \triangleq \langle \Sigma, \Delta, Q, \delta, q_0, F \rangle$*, where* Σ*,* Δ*, and* Q *are non-empty finite sets of* input symbols, directions, *and* states, *respectively,* $q_0 \in Q$ *is an* initial state, F *is an* acceptance condition *to be defined later, and* $\delta : Q \times \Sigma \to B^+(\Delta \times Q)$ *is an* alternating transition function *that maps each pair of states and input symbols to a positive Boolean combination on the set of propositions of the form* $(d, q) \in \Delta \times Q$*, a.k.a.* moves.

A *nondeterministic tree automaton* (NTA, for short) is a special ATA in which each conjunction in the transition function δ has exactly one move (d, q) associated with each direction d. In addition, a *universal tree automaton* (UTA, for short) is a special ATA in which all the Boolean combinations that appear in δ are only conjunctions of moves.

The semantics of the ATAs is now given through the following concept of run.

Definition 3.15 (ATA **Run**) *A* run *of an* ATA $\mathcal{A} = \langle \Sigma, \Delta, Q, \delta, q_0, F \rangle$ *on a* Σ*-labeled* Δ*-tree* $\mathcal{T} = \langle T, v \rangle$ *is a* $(Q \times T)$*-labeled* $\mathbb{N}$*-tree* $\mathcal{R} \triangleq \langle R, r \rangle$ *such that* (i) $r(\varepsilon) = (q_0, \varepsilon)$ *and* (ii) *for all nodes* $y \in R$ *with* $r(y) = (q, x)$*, there is a set of moves* $S \subseteq \Delta \times Q$ *with* $S \models \delta(q, v(x))$ *such that, for all* $(d, q') \in S$*, there is an index* $j \in [0, |S|[$ *for which it holds that* $y \cdot j \in R$ *and* $r(y \cdot j) = (q', x \cdot d)$.

In the following, we consider ATAs along with the *parity* $F = (F_1, \ldots, F_k) \in (2^Q)^+$ with $F_1 \subseteq \ldots \subseteq F_k = Q$ (APT, for short) acceptance condition (see [Kupferman et al. (2000)], for more). The number k of sets in F is called the *index* of the automaton. We also use ATAs with the *Co-Büchi* acceptance condition $F \subseteq Q$ (ACT, for short) that are APTs of index 2 in which the set of final states is represented by F_1.

Let $\mathcal{R} = \langle R, r \rangle$ be a run of an ATA $\mathcal{A}$ on a tree $\mathcal{T}$ and $R' \subseteq R$ one of its branches. Then, by $\mathsf{inf}(R') \triangleq \{q \in Q : |\{y \in R' : r(y) = q\}| = \omega\}$ we denote the set of states that occur infinitely often as labeling of the nodes in the branch R'. We say that a branch R' of $\mathcal{T}$ satisfies the parity acceptance condition $F = (F_1, \ldots, F_k)$ iff the least index $i \in [1, k]$ for which $\mathsf{inf}(R') \cap F_i \neq \emptyset$ is even.

At this point, we can define the concept of language accepted by an ATA.

Definition 3.16 (ATA **Acceptance**) *An* ATA $\mathcal{A} = \langle \Sigma, \Delta, Q, \delta, q_0, F \rangle$ accepts *a* Σ*-labeled* Δ*-tree* $\mathcal{T}$ *iff is there exists a run* $\mathcal{R}$ *of* $\mathcal{A}$ *on* $\mathcal{T}$ *such that all its infinite branches satisfy the acceptance condition* F*, where the concept of satisfaction is dependent from the definition of* F.

By $L(\mathcal{A})$ we denote the language accepted by the ATA $\mathcal{A}$, i.e., the set of trees $\mathcal{T}$ accepted by $\mathcal{A}$. Moreover, $\mathcal{A}$ is said to be *empty* if $L(\mathcal{A}) = \emptyset$. The *emptiness problem* for $\mathcal{A}$ is to decide whether $L(\mathcal{A}) = \emptyset$ or not.

3.7 Model Checking

In this section, we study the model-checking problem for SL under the elementary semantics and show that it is decidable and 2EXPTIME-COMPLETE, as for ATL*.

The lower bound immediately follows from ATL*, which SL properly includes. For the upper bound, we follow an *automata-theoretic approach* (Kupferman et al. 2000), reducing the decision problem for the logic of interest to the emptiness problem of automata.

We recall that an approach with tree automata to model checking is only possible once the logic satisfies invariance under unwinding. In fact, this property holds for SL as we have proved in Item 2 Theorem 3.2. By the size of the automaton and the complexity required for checking its emptiness, we get the desired 2EXPTIME upper bound.

We now proceed with the model-checking algorithm for SL. As for ATL*, we use a bottom-up model-checking algorithm, in which we start with the innermost sub-sentences and terminate with the sentence under checking. At each step, we label each state of the model with all the sub-sentences that are satisfied on it. The procedure we propose here extends that used for ATL* in Alur et al. (2002) by means of a richer structure of the automata involved in.

First, we introduce some extra notation. A *principal sentence* φ is a sentence of the form $\mathsf{Qn}_1 x_1 \cdots \mathsf{Qn}_k x_k\ \psi_\varphi$, where $\mathsf{Qn}_i x_i \in \{\langle\!\langle x_i \rangle\!\rangle, [\![x_i]\!]\}$ and the *matrix* ψ_φ is an agent-closed formula, with $\mathsf{free}(\psi_\varphi) = \{x_1, \ldots, x_k\}$, such that it does not contain any quantification. For the sake of space and clarity of exposition, we only discuss the model checking of principal formulas. By a slight variation of both the notion of principal formulas and our procedure, we can also address the full SL. We also need the notion of atom. An *atom* ψ is an agent-closed formula of the form $(\alpha_1, y_1) \cdots (\alpha_n, y_n)\psi'$, where $\mathrm{Ag} = \{\alpha_1, \ldots, \alpha_n\}$, $y_1, \ldots, y_n$ are possible equal variables and either *(i)* ψ' does not contain any quantification and binding, i.e., it is an LTL formula, or *(ii)* the derived formula $\widehat{\psi}'$ does not contain any quantification and binding at all, where $\widehat{\psi}'$ is obtained by ψ' substituting its sub-atoms with fresh atomic propositions. W.l.o.g., we assume that each principal sentence has a matrix that is a Boolean combination of atoms. $\mathrm{Atm}(\varphi)$ denotes the set of all sub-formulas of φ that are atoms.

The core idea behind our model-checking procedure is the following. Let $\mathcal{G} = \langle \mathrm{AP}, \mathrm{Ag}, \mathrm{Ac}, \mathrm{St}, \lambda, \tau, s_0 \rangle$ be a CGS and φ be an SL principal sentence over the set $\mathrm{Ag} = \{\alpha_1, \ldots, \alpha_n\}$ of n different agents, for which we want to check if $\mathcal{G} \models \varphi$ holds or not. We first build an NPT $\mathcal{D}_\mathcal{G}$ recognizing the unwinding $\mathcal{G}^\mathcal{U}$ of $\mathcal{G}$. Then, we build an APT $\mathcal{A}'_{\mathcal{G},\varphi}$ accepting all prunings of $\mathcal{G}^\mathcal{U}$ that are coherent with the strategy quantification of φ. Such prunings are done by properly labeling its paths with elements from the set $\mathrm{Z} \triangleq \mathrm{Atm}(\varphi) \times \{start, pass\}$ of atoms associated with a flag in $\{start, pass\}$, in a way similar as it has been done for ATL* satisfiability in Schewe (2008). The $start$ and $pass$ flags are used to indicate whether a path guessed to satisfy at a specific state an atom $\psi \in \mathrm{Atm}(\varphi)$, starts or passes through that state, respectively. Namely, the unlabeled paths are the pruned ones that are not needed in order to satisfy the formula. Hence, $\mathcal{A}'_{\mathcal{G},\varphi}$ accepts $\mathcal{G}^\mathcal{U}$ with this additional labeling. The automata $\mathcal{D}_\mathcal{G}$ and $\mathcal{A}'_{\mathcal{G},\varphi}$ have index 2 and a number of states polynomial in the size of $\mathcal{G}$ and φ, respectively. With more details, they are both safety automata.[2] Finally, we build an APT $\mathcal{A}''_{\varphi}$ that checks that all paths of a pruned model accepted by $\mathcal{A}'_{\mathcal{G},\varphi}$, i.e., all labeled paths, satisfy the atoms of φ. The automaton $\mathcal{A}''_{\varphi}$ has index 2 and a number of states exponential in φ.

Now, recall that APTs are linearly closed under intersection. More precisely, two APTs having n_1 and n_2 states and k_1 and k_2 as indexes, respectively, can be intersected in an APT with $n_1 + n_2$ states and index $\max\{k_1, k_2\}$ (Muller and Schupp 1995). So, we can build an APT $\mathcal{A}_{\mathcal{G},\varphi}$ such that $\mathrm{L}(\mathcal{A}_{\mathcal{G},\varphi}) = \mathrm{L}(\mathcal{A}'_{\mathcal{G},\varphi}) \cap \mathrm{L}(\mathcal{A}''_{\varphi})$, having in particular index 2. Also, by Muller and Schupp (1995), we can translate an APT with n states and index k in an equivalent NPT having $n^{\mathrm{O}(n)}$ states and index $\mathrm{O}(n)$. Hence, we can transform $\mathcal{A}_{\mathcal{G},\varphi}$ in an NPT $\mathcal{N}_{\mathcal{G},\varphi}$ with a number of states double exponential in φ and an index exponential in φ. It is well known that an NPT having n states and index k and a safety automaton with m states can be intersected in an NPT with $n \cdot m$ states and index k. Hence, by intersecting $\mathcal{D}_{\mathcal{G}}$ with $\mathcal{N}_{\mathcal{G},\varphi}$, we get an NPT $\mathcal{N}'_{\mathcal{G},\varphi}$ such that $\mathrm{L}(\mathcal{N}'_{\mathcal{G},\varphi}) = \mathrm{L}(\mathcal{D}_{\mathcal{G}}) \cap \mathrm{L}(\mathcal{N}_{\mathcal{G},\varphi})$. At this point, it is possible to prove that $\mathcal{G} \models \varphi$ iff $\mathrm{L}(\mathcal{N}'_{\mathcal{G},\varphi}) \neq \emptyset$. Observe that $\mathcal{N}'_{\mathcal{G},\varphi}$ has a number of states double exponential in φ and polynomial in $\mathcal{G}$, while it has an index exponential in φ, but independent from $\mathcal{G}$. Moreover, the automata run over the alphabet $\Sigma = \{\sigma \subseteq \mathrm{AP} \cup \mathrm{St} \cup \mathrm{Z} : |\sigma \cap \mathrm{St}| = 1\}$, where $|\mathrm{Z}| = \mathrm{O}(|\mathcal{G}| \times 2^{|\varphi|})$. Since the emptiness of an NPT with n states, index k, and alphabet size h can be checked in time $\mathrm{O}(h \cdot n^k)$ (Kupferman and Vardi 1998), we get that to check whether $\mathcal{G} \models \varphi$ can be done in time double exponential in φ and polynomial in $\mathcal{G}$. More precisely, the algorithm runs in $|\mathcal{G}|^{2^{\mathrm{O}(|\varphi|)}}$. The details of the automata construction follow.

The NPT $\mathcal{D}_{\mathcal{G}} = \langle \Sigma, \mathrm{St}, \mathrm{St}, \delta, s_0, (\emptyset, \mathrm{St})\rangle$ has the set of directions and states formed by the states of $\mathcal{G}$ that are used to build its unwinding. Moreover, the transition function is defined as follows. At the state $s \in \mathrm{St}$, the automaton first checks that the labeling of the node of the input tree corresponds to the union of $\{s\}$ and its labeling $\lambda(s)$ in $\mathcal{G}$. Then, it sends all successors of s in the relative directions. Formally, $\delta(s, \sigma)$ is set to $\mathfrak{f}$ (false) if $\lambda(s) \cup \{s\} \neq \sigma \cap (\mathrm{AP} \cup \mathrm{St})$ and to $\bigwedge_{s' \in \{\tau(s,\mathsf{d}) : \mathsf{d} \in \mathrm{Dc}\}} (s', s')$ otherwise. Note that $|\mathcal{D}_{\mathcal{G}}| = \mathrm{O}(|\mathcal{G}|)$.

The APT $\mathcal{A}'_{\mathcal{G},\varphi} = \langle \Sigma, \mathrm{St}, \{q_0\} \cup \mathrm{Atm}(\varphi), \delta, q_0, (\emptyset, \{q_0\} \cup \mathrm{Atm}(\varphi))\rangle$ has the set of states formed by a distinguished state q_0, which is also initial, and from the atoms in $\mathrm{Atm}(\varphi)$ that are used to verify the correctness of the additional labeling Z. Moreover, the transition function is defined as follows. $\delta(\psi, \sigma)$ is equal to $\mathfrak{t}$ (true) if $(\psi, pass) \in \sigma \cap \mathrm{Z}$ and to $\mathfrak{f}$ (false) otherwise. The automaton at state q_0 sends the same state in all the directions individuated by the quantification, together with the control state ψ. It is important to note that the quantification here is reproduced by conjunctions and disjunctions on all possible actions of $\mathcal{G}$. Formally, $\delta(q_0, \sigma)$ is set to $\mathsf{Op}_{\,1\, c_1 \in \mathrm{Ac}} \cdots \mathsf{Op}_{\,k\, c_k \in \mathrm{Ac}} \bigwedge_{(\psi,\star) \in \sigma \cap \mathrm{Z}} (\tau(s, \mathsf{d}), q_0) \wedge (\tau(s, \mathsf{d}), \psi)$, where $\mathsf{Op}_{\,i\, c_i \in \mathrm{Ac}}$ is a disjunction if $\mathsf{Qn}_{\,i} x_i = \langle x_i \rangle$ and a conjunction if $\mathsf{Qn}_{\,i} x_i = [x_i]$, $\{s\} = \sigma \cap \mathrm{St}$, and $\mathsf{d}(\alpha_i) = c_j$ iff in the atom ψ the binding (α_i, x_j) appears. Note that $|\mathcal{A}'_{\mathcal{G},\varphi}| = \mathrm{O}(\mathsf{lng}(\varphi))$.

---

[2] A safety condition is the special parity condition $(\emptyset, \mathrm{Q})$ of index 2.

Finally, we build the APT $\mathcal{A}''_{\varphi}$. Let $\widehat{\psi}$ be the LTL formula obtained by replacing in $\psi \in \mathrm{Atm}(\varphi)$ all the occurrences of each other atom $\psi' \in \mathrm{Atm}(\psi)$ with the fresh atomic proposition $(\psi', start)$. By using a slight variation of the procedure developed in Vardi and Wolper (1986a), we can translate $\widehat{\psi}$ into a universal co-Büchi word automaton[3] $\mathcal{U}_{\psi} = \langle \Sigma, \mathrm{Q}_{\psi}, \delta_{\psi}, \mathrm{Q}_{0\psi}, \mathrm{F}_{\psi} \rangle$, with a number of states at most exponential in $\mathsf{lng}(\psi)$, accepting the infinite words on Σ that are models of $\widehat{\psi}$. At this point, we can construct the automaton $\mathcal{A}''_{\varphi}$ that recognizes the trees whose paths, labeled with the flags $(\psi, \star)$, for $\star \in \{start, pass\}$, and starting with the label $(\psi, start)$, satisfy the LTL formula $\widehat{\psi}$, for all $\psi \in \mathrm{Atm}(\varphi)$. Formally, $\mathcal{A}''_{\varphi} = \langle \Sigma, \mathrm{St}, \{q_0, q_c\} \cup \mathrm{Q}, \delta, q_0, (\mathrm{F}, \{q_0, q_c\} \cup \mathrm{Q}) \rangle$ is built as follows. $\mathrm{Q} = \bigcup_{\psi \in \mathrm{Atm}(\varphi)} \{\psi\} \times \mathrm{Q}_{\psi}$ and $\mathrm{F} = \bigcup_{\psi \in \mathrm{Atm}(\varphi)} \{\psi\} \times \mathrm{F}_{\psi}$ are, respectively, the disjoint union of the set of states and final states of the word automata $\mathcal{U}_{\psi}$, for every atom $\psi \in \mathrm{Atm}(\varphi)$. q_0 is the *initial state* used to verify that the formula ψ_{φ} (the matrix of φ) holds at the root of the tree in input, by checking whether the labeling of the root contains all the propositions required by ψ_{φ} to hold. If the checking succeeds, q_0 behaves as the state q_c. Formally, let ψ_{φ} be considered as a boolean formula on the set of atoms $\mathrm{Atm}(\varphi)$ in which we assume $\psi = (\psi, start)$, for all $\psi \in \mathrm{Atm}(\varphi)$. Then, $\delta(q_0, \sigma)$ is set to $\delta(q_c, \sigma)$, if $\sigma \cap \mathrm{Z} \models \psi_{\varphi}$ and to $\mathfrak{f}$ (false), otherwise. q_c is the *checking state* used to start the verification of the atoms ψ in every node of the input tree that contains the flag $(\psi, start)$, which indicates the existence of a path starting in that node that satisfies ψ. To do this, q_c sends in all the directions *(i)* a copy of the state itself, to continue the control on the remaining part of the tree, and *(ii)* the states derived by all initial states of the automata $\mathcal{U}_{\psi}$, for all the atoms ψ for which a flag $(\psi, start)$ appears in the labeling σ. Formally, $\delta(q_c, \sigma)$ is $\bigwedge_{s \in \mathrm{St}} (s, q_c) \wedge \bigwedge_{(\psi, start) \in \sigma \cap \mathrm{Z}} \bigwedge_{q \in \mathrm{Q}_{0\psi}} \bigwedge_{q' \in \delta_{\psi}(q, \sigma \cap \mathrm{AP})} (s, (\psi, q'))$. The states of the form (ψ, q) are used to run $\mathcal{U}_{\psi}$ on all paths labeled by the related flags $(\psi, pass)$. Formally, $\delta((\psi, q), \sigma)$ is set to $\mathfrak{t}$ (true) if $(\psi, pass) \notin \sigma \cap \mathrm{Z}$ and to $\bigwedge_{s \in \mathrm{St}} \bigwedge_{q' \in \delta_{\psi}(q, \sigma \cap \mathrm{AP})} (s, (\psi, q'))$ otherwise. Note that $|\mathcal{A}''_{\varphi}| = \mathrm{O}(2^{\mathsf{lng}(\varphi)})$.

By a simple calculation, it follows that the overall procedure results in an algorithm that is in PTIME w.r.t the size of $\mathcal{G}$ and in 2EXPTIME w.r.t. the size of φ. Hence, by getting the lower bound from ATL*, the following result holds.

Theorem 3.6 (SL **Model Checking**) *The* SL *model-checking problem under the elementary semantics is* PTIME-COMPLETE *w.r.t. the size of the model and* 2EXPTIME-COMPLETE *w.r.t the size of the specification.*

We conclude this section by pointing out that the model checking procedure described above for SL is completely different from that one used in Chatterjee et al. (2007) for CHP-SL. Indeed in Chatterjee et al. (2007), the authors use a top-down approach and, most important, for every quantification in the formula, they make a projection of the automaton they build at each stage (one for each quantification). Since at each projection they have an exponential blow-up, at the end their procedure results in a non-elementary one, both in the size of the system and the formula. Our

---

[3] Word automata can be seen as tree automata in which the tree has just one path. A universal word automaton is a particular case of alternating automata in which there is no nondeterminism. A co-Büchi acceptance condition $\mathrm{F} \subseteq \mathrm{Q}$ is the special parity condition (F, Q) of index 2.

iterative approach, instead, does not make use of any projection, since we reduce strategy quantifications to action quantifications, which, as we have stated, can be handled locally on each state of the model.

3.8 Satisfiability

In this section, we show the undecidability of the satisfiability problem for SL through a reduction of the *recurrent domino problem*. In particular, as we discuss later, the reduction also holds for CHP-SL under the concurrent game semantics.

The *domino problem*, proposed for the first time by Wang (1961), consists of placing a given number of tile types on an infinite grid, satisfying a predetermined set of constraints on adjacent tiles. One of its standard versions asks for a compatible tiling of the whole plane $\mathbb{N} \times \mathbb{N}$. The *recurrent domino problem* further requires the existence of a distinguished tile type that occurs infinitely often in the first row of the grid. This problem was proved to be highly undecidable by Harel, and in particular, Σ_1^1-COMPLETE (Harel 1984). The formal definition follows.

Definition 3.17 (Recurrent Domino System) *An* $\mathbb{N} \times \mathbb{N}$ recurrent domino system $\langle \mathscr{D}, t^* \rangle$ *with* $\mathscr{D} = \langle \mathrm{D}, H, V \rangle$ *consists of a finite non-empty set* D *of* domino types, *two* horizontal *and* vertical matching relations $H, V \subseteq \mathrm{D} \times \mathrm{D}$, *and a distinguished tile type* $t^* \in \mathrm{D}$. *The recurrent domino problem asks for an* admissible tiling *of* $\mathbb{N} \times \mathbb{N}$, *which is a* solution mapping $\partial : \mathbb{N} \times \mathbb{N} \to \mathrm{D}$ *such that, for all* $x, y \in \mathbb{N}$, *it holds that* (i) $(\partial(x, y), \partial(x + 1, y)) \in H$, (ii) $(\partial(x, y), \partial(x, y + 1)) \in V$, *and* (iii) $|\{x \in \mathbb{N} : \partial(x, 0) = t^*\}| = \omega$.

By showing a reduction from the recurrent domino problem, we prove that the satisfiability problem for SL is Σ_1^1-HARD, which implies that it is even not computably enumerable. We achieve this reduction by describing how a given recurrent tiling system $\langle \mathscr{D}, t^* \rangle$ with $\mathscr{D} = \langle \mathrm{D}, H, V \rangle$ can be "*embedded*" into a model of a particular sentence $\varphi^{dom} \triangleq \varphi^{grd} \wedge \varphi^{til} \wedge \varphi^{rec}$ over $\mathrm{AP} \triangleq \{p\} \cup \mathrm{D}$ and $\mathrm{Ag} \triangleq \{\alpha, \beta\}$, where $p \notin \mathrm{D}$, in such a way that φ^{dom} is satisfiable iff $\mathscr{D}$ allows an admissible tiling. For the sake of clarity, we split the reduction into three tasks where we explicit the sentences φ^{grd}, φ^{til}, and φ^{rec}.

Grid specification Consider the sentence $\varphi^{grd} \triangleq \bigwedge_{a \in \mathrm{Ag}} \varphi_a^{ord}$, where $\varphi_a^{ord} = \varphi_a^{unb} \wedge \varphi_a^{trn}$ are the *order sentences* and φ_a^{exs} and φ_a^{trn} are the *unboundedness* and *transitivity* strategy requirements for agents α and β defined, similarly to Definition 3.7, as follows:

(1) $\varphi_a^{unb} \triangleq [[z_1]]\langle\langle z_2\rangle\rangle\, z_1 <_a z_2$;
(2) $\varphi_a^{trn} \triangleq [[z_1]][[z_2]][[z_3]]\, (z_1 <_a z_2 \wedge z_2 <_a z_3) \to z_1 <_a z_3$;

where $x_1 <_\alpha x_2 \triangleq \langle\langle y\rangle\rangle\, (\beta, y)((\alpha, x_1)(\mathsf{X} p) \wedge (\alpha, x_2)(\mathsf{X} \neg p))$ and $y_1 <_\beta y_2 \triangleq \langle\langle x\rangle\rangle\, (\alpha, x)((\beta, y_1)(\mathsf{X} \neg p) \wedge (\beta, y_2)(\mathsf{X} p))$ are the two *partial order* formulas on strategies of α and β, respectively. Intuitively, $<_\alpha$ and $<_\beta$ correspond to the horizontal and vertical ordering of the positions in the grid, respectively.

It is easy to see that φ^{grd} is satisfiable, as it follows by the use of the same candidate model $\mathcal{G}^\star$ (see Fig. 3.5) and of a proof argument similar to that proposed in Lemma 3.1 for the simpler order sentence.

Lemma 3.7 (Grid Ordering Satisfiability) *The* SL *sentence* φ^{grd} *is satisfiable.*

Moreover, is is also immediate to see that φ^{grd} cannot have turn-based models, by using the same proof of Lemma 3.2.

Lemma 3.8 (Grid Ordering Turn-Based Unsatisfiability) *The* SL *sentence* φ^{grd} *is unsatisfiable over turn-based* CGSs.

Consider now a model $\mathcal{G} = \langle \mathrm{AP}, \mathrm{Ag}, \mathrm{Ac}, \mathrm{St}, \lambda, \tau, s_0 \rangle$ of φ^{grd} and, for all agents $a \in \mathrm{Ag}$, the relation $r_a^< \subseteq \mathrm{Str}(\mathcal{G}, s_0) \times \mathrm{Str}(\mathcal{G}, s_0)$ between s_0-total strategies defined as follows: $r_a^<(\mathsf{f}_1, \mathsf{f}_2)$ holds iff $\mathcal{G}, \chi, s_0 \models z_1 <_a z_2$, where $\chi(z_1) = \mathsf{f}_1$ and $\chi(z_2) = \mathsf{f}_2$, for all strategies $\mathsf{f}_1, \mathsf{f}_2 \in \mathrm{Str}(\mathcal{G}, s_0)$ and assignments $\chi \in \mathrm{Asg}(\mathcal{G}, \{z_1, z_2\}, s_0)$. By using a proof similar to that of Lemma 3.3, it is possible to see that $r_a^<$ is a *strict partial order without maximal element* on $\mathrm{Str}(\mathcal{G}, s_0)$. Now, to apply the desired reduction, we need to transform $r_a^<$ into a total order over strategies, by using the following two lemmas.

Lemma 3.9 (Strategy Equivalence) *Let* $r_a^\equiv \subseteq \mathrm{Str}(\mathcal{G}, s_0) \times \mathrm{Str}(\mathcal{G}, s_0)$, *with* $a \in \mathrm{Ag}$, *be the relation between strategies such that* $r_a^\equiv(\mathsf{f}_1, \mathsf{f}_2)$ *holds iff neither* $r_a^<(\mathsf{f}_1, \mathsf{f}_2)$ *nor* $r_a^<(\mathsf{f}_2, \mathsf{f}_1)$ *holds, for all* $\mathsf{f}_1, \mathsf{f}_2 \in \mathrm{Str}(\mathcal{G}, s_0)$. *Then* $r_a^\equiv$ *is an* equivalence relation.

Proof It is immediate to see that the relation $r_a^\equiv$ is reflexive, since $r_a^<$ is not reflexive, and symmetric, by definition. Moreover, due to the definition of the partial order formula $<_a$, it is also transitive and, thus, $r_a^\equiv$ is an *equivalence relation*. Indeed, if both $r_\alpha^\equiv(\mathsf{f}_1, \mathsf{f}_2)$ and $r_\alpha^\equiv(\mathsf{f}_2, \mathsf{f}_3)$ hold, we have that either $\mathcal{G}, \chi_1, \varepsilon \models (\beta, y)((\alpha, x_1)(\mathsf{X} p) \wedge (\alpha, x_2)(\mathsf{X} p))$ or $\mathcal{G}, \chi_1, \varepsilon \models (\beta, y)((\alpha, x_1)(\mathsf{X} \neg p) \wedge (\alpha, x_2)(\mathsf{X} \neg p))$ holds and either $\mathcal{G}, \chi_2, \varepsilon \models (\beta, y)((\alpha, x_2)(\mathsf{X} p) \wedge (\alpha, x_3)(\mathsf{X} p))$ or $\mathcal{G}, \chi_2, \varepsilon \models (\beta, y)((\alpha, x_2)(\mathsf{X} \neg p) \wedge (\alpha, x_3)(\mathsf{X} \neg p))$ holds, for all assignments $\chi_1 \in \mathrm{Asg}(\mathcal{G}, \{x_1, x_2\}, s_0)$ and $\chi_2 \in \mathrm{Asg}(\mathcal{G}, \{x_2, x_3\}, s_0)$ such that $\chi_1(\alpha, x_1) = \mathsf{f}_1$, $\chi_1(\alpha, x_2) = \chi_2(\alpha, x_2) = \mathsf{f}_2$, and $\chi_2(\alpha, x_3) = \mathsf{f}_3$. Hence, we have also that either $\mathcal{G}, \chi_3, \varepsilon \models (\beta, y)((\alpha, x_1)(\mathsf{X} p) \wedge (\alpha, x_3)(\mathsf{X} p))$ or $\mathcal{G}, \chi_3, \varepsilon \models (\beta, y)((\alpha, x_1)(\mathsf{X} \neg p) \wedge (\alpha, x_3)(\mathsf{X} \neg p))$ holds, for all assignments $\chi_3 \in \mathrm{Asg}(\mathcal{G}, \{x_1, x_3\}, s_0)$ such that $\chi_3(\alpha, x_1) = \mathsf{f}_1$ and $\chi_3(\alpha, x_3) = \mathsf{f}_3$, i.e., for all strategies of β assigned to y. Thus, $r_\alpha^\equiv(\mathsf{f}_1, \mathsf{f}_3)$ holds, too. The same reasoning applies to $r_\beta^\equiv$. □

Let $\mathrm{Str}_a^\equiv = (\mathrm{Str}(\mathcal{G}, s_0)/r_a^\equiv)$ be the quotient set of $\mathrm{Str}(\mathcal{G}, s_0)$ w.r.t. $r_a^\equiv$, for $a \in \mathrm{Ag}$, i.e., the set of the related equivalence classes over s_0-total strategies. Then, the following holds.

Lemma 3.10 (Strategy Total Order) *Let* $s_a^< \subseteq \mathrm{Str}_a^\equiv \times \mathrm{Str}_a^\equiv$, *with* $a \in \mathrm{Ag}$, *be the relation between classes of strategies such that* $s_a^<(\mathrm{F}_1, \mathrm{F}_2)$ *holds iff* $r_a^<(\mathsf{f}_1, \mathsf{f}_2)$ *holds, for all* $\mathsf{f}_1 \in \mathrm{F}_1$, $\mathsf{f}_2 \in \mathrm{F}_2$, *and* $\mathrm{F}_1, \mathrm{F}_2 \in \mathrm{Str}_a^\equiv$. *Then* $s_a^<$ *is a* strict total order with minimal element but no maximal element.

Proof The fact that $s_a^<$ is a *strict partial order without maximal element* derives directly from the same property of $r_a^<$. Indeed, due to the definition of the partial order formula $<_a$, if $r_a^\equiv(f', f'')$ and $r_a^<(f', f)$ (resp., $r_a^<(f, f')$) hold, we obtain that $r_a^<(f'', f)$ (resp., $r_a^<(f, f'')$) holds too. Hence, if there are $f_1 \in F_1$ and $f_2 \in F_2$ such that $r_a^<(f_1, f_2)$ holds, we directly obtain that $s_a^<(F_1, F_2)$ holds as well, for all $F_1, F_2 \in Str_a^\equiv$ and $a \in Ag$.

Moreover, $s_a^<$ is total, since $r_a^\equiv$ is an equivalence relation that cluster together all strategies of the agent a that are not in relation w.r.t. either $r_a^<$ or its inverse $(r_a^<)^{-1}$. Indeed, suppose by contradiction that there are two different classes $F_1, F_2 \in Str_a^\equiv$ such that neither $s_a^<(F_1, F_2)$ nor $s_a^<(F_2, F_1)$ holds. This means that, for all $f_1 \in F_1$ and $f_2 \in F_2$, neither $r_a^<(f_1, f_2)$ nor $r_a^<(f_2, f_1)$ holds, so $r_a^\equiv(f_1, f_2)$. However, this contradict the fact that F_1 and F_2 are different equivalences classes.

Finally, it is important to note that in $Str_a^\equiv$ there is also a minimal element w.r.t. $s_a^<$. Indeed, for a strategy $f \in Str(\mathcal{G}, s_0)$ for α (resp., for β) that forces the play to reach only nodes labeled with p (resp., $\neg p$), as successor of the root in $\mathcal{G}$, independently from the strategy of β (resp., α), the relation $r_\alpha^<(f', f)$ (resp., $r_\beta^<(f', f)$) does not hold, for any $f' \in Str(\mathcal{G}, s_0)$. $\qquad\square$

By a classical result on first order logic model theory (Ebbinghaus and Flum 1995), the relation $s_a^<$ cannot be defined on a finite set. Hence, $|Str_a^\equiv| = \omega$, for all $a \in Ag$. Now, let $s_a^\prec$ be the *successor* relation on $Str_a^\equiv$ compatible with the strict total order $s_a^<$, i.e., such that $s_a^\prec(F_1, F_2)$ holds iff *(i)* $s_a^<(F_1, F_2)$ holds and *(ii)* there is no $F_3 \in Str_a^\equiv$ for which both $s_a^<(F_1, F_3)$ and $s_a^<(F_3, F_2)$ hold, for all $F_1, F_2 \in Str_a^\equiv$. Then, we can write the two sets of classes $Str_\alpha^\equiv$ and $Str_\beta^\equiv$ as the infinite ordered lists $\{F_0^\alpha, F_1^\alpha, \ldots\}$ and $\{F_0^\beta, F_1^\beta, \ldots\}$, respectively, such that $s_a^\prec(F_i^a, F_{i+1}^a)$ holds, for all indexes $i \in \mathbb{N}$. Note that F_0^a is the class of minimal strategies w.r.t the relation $s_a^<$.

At this point, we have all the machinery to build an embedding of the plane $\mathbb{N} \times \mathbb{N}$ into the model $\mathcal{G}$ of φ^{grd}. In particular, we are able to construct a *bijective map* $\aleph : \mathbb{N} \times \mathbb{N} \to Str_\alpha^\equiv \times Str_\beta^\equiv$ such that $\aleph(i, j) = (F_i^\alpha, F_j^\beta)$, for all $i, j \in \mathbb{N}$.

Compatible tiling Given the grid structure built on the model $\mathcal{G}$ of φ^{grd} through the bijective map $\aleph$, we can express that a tiling of the grid is admissible by making use of the formula $z_1 \prec_a z_2 \triangleq z_1 <_a z_2 \land \neg\langle\!\langle z_3 \rangle\!\rangle z_1 <_a z_3 \land z_3 <_a z_2$ corresponding to the successor relation $s_a^\prec$, for all $a \in Ag$. Indeed, it is not hard to see that $\mathcal{G}, \chi, \varepsilon \models z_1 \prec_a z_2$ iff $\chi(z_1) \in F_i^a$ and $\chi(z_2) \in F_{i+1}^a$, for all assignments $\chi \in Asg(\mathcal{G}, \{z_1, z_2\}, s_0)$ and indexes $i \in \mathbb{N}$. The idea here is to associate to each domino type $t \in D$ a corresponding atomic proposition $t \in AP$ and to express the horizontal and vertical matching conditions via suitable object labeling. In particular, we can express, respectively, that the tiling is locally compatible, that the horizontal neighborhood of a tile satisfies the H requirement, and that also its vertical neighborhood satisfies the V requirement, all through the following three agent-closed formulas:

(1) $\varphi^{t,loc}(x, y) \triangleq (\alpha, x)(\beta, y)(X(t \land \bigwedge_{t' \in D}^{t' \neq t} \neg t'))$;

(2) $\varphi^{t,hor}(x, y) \triangleq \bigvee_{(t,t') \in H} [[x']] x \prec_\alpha x' \to (\alpha, x')(\beta, y)(Xt')$;

(3) $\varphi^{t,ver}(x, y) \triangleq \bigvee_{(t,t')\in V} [[y']]\, y \prec_\beta y' \rightarrow (\alpha, x)(\beta, y')(\mathsf{X} t')$.

Informally, we have the following: $\varphi^{t,loc}(x, y)$ asserts that t is the only domino type labeling the successors of the root of the model $\mathcal{G}$ that can be reached using the strategies related to the variables x and y; $\varphi^{t,hor}(x, y)$ asserts that the tile t' labeling the successors of the root reachable through the strategies x' and y is compatible with t w.r.t. the horizontal requirement H, for all strategies x' that immediately follow that related to x w.r.t. the order $r_\alpha^<$; $\varphi^{t,ver}(x, y)$ asserts that the tile t' labeling the successors of the root reachable through the strategies x and y' is compatible with t w.r.t. the vertical requirement V, for all strategies y' that immediately follow that related to y w.r.t. the order $r_\beta^<$.

Finally, to express that the whole grid has an admissible tiling, we use the sentence $\varphi^{til} \triangleq [[x]][[y]] \bigvee_{t\in \mathrm{D}} \varphi^{t,loc}(x, y) \wedge \varphi^{t,hor}(x, y) \wedge \varphi^{t,ver}(x, y)$ that asserts the existence of a domino type t satisfying the three conditions mentioned above, for every point individuated by the strategies x and y.

Recurrent tile As last task, we impose that the grid embedded into $\mathcal{G}$ has the distinguished domino type t^* occurring infinitely often in its first row. To do this, we first use two formulas that determine if a row or a column is the first one w.r.t. the orders $s_\alpha^<$ and $s_\beta^<$, respectively. Formally, we use $0_a(z) \triangleq \neg\langle\langle z'\rangle\rangle z' <_a z$, for $a \in \mathrm{Ag}$. One can easily prove that $\mathcal{G}, \chi, \varepsilon \models 0_\alpha(z)$ iff $\chi(z) \in \mathrm{F}_0^a$, for all assignments $\chi \in \mathrm{Asg}(\mathcal{G}, \{z\}, s_0)$. Now, the infinite occurrence requirement on t^* can be expressed with the following sentence: $\varphi^{rec} \triangleq [[x]][[y]](0_\beta(y) \wedge (0_\alpha(x) \vee (\alpha, x)(\beta, y)(\mathsf{X} t^*))) \rightarrow \langle\langle x'\rangle\rangle\, x <_\alpha x' \wedge (\alpha, x')(\beta, y)(\mathsf{X} t^*)$. Informally, φ^{rec} asserts that, when we are on the first row individuated by the variable y and at a column individuated by x such that it is the first column or the node of the "intersection" between x and y is labeled by t^*, we have that there exists a greater column individuated by x' such that its "intersection" with y is labeled by t^* as well.

Construction correctness At this point, we have all the tools to formally prove the correctness of the undecidability reduction, by showing the equivalence between finding the solution of the recurrent tiling problem and the satisfiability of the sentence φ^{dom}. In particular, one can note that in the reduction we propose, only the CHP-SL fragment of SL is involved. Thus, we prove that CHP-SL under the concurrent semantics has an highly undecidable satisfiability problem, while it remains an open question whether this problem is undecidable in the turned-based framework too, since the proof we propose cannot be applied to this case, as reported in Lemma 3.8.

Theorem 3.7 (SL **Satisfiability**) *The satisfiability problem for* SL *and* CHP-SL, *under the concurrent semantics, is highly undecidable. In particular, it is* Σ_1^1-HARD.

Proof Assume, for the direct reduction, that there exists a solution mapping $\partial : \mathbb{N} \times \mathbb{N} \rightarrow \mathrm{D}$ for the given recurrent domino system $\langle \mathcal{D}, t^*\rangle$ with $\mathcal{D} = \langle \mathrm{D}, H, V\rangle$. Then, we can build a CGS $\mathcal{G}_\partial^\star \triangleq \langle \mathrm{AP}, \mathrm{Ag}, \mathrm{Ac}, \mathrm{St}, \lambda, \tau, s_0\rangle$ similar to that used in Lemma 3.1 and satisfying the sentence φ^{dom} in the following way: *(i)* $\mathrm{Ac} \triangleq \mathbb{N}$; *(ii)* there are $2 \cdot |\mathrm{D}| + 1$ different states $\mathrm{St} \triangleq \{s_0\} \cup (\{p, \neg p\} \times \mathrm{D})$ such that $\lambda(s_0) \triangleq \emptyset$, $\lambda((p, t)) \triangleq \{p, t\}$, and $\lambda((\neg p, t)) \triangleq \{t\}$, for all $t \in \mathrm{D}$; *(iii)* each state $(z, t) \in \{p, \neg p\} \times \mathrm{D}$ has only

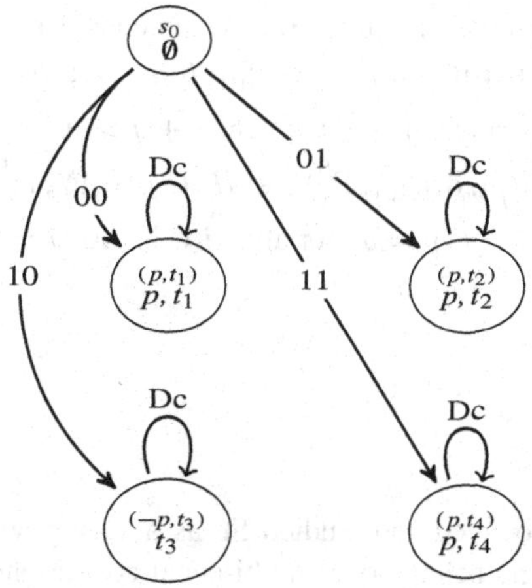

Fig. 3.6 Part of the CGS $\mathcal{G}^{\star}_{\partial}$ model of φ^{dom}, where $\partial(0,0) = t_1$, $\partial(0,1) = t_2$, $\partial(1,0) = t_3$, and $\partial(1,1) = t_4$

loops $\tau((z,t), \mathsf{d}) \triangleq (z,t)$ on itself and the initial state s_0 is connected to (z,t) through the decision d, in symbols $\tau(s_0, \mathsf{d}) \triangleq (z,t)$, iff *(iii.i)* $t = \partial(\mathsf{d}(\alpha), \mathsf{d}(\beta))$ and *(iii.ii)* $z = p$ iff $\mathsf{d}(\alpha) \leq \mathsf{d}(\beta)$, for all $\mathsf{d} \in \mathrm{Dc}$ (see Fig. 3.6). By a simple case analysis on the subformulas of φ^{dom}, it is possible to see that $\mathcal{G}^{\star}_{\partial} \models \varphi^{dom}$.

Conversely, let $\mathcal{G} = \langle \mathrm{AP}, \mathrm{Ag}, \mathrm{Ac}, \mathrm{St}, \lambda, \tau, s_0 \rangle$ be a model of the sentence φ^{dom}, and $\aleph : \mathbb{N} \times \mathbb{N} \rightarrow \mathrm{Str}^{\equiv}_{\alpha} \times \mathrm{Str}^{\equiv}_{\beta}$ be the related bijective map built for the grid specification task. As first thing, we have to prove the existence of a coloring function $\eth : \mathrm{Str}^{\equiv}_{\alpha} \times \mathrm{Str}^{\equiv}_{\beta} \rightarrow \mathrm{D}$ such that, for all pairs of classes of strategies $(\mathrm{F}^{\alpha}, \mathrm{F}^{\beta}) \in \mathrm{Str}^{\equiv}_{\alpha} \times \mathrm{Str}^{\equiv}_{\beta}$ and assignments $\chi \in \mathrm{Asg}(\mathcal{G}, \{\alpha, \beta\}, s_0)$ with $\chi(\alpha) \in \mathrm{F}^{\alpha}$ and $\chi(\beta) \in \mathrm{F}^{\beta}$, it holds that $\mathcal{G}, \chi, s_0 \models \mathsf{X}\eth(\mathrm{F}^{\alpha}, \mathrm{F}^{\beta})$. Then, it remains to note that the solution mapping $\partial = \eth \circ \aleph$ built as a composition of the bijective map $\aleph$ and the coloring function $\eth$ is an admissible tiling of the plane $\mathbb{N} \times \mathbb{N}$.

Due to the $\varphi^{t,loc}$ formula in φ^{til}, we have that, for all assignments $\chi \in \mathrm{Asg}(\mathcal{G}, \{\alpha, \beta\}, s_0)$, there exists just one domino type $t \in \mathrm{D}$ satisfying the property $\mathcal{G}, \chi, s_0 \models \mathsf{X}t$. Let $\widehat{\eth} : \mathrm{Str}(\mathcal{G}, s_0) \times \mathrm{Str}(\mathcal{G}, s_0) \rightarrow \mathrm{D}$ be the function that returns such a type, for all pair of strategies of α and β, i.e., such that $\mathcal{G}, \chi, s_0 \models \mathsf{X}\widehat{\eth}(\chi(\alpha), \chi(\beta))$, for all assignments $\chi \in \mathrm{Asg}(\mathcal{G}, \{\alpha, \beta\}, s_0)$. It is not hard to see that, due to the formulas $\varphi^{t,hor}$ and $\varphi^{t,ver}$ in φ^{til}, it holds *(i)* $(\widehat{\eth}(\mathsf{f}_{\alpha}, \mathsf{f}_{\beta}), \widehat{\eth}(\mathsf{f}'_{\alpha}, \mathsf{f}_{\beta})) \in H$ and *(ii)* $(\widehat{\eth}(\mathsf{f}_{\alpha}, \mathsf{f}_{\beta}), \widehat{\eth}(\mathsf{f}_{\alpha}, \mathsf{f}'_{\beta})) \in V$, for all $\mathsf{f}_{\alpha} \in \mathrm{F}^{\alpha}_{i}$, $\mathsf{f}'_{\alpha} \in \mathrm{F}^{\alpha}_{i+1}$, $\mathsf{f}_{\beta} \in \mathrm{F}^{\beta}_{j}$, $\mathsf{f}'_{\beta} \in \mathrm{F}^{\beta}_{j+1}$, and $i, j \in \mathbb{N}$. Moreover, since the guess of the tile type t' adjacent to t is uniform w.r.t. the choice of the successor strategy, we have that, for all $\mathsf{f}'_{\alpha}, \mathsf{f}''_{\alpha} \in \mathrm{F}^{\alpha}_{i}$ and $\mathsf{f}'_{\beta}, \mathsf{f}''_{\beta} \in \mathrm{F}^{\beta}_{j}$ with $i, j \in \mathbb{N}$ and $i + j > 0$, it holds that $\widehat{\eth}(\mathsf{f}'_{\alpha}, \mathsf{f}'_{\beta}) = \widehat{\eth}(\mathsf{f}''_{\alpha}, \mathsf{f}''_{\beta})$. This fact, is not necessarily true for strategies that both belong to the minimal classes F^{α}_{0} and F^{β}_{0}, since the sentence φ^{dom} does not contain a relative requirement. However, every domino type $\widehat{\eth}(\mathsf{f}_{\alpha}, \mathsf{f}_{\beta})$, with $\mathsf{f}_{\alpha} \in \mathrm{F}^{\alpha}_{0}$ and $\mathsf{f}_{\beta} \in \mathrm{F}^{\beta}_{0}$, can be used to label the origin of the plane $\mathbb{N} \times \mathbb{N}$

in order to obtain an admissible tiling. So, we can consider a function $\eth$, defined as follows: *(i)* $\eth(\mathsf{F}_0^\alpha, \mathsf{F}_0^\beta) \in \{\widehat{\eth}(\mathsf{f}_\alpha, \mathsf{f}_\beta) : \mathsf{f}_\alpha \in \mathsf{F}_0^\alpha \wedge \mathsf{f}_\beta \in \mathsf{F}_0^\beta\}$; *(ii)* $\eth(\mathsf{F}_i^\alpha, \mathsf{F}_j^\beta) = \widehat{\eth}(\mathsf{f}_\alpha, \mathsf{f}_\beta)$, for all $\mathsf{f}_\alpha \in \mathsf{F}_i^\alpha$, $\mathsf{f}_\beta \in \mathsf{F}_j^\beta$, and $i, j \in \mathbb{N}$ with and $i + j > 0$.

Clearly, *(i)* $(\eth(\mathsf{F}_i^\alpha, \mathsf{F}_j^\beta), \eth(\mathsf{F}_{i+1}^\alpha, \mathsf{F}_j^\beta)) \in H$, *(ii)* $(\eth(\mathsf{F}_i^\alpha, \mathsf{F}_j^\beta), \eth(\mathsf{F}_i^\alpha, \mathsf{F}_{j+1}^\beta)) \in V$, and *(iii)* $|\{i : \eth(\mathsf{F}_i^\alpha, \mathsf{F}_0^\beta) = t^*\}| = \omega$, for all $i, j \in \mathbb{N}$. So, $\partial = \eth \circ \aleph$ is an admissible tiling. □

3.9 Discussion

In this paper, we introduced and studied SL as a very powerful logic formalism to reason about strategic behaviors of multi-agent concurrent games. In particular, we proved that it subsumes numerous classical temporal and game logics not using explicit fix-points. As main results about SL, we shown that the relative model-checking problem under a reasonable semantics is decidable in 2EXPTIME, while its satisfiability problem is highly-undecidable.

From a theoretical point of view, we are convinced that our framework can be used as a unifying basis for logic reasonings about strategic behaviors in multi-agent scenarios and their relationships. In particular, it can be used to study fragments of SL in a way similar as it has been done in the literature for ATL* and other formalisms. For example, it could be interesting to investigate memoryful SL, by inheriting and extending the "memoryful" concept used for ATL* and CHP-SL and investigated in Mogavero et al. (2010b) and Fisman et al. (2010), respectively. As another example, it would be interesting to investigate the graded extension of SL, in a way similar as it has been done in Bianco et al. (2009, 2010, 2012) and Kupferman et al. (2002); Bonatti et al. (2008) for CTL and μCALCULUS, respectively. We recall that graded quantifiers in branching-time temporal logics allow to count how many equivalent classes of paths satisfy a given property. This concept in SL would further allow the counting of strategies and so to succinctly check the existence of more than one nonequivalent winning strategy for a given agent, in one shot.

Chapter 4
Relentful Strategic Reasoning

Abstract Temporal logics are a well investigated formalism for the specification, verification, and synthesis of reactive systems. Within this family, *Alternating-Time Temporal Logic* (ATL*, for short) has been introduced as a useful generalization of classical linear- and branching-time temporal logics, by allowing temporal operators to be indexed by coalitions of agents. Classically, temporal logics are memoryless: once a path in the computation tree is quantified at a given node, the computation that has led to that node is forgotten. Recently, mCTL* has been defined as a memoryful variant of CTL*, where path quantification is memoryful. In the context of multi-agent planning, memoryful quantification enables agents to "relent" and change their goals and strategies depending on their history. In this work, we define mATL*, a memoryful extension of ATL*, in which a formula is satisfied at a certain node of a path by taking into account both the future and the past. We study the expressive power of mATL*, its succinctness, as well as related decision problems. We also investigate the relationship between memoryful quantification and past modalities and show their equivalence. We show that both the memoryful and the past extensions come without any computational price; indeed, we prove that both the satisfiability and the model-checking problems are 2EXPTIME-COMPLETE, as they are for ATL*.

4.1 Introduction

Multi-agent concurrent systems recently emerged as a new paradigm for better understanding distributed systems (Fagin et al. 1995; Wooldridge 2001). In this kind of systems, different processes can have different goals and the interactions between them may be adversarial or cooperative. Thus the latter can be seen as games in the classical framework of game theory, with adversarial coalitions (Osborne and Rubinstein 1994). Classical branching-time temporal logics, such as CTL*(Emerson and Halpern 1986), turn out to be of very limited power when applied to multi-agent systems. For example, consider the property p: "processes 1 and 2

F. Mogavero, *Logics in Computer Science*, Atlantis Studies in Computing 3, DOI: 10.2991/978-94-91216-95-4_4,

cooperate to ensure that a system (having more than two processes) never enters a failure state". It is well known that CTL* cannot express p (Alur et al. 2002). Rather, CTL* can only say whether the set of all agents can or cannot prevent the system from failing.

In order to allow the temporal-logic framework to work within the setting of multi-agent concurrent systems, Alur, Henzinger, and Kupferman introduced *Alternating-Time Temporal Logic* (ATL*, for short) (Alur et al. 2002). This is a generalization of CTL* obtained by replacing the path quantifiers, "E" (*there exists*) and "A" (*for all*), with "*cooperation modalities*" of the form $\langle\langle A\rangle\rangle$ and $[[A]]$ where A is a set of *agents*. These modalities can be used to represent the power that a coalition of agents has to achieve certain results. In particular, they can express selective quantifications over those paths that are obtained as outcomes of the infinite game between the coalition and its complement. ATL* formulas are interpreted over *concurrent game structures* (CGS, for short) (Alur et al. 2002), closely related to *systems* in (Fagin et al. 1995), which model a set of interacting processes. Given a CGS $\mathscr{G}$ and a set A of agents, the ATL* formula $\langle\langle A\rangle\rangle\psi$ is satisfied at a state s of $\mathscr{G}$ iff there exists a *strategy* for the agents in A such that, no matter the strategy that is executed by agents not in A, the resulting outcome of the interaction in $\mathscr{G}$ satisfies ψ at s. Coming back to the previous example, one can see that the property p can be expressed by the ATL* formula $\langle\langle\{1,2\}\rangle\rangle \mathsf{G}\neg \mathit{fail}$, where G is the classic LTL temporal operator "*globally*".

Traditionally, temporal logics are *memoryless*: once a path in the underlying structure (usually a computation tree) is quantified at a given state, the computation that led to that state is forgotten (Kupferman and Vardi 2006). In the case of ATL*, we have even more: the logic is also "relentless", in the sense that the agents are not able to formulate their strategies depending on the history of the computation; when $\langle\langle A\rangle\rangle\psi$ is asserted in a state s, its truth is independent of the path that led to s. Inspired by a work on *strong cyclic planning* (Daniele et al. 2000), Pistore and Vardi proposed a logic that can express the spectrum between strong goal $\mathsf{A}\psi$ and the weak goal $\mathsf{E}\psi$ in planning (Pistore and Vardi 2007). A novel aspect of the Pistore-Vardi logic is that it is "*memoryful*", in the sense that the satisfiability of a formula at a state s depends on the future as well as on the past, i.e., the trace starting from the initial state and leading to s. Nevertheless, this logic does not have a standard temporal logical syntax (for example, it is not closed under conjunction and disjunction). Also, it is less expressive than CTL*. This has lead Kupferman and Vardi (2006) to introduce a memoryful variant of CTL* (mCTL*, for short), which unifies in a common framework both CTL* and the Pistore-Vardi logic. Syntactically, mCTL* is obtained from CTL* by simply adding a special proposition *present*, which is needed to emulate the ability of CTL* to talk about the "present" time. Semantically, mCTL* is obtained from CTL* by reinterpreting the path quantifiers of the logic to be memoryful.

Recently, ATL* has become a popular specification logic in the context of multi-agent system planning (Jamroga 2004; van der Hoek and Wooldridge 2002). In such a framework, a memoryful enhancement of ATL* enables "relentful" planning, that is, agents can relent and change their goals, depending on their history.[1] That is, when a specific goal at a certain state is checked, agents may learn from the past to change their goals. Note that this does not mean that agents change their strategy, but that they can choose a strategy that allows them to change their goals. For example, consider the ATL* formula $\langle\langle\emptyset\rangle\rangle \mathsf{G} \langle\langle A\rangle\rangle \psi$. In the memoryful framework, this formula is satisfied by a CGS $\mathcal{G}$ (at its starting node) iff for each possible trace (history) ρ the agents in A can ensure that the evolution of $\mathcal{G}$ that extends ρ satisfies ψ from the start state.

In this work, we introduce and study the logic mATL*, a memoryful extension of ATL*. Thus, mATL* can be thought of as a fusion of mCTL* and ATL* in a common framework. Similarly to mCTL*, the syntax of mATL* is obtained from ATL* by simply adding a special proposition *present*. Semantically, mATL* is obtained from ATL* by reinterpreting the path quantifiers of the logic to be memoryful. More specifically, for a CGS $\mathcal{G}$, the mATL* formula $\langle\langle A\rangle\rangle \psi$ holds at a state s of $\mathcal{G}$ if there is a strategy for agents in A such that, no matter which is the strategy of the agents not in A, the resulting outcome of the game, obtained by *extending* the execution trace of the system ending in s, satisfies ψ. As an example of the usefulness of the relentful reasoning, consider the situation in which the agents in a set A have the goal to eventually satisfy q and, if they see r, they can also change their goal to eventually satisfy v. It is easy to formalize this property in ATL* with the formula $\langle\langle A\rangle\rangle(\mathsf{F}(q \vee r) \wedge \mathsf{G} f)$, where f is $r \rightarrow \langle\langle A\rangle\rangle(\mathsf{F} v)$. Consider, instead, the situation in which the agents in A have the goal to satisfy p until q holds, unless they see r in which case they change their goal to satisfy u until v holds from the *start* of the computation. This cannot be easily handled in ATL*, since the specification depends on the past. On the other hand, it can be handled in mATL*, with the formula $\langle\langle A\rangle\rangle((p \mathsf{U}(q \vee r)) \wedge \mathsf{G} f)$, where f is $r \rightarrow \langle\langle A\rangle\rangle(u \mathsf{U} v)$.

We also consider an extension of mATL* with *past operators* (mpATL*, for short). As for classical temporal and modal logics, past operators allow reasoning about the past in a computation (Bonatti et al. 2008, 2006; Lichtenstein et al. 1985; Mogavero 2007; Vardi 1998). In mpATL*, we can further require that coalitions of agents had a memoryful goal in the past. In more details, we can write a formula whose satisfaction, at a state s, depends on the trace starting from the initial state and leading to a state s' occurring before s. Coming back to the previous example, by using P as the dual of F, we can change the alternative goal f of agents in A to be $r \rightarrow \mathsf{P}(h \wedge \langle\langle A\rangle\rangle(u\ \mathsf{U} v))$, which requires that once r occurs at a state s, at a previous state s' of s in which h holds, the subformula u until v from the start of the computation must be true.

As a direct consequence and important contribution of this work, we show for the first time a clear and complete picture of the relationships among ATL* and its various extensions with memoryful quantification and past modalities, which goes beyond the expressiveness results obtained in (Kupferman and Vardi 2006) for mCTL*. Since memoryfulness refers to behavior from the start of the computation, which occurred

[1] In Middle English to relent means to melt. In modern English it is used only in the combination of "relentless".

in the past, memoryfulness is intimately connected to the past. Indeed, we prove this formally. We study the expressive power and the succinctness of mATL* w.r.t ATL*, as well as the memoryless fragment of mpATL* (i.e., the extension of ATL* with past modalities), which we call pATL*. We show that the three logics have the same expressive power, but both mATL* and pATL* are at least exponentially more succinct than ATL*. As for m$^-$ATL* (where the minus stands for the variant of the logic without the "present" proposition, but the path interpretation is still memoryful), we prove that it is strictly less expressive than ATL*. On the other hand, we prove that pATL* is equivalent to p$^-$ATL*, but exponentially more succinct.

From an algorithmic point of view, we examine, for mpATL*, the two classical decision problems: *model checking* and *satisfiability*. We show that model checking is not easier than satisfiability and in particular that both are 2EXPTIME-COMPLETE, as for ATL*. We recall that this is not the case for mCTL*, where the model checking is EXPSPACE-COMPLETE, while satisfiability is 2EXPTIME-COMPLETE. For the upper bounds, we follow an *automata-theoretic approach* (Kupferman et al. 2000). In order to develop a decision procedure for a logic with the *tree-model property*, one first develops an appropriate notion of tree automata and studies their emptiness problem. Then, the decision problem for the logic can be reduced to the emptiness problem of such automata. To this aim, we introduce a new automaton model, the complex *symmetric alternating tree automata with satellites* (SATAS, for short), which extends both *automata over concurrent game structures* in Schewe and Finkbeiner (2006) and *alternating automata with satellites* in Kupferman and Vardi (2006), in a common setting. For technical convenience, the states of the whole automaton are partitioned into states regarding the satellite and those regarding the rest of the automaton, which we call the *main automaton*. The complexity results then come from the fact that mpATL* formulas can be translated into a SATAS with an exponential number of states for the main automaton and doubly exponential number of states for the satellite, and from the fact that the emptiness problem for this kind of automata is solvable in EXPTIME w.r.t. both the size of the main automaton and the logarithm of the size of the satellite.

Outline

In Sect. 4.2, we recall the basic notions regarding concurrent game structures and trees, tracks and plays, strategies, plays, and unwinding. Then, we have Sect. 4.3, in which we introduce mATL* and define its syntax and semantics, followed by Sect. 4.4, in which it is defined the extension mpATL* and there are studied the expressiveness and succinctness relationship of both the logics. In Sect. 4.5, we introduce the SATAS automaton model. Finally, in Sect. 4.6 we describe how to solve the satisfiability and model-checking problems for both mATL* and mpATL*. Note that, in the accompanying Appendix , we recall standard mathematical notation and some basic definitions that are used in the work.

4.2 Preliminaries

Decisions and Counterdecisions Given a set $A \subseteq \mathrm{Ag}$ of agents, a *decision* and a *counterdecision* for A are, respectively, two functions $\mathsf{d}_A \in \mathrm{Ac}^A$ and $\mathsf{d}^c_A \in \mathrm{Ac}^{\mathrm{Ag}\setminus A}$. By $\mathsf{d} \triangleq (\mathsf{d}_A, \mathsf{d}^c_A) \in \mathrm{Dc}$ we denote the *composition* of d_A and d^c_A, i.e., the total decision such that $\mathsf{d}_{\restriction A} = \mathsf{d}_A$ and $\mathsf{d}_{\restriction(\mathrm{Ag}\setminus A)} = \mathsf{d}^c_A$.

Strategies A *strategy* for $\mathcal{G}$ w.r.t. a set of agents $A \subseteq \mathrm{Ag}$ is a partial function $\mathsf{f}_A : \mathrm{Trk} \rightharpoonup \mathrm{Ac}^A$ that maps a non-empty trace ρ in its domain to a decision $\mathsf{f}_A(\rho)$ of agents in A. Intuitively, a strategy for agents in A is a *combined plan* that contains all choices of moves as a function of the history of the current outcome. For a state s, we say that f_A is *s-total* iff it is defined on all non-trivial tracks starting in s that are reachable through f_A itself, i.e., $\rho \cdot s' \in \mathsf{dom}(\mathsf{f}_A)$, with $\rho \in \mathsf{dom}(\mathsf{f}_A)$, iff $\mathsf{fst}(\rho) = s$ and there is a counterdecision $\mathsf{d}^c_A \in \mathrm{Ac}^{\mathrm{Ag}\setminus A}$ for A such that $\tau(\mathsf{lst}(\rho), (\mathsf{f}_A(\rho), \mathsf{d}^c_A)) = s'$. We use $\mathrm{Str}(A)$ (resp., $\mathrm{Str}(A, s)$ with $s \in \mathrm{St}$) to indicate the set of all the (resp., s-total) strategies of agents in A.

Plays A path π in $\mathcal{G}$ starting at a state s is a *play* w.r.t. an s-total strategy f_A (*f_A-play*, for short) iff, for all $i \in \mathbb{N}$, there is a counterdecision $\mathsf{d}^c_A \in \mathrm{Ac}^{\mathrm{Ag}\setminus A}$ such that $\pi_{i+1} = \tau(\pi_i, \mathsf{d})$, where $\mathsf{d} = (\mathsf{f}_A(\pi_{\leq i}), \mathsf{d}^c_A)$. Observe that π is an f_A-play iff $\pi_{\leq i} \in \mathsf{dom}(\mathsf{f}_A)$, for all $i \in \mathbb{N}$. Intuitively, a play is the outcome of the game determined by all the agents participating to it. By $\mathrm{Play}(\mathsf{f}_A)$ we denote the set of all f_A-plays.

Decision trees A *decision tree* (DT, for short) is a CGT $\mathcal{T}$ where *(i)* $\mathrm{St} = \mathrm{Dc}^*$ and *(ii)* if $t \cdot \mathsf{d} \in \mathrm{St}$ then $\tau(t, \mathsf{d}) = t \cdot \mathsf{d}$, for all $t \in \mathrm{St}$ and $\mathsf{d} \in \mathrm{Dc}$.

Unwinding Given a CGS $\mathcal{G}$, its *unwinding* is the DT $\mathcal{G}_U \triangleq \langle \mathrm{AP}, \mathrm{Ag}, \mathrm{Ac}, \mathrm{Dc}^*, \lambda', \tau', \varepsilon \rangle$ for which there is a surjective function $\mathsf{unw} : \mathrm{Dc}^* \rightarrow \mathrm{St}$, called *unwinding function*, such that *(i)* $\mathsf{unw}(\varepsilon) = s_0$, *(ii)* $\mathsf{unw}(\tau'(t, \mathsf{d})) = \tau(\mathsf{unw}(t), \mathsf{d})$, and *(iii)* $\lambda'(t) = \lambda(\mathsf{unw}(t))$, for all $t \in \mathrm{Dc}^*$ and $\mathsf{d} \in \mathrm{Dc}$.

From now on, we use the name of a CGSas a subscript to extract the components from its tuple-structure. Accordingly, if $\mathcal{G} = \langle \mathrm{AP}, \mathrm{Ag}, \mathrm{Ac}, \mathrm{St}, \lambda, \tau, s_0 \rangle$, we have $\mathrm{Ac}_{\mathcal{G}} = \mathrm{Ac}$, $\lambda_{\mathcal{G}} = \lambda$, $s_{0\mathcal{G}} = s_0$, and so on. Also, we use the same notational concept to make explicit to which CGSthe sets Dc, Trk, Pth, etc. are related to. Note that, we omit the subscripts if the structure can be unambiguously individuated from the context.

4.3 Memoryful Alternating-Time Temporal Logic

In this section, we introduce an extension of classic alternating-time temporal logic ATL* (Alur et al. 2002), obtained by allowing the use of memoryful quantification over paths, in a similar way it has been done for the memoryful branching-time temporal logic mCTL*(Kupferman and Vardi 2006).

4.3.1 Syntax

The *memoryful alternating-time temporal logic* (mATL*, for short) inherits from ATL* the existential $\langle\langle A\rangle\rangle$ and the universal $[[A]]$ *strategy quantifiers*, where A denotes a set of agents. We recall that these two quantifiers can be read as *"there exists a collective strategy for agents in A"* and *"for all collective strategies for agents in A"*, respectively. The syntax of mATL* is similar to that for ATL*: there are two types of formulas, *state* and *path formulas*. Strategy quantifiers can prefix an assertion composed of an arbitrary Boolean combination and nesting of the linear-time operators X *"next"*, U *"until"*, and R *"release"*. The only syntactical difference between the two logics is that mATL* formulas can refer to a special proposition present, which enables us to refer to the present time. Readers familiar with mCTL* can see mATL* as mCTL* where strategy quantifiers substitute path quantifiers. The formal syntax of mATL* follows.

Definition 4.1 *(*mATL* **Syntax***)* mATL*state *(φ) and* path *(ψ)* formulas *are built inductively from the sets of atomic propositions* AP *and agents* Ag *in the following way, where $p \in$ AP and $A \subseteq$ Ag:*

(1) $\varphi ::= \mathsf{present} \mid p \mid \neg\varphi \mid \varphi \wedge \varphi \mid \varphi \vee \varphi \mid \langle\langle A\rangle\rangle\psi \mid [[A]]\psi$;
(2) $\psi ::= \varphi \mid \neg\psi \mid \psi \wedge \psi \mid \psi \vee \psi \mid \mathsf{X}\psi \mid \psi\mathsf{U}\psi \mid \psi\mathsf{R}\psi$.

mATL* *is the set of all state formulas generated by the above grammar, in which the occurrences of the special proposition* present *is in the scope of a strategy quantifier.*

We now introduce some auxiliary syntactical notation.

For a formula φ, we define the *length* $\mathsf{lng}(\varphi)$ of φ as for ATL*. Formally, *(i)* $\mathsf{lng}(p) \triangleq 1$, for $p \in \mathrm{AP} \cup \{\mathsf{present}\}$, *(ii)* $\mathsf{lng}(\mathsf{Op}\ \psi) \triangleq 1 + \mathsf{lng}(\psi)$, for all $\mathsf{Op} \in \{\neg, \mathsf{X}\}$, *(iii)* $\mathsf{lng}(\psi_1 \mathsf{Op}\ \psi_2) \triangleq 1 + \mathsf{lng}(\psi_1) + \mathsf{lng}(\psi_2)$, for all $\mathsf{Op} \in \{\wedge, \vee, \mathsf{U}, \mathsf{R}\}$, and *(iv)* $\mathsf{lng}(\mathsf{Qn}\ \psi) \triangleq 1 + \mathsf{lng}(\psi)$, for all $\mathsf{Qn} \in \{\langle\langle A\rangle\rangle, [[A]]\}$.

We also use $\mathsf{cl}(\psi)$ to denote a variation of the classical Fischer-Ladner *closure* (Fischer and Ladner 1979) of ψ defined recursively as for ATL* in the following way: $\mathsf{cl}(\varphi) \triangleq \{\varphi\} \cup \mathsf{cl}'(\varphi)$, for all *basic formulas* $\varphi = \mathsf{Qn}\ \psi$, with $\mathsf{Qn} \in \{\langle\langle A\rangle\rangle, [[A]]\}$, and $\mathsf{cl}(\psi) \triangleq \mathsf{cl}'(\psi)$, in all other cases, where *(i)* $\mathsf{cl}'(p) \triangleq \emptyset$, for $p \in \mathrm{AP} \cup \{\mathsf{present}\}$, *(ii)* $\mathsf{cl}'(\mathsf{Op}\ \psi) \triangleq \mathsf{cl}(\psi)$, for all $\mathsf{Op} \in \{\neg, \mathsf{X}\}$, *(iii)* $\mathsf{cl}'(\psi_1 \mathsf{Op}\ \psi_2) \triangleq \mathsf{cl}(\psi_1) \cup \mathsf{cl}(\psi_2)$, for all $\mathsf{Op} \in \{\wedge, \vee, \mathsf{U}, \mathsf{R}\}$, and *(iv)* $\mathsf{cl}'(\mathsf{Qn}\ \psi) \triangleq \mathsf{cl}(\psi)$, for all $\mathsf{Qn} \in \{\langle\langle A\rangle\rangle, [[A]]\}$. Intuitively, $\mathsf{cl}(\varphi)$ is the set of all basic formulas that are subformulas of φ.

Finally, by $\mathsf{rcl}(\psi)$ we denote the *reduced closure* of ψ, i.e., the set of maximal basic formulas contained in ψ. Formally, (i) $\mathsf{rcl}(\varphi) \triangleq \{\varphi\}$, for all basic formulas $\varphi = \mathsf{Qn}\ \psi$, with $\mathsf{Qn} \in \{\langle\langle A\rangle\rangle, [[A]]\}$, *(ii)* $\mathsf{rcl}(\mathsf{Op}\ \psi) \triangleq \mathsf{rcl}(\psi)$ when $\mathsf{Op}\ \psi$ is a path formula, for all $\mathsf{Op} \in \{\neg, \mathsf{X}\}$, and *(iii)* $\mathsf{rcl}(\psi_1 \mathsf{Op}\ \psi_2) \triangleq \mathsf{rcl}(\psi_1) \cup \mathsf{rcl}(\psi_2)$ when $\psi_1 \mathsf{Op}\ \psi_2$ is a path formula, for all $\mathsf{Op} \in \{\wedge, \vee, \mathsf{U}, \mathsf{R}\}$. It is immediate to see that $\mathsf{rcl}(\psi) \subseteq \mathsf{cl}(\psi)$ and $|\mathsf{cl}(\psi)| = \mathsf{O}(\mathsf{lng}(\psi))$.

4.3.2 Semantics

As for ATL*, the semantics of mATL* is defined w.r.t. concurrent game structures. However, the two logics differ on interpreting state formulas. First, in mATL* the satisfaction of a state formula is related to a specific track, while in ATL* it is related only to a state. Moreover, a path quantification in mATL* ranges over paths that start at the initial state and contain as prefix the track that lead to the present state. We refer to this track as the *present track*. The whole concept is what we name *memoryful quantification*. On the contrary, in ATL*, path quantifications range over paths that start at the present state. For example, consider the formula $\varphi = [[A]]\mathsf{G}\langle\langle B\rangle\rangle\psi$. Considered as an ATL* formula, φ holds in the initial state of a structure if the agents in B can force a path satisfying ψ from every state that can be reached by a strategy of the agents in A. In contrast, considered as an mATL* formula, φ holds in the initial state of the structure if the agents in B can extend to a path satisfying ψ every track generated by a strategy of the agent in A. Thus, when evaluating path formulas in mATL* one cannot ignore the past, and satisfaction may depend on the events that preceded the point of quantification. In ATL*, state and path formulas are evaluated w.r.t. states and paths in the structure, respectively. In mATL*, instead, we add an additional parameter, the *present track*, which is the track that led from the initial state to the point of quantification. Path formulas are again evaluated w.r.t. paths, but state formulas are now evaluated w.r.t. tracks, which are viewed as partial executions.

We now formally define mATL* semantics w.r.t. a CGS $\mathcal{G}$. For two non-empty initial tracks $\rho, \rho_p \in \mathrm{Trk}(s_0)$, where ρ_p is the present track, we write $\mathcal{G}, \rho, \rho_p \models \varphi$ to indicate that the state formula φ holds at ρ, with ρ_p being the present. Similarly, for a path $\pi \in \mathrm{Pth}(s_0)$, a non-empty present track $\rho_p \in \mathrm{Trk}(s_0)$ and a natural number k, we write $\mathcal{G}, \pi, k, \rho_p \models \psi$ to indicate that the path formula ψ holds at the position k of π, with ρ_p being the present. The semantics of mATL* state formulas involving $\neg$, $\wedge$, and $\vee$, as well as that for mATL* path formulas, except for the state formula case, is defined as usual in ATL*. The semantics of the remaining part, which involves the memoryful feature, follows.

Definition 4.2 (mATL* **Semantics**) *Given a* CGS $\mathcal{G} = \langle \mathrm{AP}, \mathrm{Ag}, \mathrm{Ac}, \mathrm{St}, \lambda, \tau, s_0\rangle$, *two initial traces* $\rho, \rho_p \in \mathrm{Trc}(s_0)$, *a path* $\pi \in \mathrm{Pth}(s_0)$, *and a number* $k \in \mathbb{N}$, *it holds that:*

(1) $\mathcal{G}, \rho, \rho_p \models \mathsf{present}$ *if* $\rho = \rho_p$;
(2) $\mathcal{G}, \rho, \rho_p \models p$ *if* $p \in \lambda(\mathsf{lst}(\rho))$, *with* $p \in \mathrm{AP}$;
(3) $\mathcal{G}, \rho, \rho_p \models \langle\langle A\rangle\rangle\psi$ *if there exists a* $\mathsf{lst}(\rho)$*-total strategy* $\mathsf{f}_A \in \mathrm{Str}(A, \mathsf{lst}(\rho))$ *such that, for all plays* $\pi \in \mathrm{Play}(\mathsf{f}_A)$, *it holds that* $\mathcal{G}, \rho \cdot \pi_{\geq 1}, 0, \rho \models \psi$;
(4) $\mathcal{G}, \rho, \rho_p \models [[A]]\psi$ *if, for all* $\mathsf{lst}(\rho)$*-total strategies* $\mathsf{f}_A \in \mathrm{Str}(A, \mathsf{lst}(\rho))$, *there exists a play* $\pi \in \mathrm{Play}(\mathsf{f}_A)$ *such that* $\mathcal{G}, \rho \cdot \pi_{\geq 1}, 0, \rho \models \psi$;
(5) $\mathcal{G}, \pi, k, \rho_p \models \varphi$ *if* $\mathcal{G}, \pi_{\leq k}, \rho_p \models \varphi$.

Observe that the present track ρ_p is used in the above definition only at Item 1 and that formulas of the form $\langle\langle A\rangle\rangle\psi$ and $[[A]]\psi$ "reset the present", i.e., their satisfaction w.r.t. ρ and ρ_p is independent of ρ_p, and the present trace, for the path formula ψ, is set to ρ.

Let $\mathcal{G}$ be a CGS and φ be an mATL* formula. Then, $\mathcal{G}$ is a *model* for φ, in symbols $\mathcal{G} \models \varphi$, iff $\mathcal{G}, s_0, s_0 \models \varphi$, where we recall that s_0 is the initial state of $\mathcal{G}$. In this case, we also say that $\mathcal{G}$ is a model for φ on s_0. A formula φ is said *satisfiable* iff there exists a model for it. Moreover, it is an *invariant* for the two CGSs $\mathcal{G}_1$ and $\mathcal{G}_2$ iff either $\mathcal{G}_1 \models \varphi$ and $\mathcal{G}_2 \models \varphi$ or $\mathcal{G}_1 \not\models \varphi$ and $\mathcal{G}_2 \not\models \varphi$.

For all state formulas φ_1 and φ_2, we say that φ_1 *implies* φ_2, in symbols $\varphi_1 \Rightarrow \varphi_2$, iff, for all CGS $\mathcal{G}$ and non-empty traces $\rho, \rho_p \in \mathrm{Trc}(\mathcal{G}, s_0)$, it holds that if $\mathcal{G}, \rho, \rho_p \models \varphi_1$ then $\mathcal{G}, \rho, \rho_p \models \varphi_2$. Consequently, we say that φ_1 is *equivalent* to φ_2, in symbols $\varphi_1 \equiv \varphi_2$, iff both $\varphi_1 \Rightarrow \varphi_2$ and $\varphi_2 \Rightarrow \varphi_1$ hold.

W.l.o.g., in the rest of the paper, we mainly consider formulas in *existential normal form* (*enf*, for short), i.e., only existential strategy quantifiers occur. Indeed, all formulas can be linearly translated in *enf* by using De Morgan's laws together with the following equivalences, which directly follow from the semantics of the logic: $\neg \mathsf{X}\varphi \equiv \mathsf{X}\neg\varphi$, $\neg(\varphi_1 \mathsf{U} \varphi_2) \equiv (\neg\varphi_1)\mathsf{R}(\neg\varphi_2)$, and $\neg\langle\langle x\rangle\rangle\varphi \equiv [[x]]\neg\varphi$.

By induction on the syntactical structure of the sentences, it is easy to prove the following two classical results. Note that these are the basic steps towards the automata-theoretic approach we use to solve the model-checking and the satisfiability problems for mATL*.

Theorem 4.1 (mATL* **Unwinding Invariance**) mATL* *is invariant under unwinding, i.e., for each* CGS $\mathcal{G}$ *and formula* φ*, it holds that* φ *is an invariant for* $\mathcal{G}$ *and* $\mathcal{G}_U$.

Proof As first thing, let $\mathsf{unw}_{trk} : \mathrm{Trk}_{\mathcal{G}_U}(\varepsilon) \rightarrow \mathrm{Trk}_{\mathcal{G}}(s_{0\mathcal{G}})$ and $\mathsf{unw}_{pth} : \mathrm{Pth}_{\mathcal{G}_U}(\varepsilon) \rightarrow \mathrm{Pth}_{\mathcal{G}}(s_{0\mathcal{G}})$ be the two functions mapping tracks and paths of the unwinding $\mathcal{G}_U$ into the corresponding ones of the original model $\mathcal{G}$, which satisfy the following properties: *(i)* $\mathsf{unw}_{trk}(\varepsilon) = s_{0\mathcal{G}}$, *(ii)* $\mathsf{unw}_{trk}(\rho \cdot t) = \mathsf{unw}_{trk}(\rho) \cdot \mathsf{unw}(t)$, for all $\rho \cdot t \in \mathrm{Trk}_{\mathcal{G}_U}(\varepsilon)$ with $t \in \mathrm{St}_{\mathcal{G}_U}$, and *(iii)* $(\mathsf{unw}_{pth}(\pi))_{\leq i} = \mathsf{unw}_{trk}((\pi)_{\leq i})$, for all $\pi \in \mathrm{Pth}_{\mathcal{G}_U}(\varepsilon)$ and $i \in \mathbb{N}$. Note that $\varepsilon \in \mathrm{Trk}_{\mathcal{G}_U}(\varepsilon)$ is not the empty track, but the track of length 1 made by the root of the tree only. Moreover, consider the following orderings between tracks and paths of $\mathcal{G}_U$: *(i)* $\rho < \rho'$ iff there exists a track $\rho'' \in \mathrm{Trk}_{\mathcal{G}_U}$ such that $\rho' = \rho \cdot \rho''$, for all $\rho, \rho' \in \mathrm{Trk}_{\mathcal{G}_U}(\varepsilon)$; *(ii)* $\rho < \pi$ iff there exists a path $\pi' \in \mathrm{Pth}_{\mathcal{G}_U}$ such that $\pi = \rho \cdot \pi'$, for all $\rho \in \mathrm{Trk}_{\mathcal{G}_U}(\varepsilon)$ and $\pi \in \mathrm{Pth}_{\mathcal{G}_U}(\varepsilon)$. Observe that $<$ forms a partial order on tracks.

At this point, we prove the statement by showing that, for all state formulas φ and path formulas ψ, it holds that *(i)* $\mathcal{G}_U, \rho, \rho_p \models \varphi$ iff $\mathcal{G}, \mathsf{unw}_{trk}(\rho), \mathsf{unw}_{trk}(\rho_p) \models \varphi$, for all $\rho, \rho_p \in \mathrm{Trk}_{\mathcal{G}_U}(\varepsilon)$, such that either $\rho < \rho_p$ or $\rho = \rho_p$ or $\rho_p < \rho$, and *(ii)* $\mathcal{G}_U, \pi, k, \rho_p \models \psi$ iff $\mathcal{G}, \mathsf{unw}_{pth}(\pi), k, \mathsf{unw}_{trk}(\rho_p) \models \psi$, for all $\pi \in \mathrm{Pth}_{\mathcal{G}_U}(\varepsilon)$, $k \in \mathbb{N}$, and $\rho_p \in \mathrm{Trk}_{\mathcal{G}_U}(\varepsilon)$, such that $\rho_p < \pi$.

We now prove, by induction on the structure of formulas, the three cases of special proposition present, atomic proposition p, and existential quantifier $\langle\langle \mathrm{A}\rangle\rangle\psi$. The remaining cases are immediate or easily derivable by the former ones.

- ($\varphi = \mathsf{present}$)
By definition of semantics, we have that $\mathcal{G}_U, \rho, \rho_p \models \mathsf{present}$ iff $\rho = \rho_p$ and $\mathcal{G}, \mathsf{unw}_{trk}(\rho), \mathsf{unw}_{trk}(\rho_p) \models \mathsf{present}$ iff $\mathsf{unw}_{trk}(\rho) = \mathsf{unw}_{trk}(\rho_p)$. Now, by the hypothesis $\rho < \rho_p$ or $\rho = \rho_p$ or $\rho_p < \rho$ on the tracks ρ and ρ_p, we have that $\rho = \rho_p$ iff $\mathsf{unw}_{trk}(\rho) = \mathsf{unw}_{trk}(\rho_p)$. Therefore, $\mathcal{G}_U, \rho, \rho_p \models \mathsf{present}$ iff $\mathcal{G}, \mathsf{unw}_{trk}(\rho), \mathsf{unw}_{trk}(\rho_p) \models \mathsf{present}$.
- ($\varphi = p$)
By definition of unw_{trk}, we have that $\mathsf{lst}(\mathsf{unw}_{trk}(\rho)) = \mathsf{unw}(\mathsf{lst}(\rho))$. Thus, by definition of the unwinding function unw, it holds that $\lambda_{\mathcal{G}}(\mathsf{lst}(\mathsf{unw}_{trk}(\rho))) = \lambda_{\mathcal{G}_U}(\mathsf{lst}(\rho))$. At this point, we derive that $\mathcal{G}_U, \rho, \rho_p \models p$ iff $p \in \lambda_{\mathcal{G}_U}(\mathsf{lst}(\rho))$ iff $p \in \lambda_{\mathcal{G}}(\mathsf{lst}(\mathsf{unw}_{trk}(\rho)))$ iff $\mathcal{G}, \mathsf{unw}_{trk}(\rho), \mathsf{unw}_{trk}(\rho_p) \models p$. Therefore, $\mathcal{G}_U, \rho, \rho_p \models p$ iff $\mathcal{G}, \mathsf{unw}_{trk}(\rho), \mathsf{unw}_{trk}(\rho_p) \models p$.
- ($\varphi = \langle\langle A \rangle\rangle \psi, \Rightarrow$)
Suppose that $\mathcal{G}_U, \rho, \rho_p \models \langle\langle A \rangle\rangle \psi$ and let $s \triangleq \mathsf{lst}(\rho) \in \mathrm{St}_{\mathcal{G}_U}$ and $s' \triangleq \mathsf{unw}(s) = \mathsf{lst}(\mathsf{unw}_{trk}(\rho)) \in \mathrm{St}_{\mathcal{G}}$. Then, by definition of semantics, we have that there exists an s-total strategy $\mathsf{f}_A \in \mathrm{Str}_{\mathcal{G}_U}(A, s)$ such that, for all plays $\pi \in \mathrm{Play}_{\mathcal{G}_U}(\mathsf{f}_A)$, it holds that $\mathcal{G}_U, \rho \cdot \pi_{\geq 1}, 0, \rho \models \psi$. Moreover, by the inductive hypothesis, it holds that $\mathcal{G}, \mathsf{unw}_{pth}(\rho \cdot \pi_{\geq 1}), 0, \mathsf{unw}_{trk}(\rho) \models \psi$. Now,to prove the statement, we have only to show that there exists an s'-total strategy $\mathsf{f}'_A \in \mathrm{Str}_{\mathcal{G}}(A, s')$ such that, for all plays $\pi' \in \mathrm{Play}_{\mathcal{G}}(\mathsf{f}'_A)$, there exists a play $\pi \in \mathrm{Play}_{\mathcal{G}_U}(\mathsf{f}_A)$ such that $\mathsf{unw}_{trk}(\rho) \cdot \pi'_{\geq 1} = \mathsf{unw}_{pth}(\rho \cdot \pi_{\geq 1})$. To do this, we first define an auxiliary function $\mathsf{h} : \mathrm{Trk}_{\mathcal{G}}(s') \rightharpoonup \mathrm{Trk}_{\mathcal{G}_U}(s)$ mapping back tracks of $\mathcal{G}$ into corresponding tracks of $\mathcal{G}_U$. This function, can be inductively defined by means of the following recursive properties:

(1) $s' \in \mathsf{dom}(\mathsf{h})$ and $\mathsf{h}(s') \triangleq s$;
(2) for all $\rho' \in \mathsf{dom}(\mathsf{h})$ and counterdecision $\mathsf{d}^c_A \in \mathrm{Ac}^{\mathrm{Ag} \setminus A}$, it holds that $\rho' \cdot t' \in \mathsf{dom}(\mathsf{h})$ and $\mathsf{h}(\rho' \cdot t') \triangleq \mathsf{h}(\rho') \cdot t$, where $t' \triangleq \tau_{\mathcal{G}}(\mathsf{lst}(\rho'), \mathsf{d})$, $t \triangleq \tau_{\mathcal{G}_U}(\mathsf{lst}(\mathsf{h}(\rho')), \mathsf{d})$, and $\mathsf{d} \triangleq (\mathsf{f}_A(\mathsf{h}(\rho')), \mathsf{d}^c_A)$.

At this point, we can define the strategy $\mathsf{f}'_A \in \mathrm{Str}_{\mathcal{G}}(A, s')$ as follows: $\mathsf{f}'_A(\rho') \triangleq \mathsf{f}_A(\mathsf{h}(\rho'))$, for all $\rho' \in \mathsf{dom}(\mathsf{f}'_A) \triangleq \mathsf{dom}(\mathsf{h})$. Now, by a simple induction on the length of the play π', we can prove that f'_A actually satisfies the required property. Hence, we obtain that if $\mathcal{G}_U, \rho, \rho_p \models \langle\langle A \rangle\rangle \psi$ then $\mathcal{G}, \mathsf{unw}_{trk}(\rho), \mathsf{unw}_{trk}(\rho_p) \models \langle\langle A \rangle\rangle \psi$.
- ($\varphi = \langle\langle A \rangle\rangle \psi, \Leftarrow$)
Suppose that $\mathcal{G}, \mathsf{unw}_{trk}(\rho), \mathsf{unw}_{trk}(\rho_p) \models \langle\langle A \rangle\rangle \psi$ and let $s \triangleq \mathsf{lst}(\rho) \in \mathrm{St}_{\mathcal{G}_U}$ and $s' \triangleq \mathsf{unw}(s) = \mathsf{lst}(\mathsf{unw}_{trk}(\rho)) \in \mathrm{St}_{\mathcal{G}}$. Then, by definition of semantics, we have that there exists an s'-total strategy $\mathsf{f}'_A \in \mathrm{Str}_{\mathcal{G}}(A, s')$ such that, for all plays $\pi' \in \mathrm{Play}_{\mathcal{G}}(\mathsf{f}'_A)$, it holds that $\mathcal{G}, \mathsf{unw}_{trk}(\rho) \cdot \pi'_{\geq 1}, 0, \mathsf{unw}_{trk}(\rho) \models \psi$. Now, define the strategy $\mathsf{f}_A \in \mathrm{Str}_{\mathcal{G}_U}(A, s)$ as follows: $\mathsf{f}_A(\rho) \triangleq \mathsf{f}'_A(\mathsf{unw}_{trk}(\rho))$, for all $\rho \in \mathrm{Trk}_{\mathcal{G}_U}(s)$. At this point, it is easy to see that, for all plays $\pi \in \mathrm{Play}_{\mathcal{G}_U}(\mathsf{f}_A)$, it holds that $\mathsf{unw}_{pth}(\pi) \in \mathrm{Play}_{\mathcal{G}}(\mathsf{f}'_A)$. Therefore, $\mathcal{G}, \mathsf{unw}_{trk}(\rho) \cdot \mathsf{unw}_{pth}(\pi)_{\geq 1}, 0, \mathsf{unw}_{trk}(\rho) \models \psi$, i.e., $\mathcal{G}, \mathsf{unw}_{pth}(\rho \cdot \pi_{\geq 1}), 0, \mathsf{unw}_{trk}(\rho) \models \psi$. Now, by the inductive hypothesis, it holds that $\mathcal{G}_U, \rho \cdot \pi_{\geq 1}, 0, \rho \models \psi$. Hence, we obtain that if $\mathcal{G}, \mathsf{unw}_{trk}(\rho), \mathsf{unw}_{trk}(\rho_p) \models \langle\langle A \rangle\rangle \psi$ then $\mathcal{G}_U, \rho, \rho_p \models \langle\langle A \rangle\rangle \psi$. □

As an immediate corollary, we obtain that mATL* also enjoys the tree model property.

Corollary 4.1 (mATL* **Tree Model Property**) mATL* *enjoys the tree model property.*

Proof Consider a formula φ and suppose that it is satisfiable. Then, $\mathcal{G}, \mathsf{unw}_{trk}(\rho)$, $\mathsf{unw}_{trk}(\rho_p) \models \langle\langle A\rangle\rangle\psi$ there is a CGS $\mathcal{G}$ such that $\mathcal{G} \models \varphi$. By Theorem 4.1, φ is satisfied at the root of the unwinding $\mathcal{G}_U$ of $\mathcal{G}$. Thus, since $\mathcal{G}_U$ is a CGT, we immediately have that φ is satisfied on a tree model. □

4.4 Expressiveness and Succinctness

In this section, we compare mATL* with other logics derived from it. The basic comparisons are in terms of *expressiveness* and *succinctness*.

Let L_1 and L_2 be two logics whose semantics are defined on the same kind of structure. We say that L_1 is *as expressive* L_2 iff every formula in L_2 is logically equivalent to some formula in L_1. If L_1 is as expressive as L_2, but there is a formula in L_1 that is not logically equivalent to any formula in L_2, then L_1 is *more expressive* than L_2. If L_1 is as expressive as L_2 and vice versa, then L_1 and L_2 are *expressively equivalent*. Note that, in the case L_1 is more expressive than L_2, there are two sets of structures $\mathcal{M}_1$ and $\mathcal{M}_2$ and an L_1 formula φ such that, for all $\mathcal{M}_1 \in \mathcal{M}_1$ and $\mathcal{M}_2 \in \mathcal{M}_2$, it holds that $\mathcal{M}_1 \models \varphi$ and $\mathcal{M}_2 \not\models \varphi$ and, for all L_2 formulas φ', it holds that there are two models $M_1 \in \mathcal{M}_1$ and $\mathcal{M}_2 \in \mathcal{M}_2$ such that $\mathcal{M}_1 \models \varphi'$ iff $\mathcal{M}_2 \models \varphi'$. Intuitively, each L_2 formula is not able to distinguish between two models that instead are different w.r.t. L_1.

We define now the comparison of the two logics L_1 and L_2 in terms of succinctness, which measures the necessary blow-up when translating between them. Note that comparing logics in terms of succinctness makes sense also when the logics are not expressively equivalent, by focusing on their common fragment. In fact, a logic L_1 can be more expressive than a logic L_2, but at the same time, less succinct than the latter. Formally, we say that L_1 is (at least) *exponentially more succinct* than L_2 iff there exist two infinite lists of models $\{\mathcal{M}_1, \mathcal{M}_2, \ldots\}$ and of L_1 formulas $\{\varphi_1, \varphi_2, \ldots\}$, with $\mathcal{M}_i \models \varphi_i$ and $\mathsf{lng}(\varphi_i) = \mathrm{O}(\mathsf{p}_1(i))$, where $\mathsf{p}_1(n)$ is a polynomial, i.e., $\mathsf{lng}(\varphi_i)$ is polynomial in $i \in \mathbb{N}$, such that, for all L_2 formulas φ, if $\mathcal{M}_i \models \varphi$ then $\mathsf{lng}(\varphi) \geq 2^{\mathsf{p}_2(i)}$, where $\mathsf{p}_2(n)$ is another polynomial, i.e., $\mathsf{lng}(\varphi)$ is (at least) exponential in i.

We now discuss expressiveness and succinctness of mATL* w.r.t. ATL* as well as some extensions/restrictions of mATL*. In particular, we consider the logics mpATL* and pATL* to be, respectively, mATL* and ATL* augmented with the past-time operators "*previous*" and "*since*", which dualize the future-time operators "*next*" and "*until*" as in pLTL (Licht-enstein et al. 1985) and pCTL* (Kupferman and Pnueli 1995). Note that pATL* still contains the present proposition and that, as for pCTL*, the semantics of its quantifiers is as for ATL*, where the past is considered linear, i.e., deterministic. Moreover, we consider the logics m¯ATL*, p¯ATL*, and mp¯ATL* to be, respectively, the syntactical restriction of mATL*, pATL*, and mpATL* in which the use of the atomic proposition present is not allowed. On one hand, we have that all mentioned logics are expressively equivalent, except for m¯ATL* and p¯ATL*. On the other hand, the ability to refer to the past makes all of them at least exponentially more succinct than the corresponding ones without the past. For example, a pATL* formula φ can be translated into an equivalent ATL* one φ', but φ' may require a non-elementary space in $\mathsf{lng}(\varphi)$ (shortly, we say that pATL* is non-elementary reducible to ATL*). Note that, to get a better complexity for this translation is not an easy question. Indeed, it would improve the non-elementary reduction from *first order logic* to LTL, which is an outstanding open problem (Gabbay 1987). All the discussed results are reported in the following theorem.

Theorem 4.2 (Reductions) *The following properties hold:*

(1) ATL* *(resp., pATL*) is linearly reducible to* mATL* *(resp., mpATL*);*
(2) *mpATL* (resp., mp¯ATL*) is linearly reducible to pATL* (resp., p¯ATL*);*
(3) *mpATL* (resp., mp¯ATL*) is non-elementarily reducible to* mATL* *(resp., m¯ATL*);*
(4) *pATL* is non-elementarily reducible to* ATL**;*
(5) *m¯ATL* and p¯ATL* are at least exponentially more succinct than* ATL**;*
(6) *m¯ATL* is less expressive then* ATL*.

Proof Let φ be an input formula for items 1–4.

- Items 1 and 2 follow by replacing each subformula $\langle\langle A\rangle\rangle\psi$ in φ by $\langle\langle A\rangle\rangle \mathsf{F}(\mathsf{present} \wedge \psi)$ and $\langle\langle A\rangle\rangle \mathsf{P}((\tilde{\mathsf{Y}}\mathfrak{f}) \wedge \psi)$, respectively, where $\mathsf{P}\psi'$ is the corresponding past-time operator for $\mathsf{F}\psi'$ and $\tilde{\mathsf{Y}}\psi'$ is the weak previous time operator, which is true if either ψ' is true in the previous time-step or such a time-step does not exist. Note that all the formula substitutions start from the innermost subformula.
- Item 3 follows by replacing each subformula $\langle\langle A\rangle\rangle\psi$ in φ by $\langle\langle A\rangle\rangle\psi'$, where ψ' is obtained by the Separation Theorem (see Theorem 2.4 of Gabbay (1987)), which allows to eliminate all pure-past formulas.[2] Note that, as in the above items, the substitutions start from the innermost subformula. Moreover, the non-elementary blow-up is inherited from the use of the Separation Theorem.
- Item 4 proceeds as for the translation of pCTL* into CTL* (see Lemma 3.3 and Theorem 3.4 of Kupferman and Pnueli (1995)). The only difference here is that, when we apply the Separation Theorem to obtain a path formula as a disjunction of formulas of the form $ps \wedge pr \wedge ft$, where ps, pr, and ft are respectively pure-past, pure-present (i.e., Boolean combinations of atomic propositions and basic formulas), and pure-future formulas, we need to substitute the present proposition with $\mathfrak{f}$ in ps and ft and with $\mathfrak{t}$ in pr. As for the previous item, the origin of the non-elementary blow-up resides in the Separation Theorem.

[2] A pure-past formula contains only past-time operators. In Item 4, we also consider pure-future formulas, which contain only future-time operators, and pure-present formulas, which do not contain any temporal operator at all.

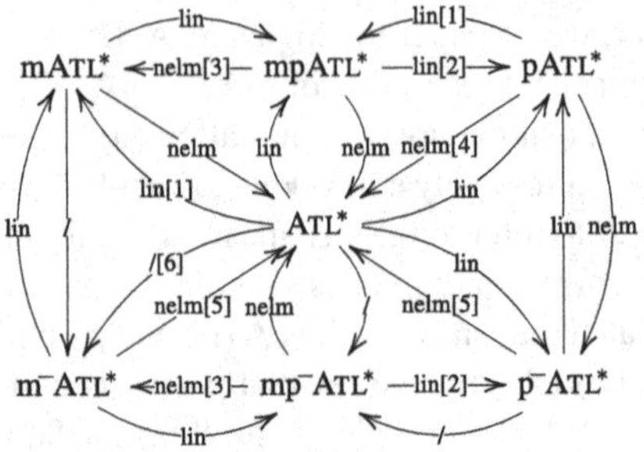

Fig. 4.1 Expressive power and succinctness hierarchy

- Item 5 follows by using the formula $\varphi \triangleq \langle\langle A\rangle\rangle \mathsf{G}(\bigwedge_{i=1}^{n}(p_i \leftrightarrow [[\emptyset]]p_i) \rightarrow (p_0 \leftrightarrow [[\emptyset]]p_0))$ (resp., $\varphi \triangleq \langle\langle A\rangle\rangle \mathsf{G}(\bigwedge_{i=1}^{n}(p_i \leftrightarrow \mathsf{P}((\tilde{\mathsf{Y}}\mathsf{f})\wedge p_i)) \rightarrow (p_0 \leftrightarrow \mathsf{P}((\tilde{\mathsf{Y}}\mathsf{f})\wedge p_0))))$, which is similar to that used to prove that pLTL is exponentially more succinct than LTL (see Theorem 3.1 of Laroussinie et al. (2002)). By using an argument similar to that used in Laroussinie et al. (2002), we obtain the desired result.
- Item 6 follows by using a proof similar to that used for m⁻CTL* (see Theorem 3.4 of Kupferman and Vardi (2006)), and so showing that the ATL formula $\varphi \triangleq \langle\langle A\rangle\rangle \mathsf{F}(([[\emptyset]]\mathsf{X}p) \wedge ([[\emptyset]]\mathsf{X}\neg p))$ has no m⁻ATL* equivalent formula. □

As an immediate consequence of combinations of the results shown into the previous theorem, it is easy to prove the following corollary.

Corollary 4.2 (Expressiveness) mATL*, *p*⁻ATL*, *p*ATL*, *and mp*ATL* *have the same expressive power of* ATL*. *m*⁻ATL* *and mp*⁻ATL* *have the same expressive power, but are less expressive than* ATL*. *Moreover, all of them are at least exponentially more succinct than* ATL*.

Figure 4.1 summarizes all the above results regarding expressiveness and succinctness. The acronym *"lin"* (resp., *"nelm"*) means that the translation exists and it is linear (resp., non-elementarily) in the size of the formula, and "/" means that such a translation is impossible. The numbers in brackets represent the item of Theorem 4.2 in which the translation is shown. We use no numbers when the translation is trivial or comes by a composition of existing ones.

4.5 Alternating Tree Automata

In this section, we briefly introduce an automaton model used to solve efficiently the satisfiability and model-checking problems for mpATL*, by reducing them, respectively, to the emptiness and membership problems of the automaton. We recall that, in general, such an approach is only possible once the logic satisfies the invariance under unwinding. In fact, this property holds for mpATL*, as it is stated in Theorem 4.1.

4.5.1 Classic Automata

Alternating tree automata (Muller and Schupp 1987) are a generalization of nondeterministic tree automata. Intuitively, while a nondeterministic automaton that visits a node of the input tree sends exactly one copy of itself to each of the successors of the node, an alternating automaton can send several copies of itself to the same successor. *Symmetric automata* (Janin and Walukiewicz 1995) are a variation of classical (asymmetric) alternating automata in which it is not necessary to specify the direction (i.e., the choice of the successors) of the tree on which a copy is sent. In fact, through two generalized directions (existential and universal moves), it is possible to send a copy of the automaton, starting from a node of the input tree, to one or all its successors. Hence, the automaton does not distinguish between directions. As a generalization of symmetric alternating automata, here we consider automata that can send copies to successor nodes, according to some entity choice. These automata are a slight variation of *automata over concurrent game structures* introduced in Schewe and Finkbeiner (2006).

We now give the formal definition of symmetric and asymmetric alternating tree automata.

Definition 4.3 (Symmetric Alternating Tree Automata) *A* symmetric alternating tree automaton *(*SATA*, for short) is a tuple* $\mathcal{A} \triangleq \langle \Sigma, \mathrm{E}, \mathrm{Q}, \delta, q_0, \mathrm{F} \rangle$*, where* Σ, E, *and* Q *are non-empty finite sets of* input symbols, entities, *and* states, *respectively,* $q_0 \in \mathrm{Q}$ *is an* initial state, F *is an* acceptance condition *to be defined later, and* $\delta : \mathrm{Q} \times \Sigma \rightarrow \mathrm{B}^+(\mathrm{D} \times \mathrm{Q})$ *is an* alternating transition function, *where* $\mathrm{D} = \{\Diamond, \Box\} \times 2^{\mathrm{E}}$ *is an extended set of* abstract directions, *which maps each pair of states and input symbols to a positive Boolean combination on the set of propositions, a.k.a.* abstract moves, *of the following form:* existential $((\Diamond, A), q)$ *and* universal $((\Box, A), q)$ *propositions, with* $A \subseteq \mathrm{E}$ *and* $q \in \mathrm{Q}$.

Definition 4.4 (Asymmetric Alternating Tree Automata) *An* asymmetric alternating tree automaton *(*AATA*, for short) is a tuple* $\mathcal{A} \triangleq \langle \Sigma, \Delta, \mathrm{Q}, \delta, q_0, \mathrm{F} \rangle$*, where* Σ, Q, q_0*, and* F *are defined as for the symmetric one,* Δ *is a non-empty finite set of* real directions, *and* $\delta : \mathrm{Q} \times \Sigma \rightarrow \mathrm{B}^+(\Delta \times \mathrm{Q})$ *is an* alternating transition function *that maps each pair of states and input symbols to a positive Boolean combination on the set of propositions of the form* $(d, q) \in \Delta \times \mathrm{Q}$*, a.k.a.* real moves.

A *nondeterministic tree automaton* (NTA, for short) is a special AATA in which each conjunction in the transition function δ has exactly one move (d, q) associated with each direction d. In addition, a *universal tree automaton* (UTA, for short) is a special AATA in which all the Boolean combinations that appear in δ are only conjunctions of moves.

In the following, we simply write ATA when we indifferently refer to its symmetric or asymmetric version.

The semantics of ATAs is now given through the following related concepts of run.

Definition 4.5 (SATA **Run**) *A* run *of an* SATA $\mathcal{A} = \langle \Sigma, \mathrm{E}, \mathrm{Q}, \delta, q_0, \mathrm{F} \rangle$ *on a* Σ*-labeled* B^{E}*-tree* $\mathcal{T} = \langle \mathrm{T}, \mathsf{v} \rangle$*, for a given set* B*, is a* $(\mathrm{Q} \times \mathrm{T})$*-labeled* $\mathbb{N}$*-tree* $\mathcal{R} \triangleq \langle \mathrm{R}, \mathsf{r} \rangle$ *such that* (i) $\mathsf{r}(\varepsilon) = (q_0, \varepsilon)$ *and* (ii) *for all nodes* $y \in \mathrm{R}$ *with* $\mathsf{r}(y) = (q, x)$*, there is a set of abstract moves* $\mathrm{S} \subseteq \Delta \times \mathrm{Q}$ *with* $\mathrm{S} \models \delta(q, \mathsf{v}(x))$ *such that, for all* $(z, q') \in \mathrm{S}$*, it holds that:*

- *if* $z = (\Diamond, A)$ *then there exists a choice* $d \in \mathrm{B}^{A}$ *such that, for all counterchoices* $d' \in \mathrm{B}^{\mathrm{E} \setminus A}$*, it holds that* $(q', x \cdot (d, d')) \in \mathsf{l}(y)$*;*
- *if* $z = (\Box, A)$ *then, for all choices* $d \in \mathrm{B}^{A}$*, there exists a counterchoice* $d' \in \mathrm{B}^{\mathrm{E} \setminus A}$ *such that* $(q', x \cdot (d, d')) \in \mathsf{l}(y)$*;*

where $(d, d') \in \mathrm{B}^{\mathrm{E}}$ *denotes composition of d and d', i.e., the function such that* $(d, d')_{\restriction A} = d$ *and* $(d, d')_{\restriction (\mathrm{E} \setminus A)} = d'$ *and* $\mathsf{l}(y) \triangleq \{\mathsf{r}(y \cdot j) : j \in \mathbb{N} \wedge y \cdot j \in \mathrm{R}\}$ *is the set of labels of successors of the node y in the run* $\mathcal{R}$*.*

Definition 4.6 (AATA **Run**) *A* run *of an* AATA $\mathcal{A} = \langle \Sigma, \Delta, \mathrm{Q}, \delta, q_0, \mathrm{F} \rangle$ *on a* Σ*-labeled* Δ*-tree* $\mathcal{T} = \langle \mathrm{T}, \mathsf{v} \rangle$ *is a* $(\mathrm{Q} \times \mathrm{T})$*-labeled* $\mathbb{N}$*-tree* $\mathcal{R} \triangleq \langle \mathrm{R}, \mathsf{r} \rangle$ *such that* (i) $\mathsf{r}(\varepsilon) = (q_0, \varepsilon)$ *and* (ii) *for all nodes* $y \in \mathrm{R}$ *with* $\mathsf{r}(y) = (q, x)$*, there is a set of real moves* $\mathrm{S} \subseteq \Delta \times \mathrm{Q}$ *with* $\mathrm{S} \models \delta(q, \mathsf{v}(x))$ *such that, for all* $(d, q') \in \mathrm{S}$*, there is an index* $j \in [0, |\mathrm{S}|[$ *for which it holds that* $y \cdot j \in \mathrm{R}$ *and* $\mathsf{r}(y \cdot j) = (q', x \cdot d)$*.*

In the following, we consider ATAs along with the *parity* $\mathrm{F} = (\mathrm{F}_1, \ldots, \mathrm{F}_k) \in (2^{\mathrm{Q}})^{+}$ with $\mathrm{F}_1 \subseteq \ldots \subseteq \mathrm{F}_k = \mathrm{Q}$ (APT, for short) acceptance condition (see Kupferman et al. 2000, for more). The number k of sets in F is called the *index* of the automaton. We also use ATAs with the *Co-Büchi* acceptance condition $\mathrm{F} \subseteq \mathrm{Q}$ (ACT, for short) that are APTs of index 2 in which the set of final states is represented by F_1.

Let $\mathcal{R} = \langle \mathrm{R}, \mathsf{r} \rangle$ be a run of an ATA $\mathcal{A}$ on a tree $\mathcal{T}$ and $\mathrm{R}' \subseteq \mathrm{R}$ one of its branches. Then, by $\mathsf{inf}(\mathrm{R}') \triangleq \{q \in \mathrm{Q} : |\{y \in \mathrm{R}' : \mathsf{r}(y) = q\}| = \omega\}$ we denote the set of states that occur infinitely often as labeling of the nodes in the branch R'. We say that a branch R' of $\mathcal{T}$ satisfies the parity acceptance condition $\mathrm{F} = (\mathrm{F}_1, \ldots, \mathrm{F}_k)$ iff the least index $i \in [1, k]$ for which $\mathsf{inf}(\mathrm{R}') \cap \mathrm{F}_i \neq \emptyset$ is even.

At this point, we can define the concept of language accepted by an ATA.

Definition 4.7 (ATA **Acceptance**) A SATA $\mathcal{A} = \langle \Sigma, \mathrm{E}, \mathrm{Q}, \delta, q_0, \mathrm{F} \rangle$ *(resp.,* AATA $\mathcal{A} = \langle \Sigma, \Delta, \mathrm{Q}, \delta, q_0, \mathrm{F} \rangle$*)* accepts *a* Σ*-labeled* B^{E}*-tree (resp.,* Δ*-tree)* $\mathcal{T}$ *iff is there exists a run* $\mathcal{R}$ *of* $\mathcal{A}$ *on* $\mathcal{T}$ *such that all its infinite branches satisfy the acceptance condition* F*, where the concept of satisfaction is dependent from of the definition of* F.

By $\mathrm{L}(\mathcal{A})$ we denote the language accepted by the ATA $\mathcal{A}$, i.e., the set of trees $\mathcal{T}$ accepted by $\mathcal{A}$. Moreover, $\mathcal{A}$ is said to be *empty* if $\mathrm{L}(\mathcal{A}) = \emptyset$. The *emptiness problem* for $\mathcal{A}$ is to decide whether $\mathrm{L}(\mathcal{A}) = \emptyset$ or not.

Now, we show how to reduce, for equivalence, a SATA to an AATA when it is known a priori the structure of the trees of interest.

Theorem 4.3 (SATA-AATA **Reduction**) *Let* $\mathscr{A} = \langle \Sigma, \mathrm{E}, \mathrm{Q}, \delta, q_0, \mathrm{F} \rangle$ *be a* SATA *and* B *be a finite set. Then there is an* AATA $\mathscr{A}' = \langle \Sigma, \mathrm{B}^{\mathrm{E}}, \mathrm{Q}, \delta', q_0, \mathrm{F} \rangle$ *such that every* Σ*-labeled* B^{E}*-tree is accepted by* $\mathscr{A}$ *iff it is accepted by* $\mathscr{A}'$.

Proof The transition function δ' of $\mathscr{A}'$ is obtained from that of $\mathscr{A}$ by substituting each existential $((\Diamond, A), q')$ and universal $((\Box, A), q')$ move with the formulas $\bigvee_{d \in \mathrm{B}^A} \bigwedge_{d' \in \mathrm{B}^{\mathrm{E} \setminus A}} ((d, d'), q')$ and $\bigwedge_{d \in \mathrm{B}^A} \bigvee_{d' \in \mathrm{B}^{\mathrm{E} \setminus A}} ((d, d'), q')$, respectively. At this point, it is immediate to see that the thesis follows directly by Definition 4.5 of SATA run. □

4.5.2 Automata with Satellite

As a generalization of ATA, here we also consider *alternating tree automata with satellites* (ATAS, for short), in a similar way it has been done in Kupferman and Vardi (2006). The satellite is used to take a bounded memory of the evaluated part of a path in a given structure and it is kept apart from the main automaton as it allows to show a tight complexity for the satisfiability problems. We use symmetric ATAS (SATAS, for short) for the solution of the satisfiability problem and asymmetric ATAS (AATAS, for short) for the model-checking problem.

We now formally define this new fundamental concept of automaton.

Definition 4.8 (Alternating Tree Automata with Satellite) *A* symmetric *(resp.,* asymmetric*)* alternating tree automaton with satellite *(*SATAS *(resp.,* AATAS*), for short) is a tuple* $\langle \mathscr{A}, \mathscr{S} \rangle$*, where* $\mathscr{A} \triangleq \langle \Sigma \times \mathrm{P}, \mathrm{E}, \mathrm{Q}, \delta, q_0, \mathrm{F} \rangle$ *(resp.,* $\mathscr{A} \triangleq \langle \Sigma \times \mathrm{P}, \Delta, \mathrm{Q}, \delta, q_0, \mathrm{F} \rangle$*) is an* SATA *(resp.,* AATA*) and* $\mathscr{S} \triangleq \langle \Sigma, \mathrm{P}, \zeta, p_0 \rangle$ *is a* deterministic safety word automaton, *a.k.a.* satellite, *where* P *is a non-empty finite set of* states, $p_0 \in \mathrm{P}$ *is an* initial states, *and* $\zeta : \mathrm{P} \times \Sigma \to \mathrm{P}$ *is a* deterministic transition function *that maps a state and an input symbol to a state. The sets* Σ *and* E *(resp.,* Δ*) are, respectively, the* alphabet *and the* entity set *(resp.,* direction sets*) of the* ATAS $\langle \mathscr{A}, \mathscr{S} \rangle$.

At this point, we can define the language accepted by an ATAS.

Definition 4.9 (ATAS **Acceptance**) *A* Σ*-labeled* B^{E}*-tree (resp.,* Δ*-tree)* $\mathscr{T}$ *is accepted by a* SATAS *(resp.,* AATAS*)* $\langle \mathscr{A}, \mathscr{S} \rangle$*, where* $\mathscr{A} \triangleq \langle \Sigma \times \mathrm{P}, \mathrm{E}, \mathrm{Q}, \delta, q_0, \mathrm{F} \rangle$ *(resp.,* $\mathscr{A} = \langle \Sigma \times \mathrm{P}, \Delta, \mathrm{Q}, \delta, q_0, \mathrm{F} \rangle$*) and* $\mathscr{S} = \langle \Sigma, \mathrm{P}, \zeta, p_0 \rangle$*, iff it is accepted by the product-automaton* $\mathscr{A}^\star \triangleq \langle \Sigma, \mathrm{E}, \mathrm{Q} \times \mathrm{P}, \delta^\star, (q_0, p_0), \mathrm{F}^\star \rangle$ *(resp.,* $\mathscr{A}^\star \triangleq \langle \Sigma, \Delta, \mathrm{Q} \times \mathrm{P}, \delta^\star, (q_0, p_0), \mathrm{F}^\star \rangle$*) with* $\delta^\star((q, p), \sigma) \triangleq \delta(q, (\sigma, p))[q' \in \mathrm{Q}/(q', \zeta(p, \sigma))]$*, where by* $f[x \in \mathrm{X}/y]$ *we denote the formula in which all occurrences of* x *in* f *are replaced by* y*, and* $\mathrm{F}^\star$ *is the acceptance condition directly derived from* F.

In words, $\delta^\star((q, p), \sigma)$ is obtained by substituting in $\delta(q, (\sigma, p))$ each occurrence of a state q' with a tuple of the form (q', p'), where $p' = \zeta(p, \sigma)$ is the new state of the satellite. By $\mathrm{L}(\langle \mathscr{A}, \mathscr{S} \rangle)$ we denote the language accepted by the ATAS $\langle \mathscr{A}, \mathscr{S} \rangle$.

In the following, we consider, in particular, ATAS along with the parity acceptance condition (APTS, for short), where $F^\star \triangleq (F_1 \times P, \dots, F_k \times P)$.

Note that satellites are just a convenient way to describe an ATA in which the state space can be partitioned into two components, one of which is deterministic, independent from the other, and that has no influence on the acceptance. Indeed, it is just a matter of technicality to see that automata with satellites inherit all the closure properties of alternating automata. In particular, we prove how to translate an AAPTS into an equivalent NPT with only an exponential blow-up in the number of states.

Theorem 4.4 (AAPTS **Nondeterminization**) *Let* $\langle \mathcal{A}, \mathcal{S} \rangle$ *be an* AAPTS, *where the main automaton* $\mathcal{A}$ *has* n *states and index* k *and the satellite* $\mathcal{S}$ *has* m *states. Then there is an* NPT $\mathcal{N}^\star$ *with* $2^{O((n \cdot k) \cdot \log(n \cdot k) + \log(m))}$ *states and index* $O(n \cdot k)$, *such that* $L(\mathcal{N}^\star) = L(\langle \mathcal{A}, \mathcal{S} \rangle)$.

Proof To deduce the thesis, we use the Muller-Schupp exponential-time nondeterminization procedure (Muller and Schupp 1995) that leads from the AAPT $\mathcal{A}$ to an NPT $\mathcal{N}$, with $2^{O((n \cdot k) \cdot \log(n \cdot k))}$ states and index $O(n \cdot k)$, such that $L(\mathcal{N}) = L(\mathcal{A})$. Since an NPT is a particular AAPT, we immediately have that $L(\langle \mathcal{N}, \mathcal{S} \rangle) = L(\langle \mathcal{A}, \mathcal{S} \rangle)$. At this point, by taking the product-automaton between $\mathcal{N}$ and the satellite $\mathcal{S}$, as described in Definition 4.9 of ATAS acceptance, we obtain a new NPT $\mathcal{N}^\star$, with $2^{O((n \cdot k) \cdot \log(n \cdot k) + \log(m))}$ states and index $O(n \cdot k)$, such that $L(\mathcal{N}^\star) = L(\langle \mathcal{N}, \mathcal{S} \rangle)$. Hence, it is evident that $L(\mathcal{N}^\star) = L(\langle \mathcal{A}, \mathcal{S} \rangle)$. □

The following theorem, directly derived by a proof idea of Kupferman and Vardi (2006), shows how the separation between $\mathcal{A}$ and $\mathcal{S}$ gives a tight analysis of the complexity of the relative emptiness problem.

Theorem 4.5 (APTS **Emptiness**) *The* emptiness problem *for an* APTS $\langle \mathcal{A}, \mathcal{S} \rangle$ *with alphabet size* h, *where the main automaton* $\mathcal{A}$ *has* n *states and index* k *and the satellite* $\mathcal{S}$ *has* m *states, can be decided in time* $2^{O(\log(h) + (n \cdot k) \cdot ((n \cdot k) \cdot \log(n \cdot k) + \log(m)))}$.

Proof The proof proceeds in two steps, the first of which is used only if $\mathcal{A}$ is a SATA, in order to translate it into an AATA. First, in order to obtain a linear translation from SATAS to AATAS, we use a bounded model theorem (see Theorem 2 of Schewe and Finkbeiner (2006)), which asserts that a SATA $\mathcal{A}$ accepts a tree iff it accepts a $|Z \times E|^{|E|}$-bounded tree, where Z is the set of abstract moves used in its transition function. Hence, by Theorem 4.3, there is an AATA $\mathcal{A}'$, with the same set of states and acceptance condition of the original automaton $\mathcal{A}$ and a set $Z \times E^E$ of directions, such that $L(\mathcal{A}') = \emptyset$ iff $L(\mathcal{A}) = \emptyset$. Hence, by definition of ATAS, we obtain that $L(\langle \mathcal{A}', \mathcal{S} \rangle) = \emptyset$ iff $L(\langle \mathcal{A}, \mathcal{S} \rangle) = \emptyset$. At this point, by Theorem 4.4, we obtain an NPT $\mathcal{N}^\star$, with $2^{O((n \cdot k) \cdot \log(n \cdot k) + \log(m))}$ states and index $O(n \cdot k)$, such that $L(\mathcal{N}^\star) = L(\langle \mathcal{A}', \mathcal{S} \rangle)$. Now, the emptiness of $\mathcal{N}^\star$ can be checked in polynomial running-time in its number of states, exponential in its index, and linear in the alphabet size (see Theorem 5.1 of Kupferman and Vardi (1998)). Overall, with this procedure, we obtain that the emptiness problem for an APTS is solvable in time $2^{O(\log(h) + (n \cdot k) \cdot ((n \cdot k) \cdot \log(n \cdot k) + \log(m)))}$. □

Finally, we show how much costs to verify if a given tree language, represented by a safety NPT, is recognized by an APTS.

Theorem 4.6 (APTS-NTA **Intersection Emptiness**) *The* emptiness problem *for the intersection of an* APTS $\langle \mathcal{A}, \mathcal{S} \rangle$ *with alphabet size h, where the main automaton* $\mathcal{A}$ *has n states and index k and the satellite* $\mathcal{S}$ *has m states, and a safety* NTA $\mathcal{N}$ *with n′ states, both running over* B^{E}*-trees, can be decided in time* $n'^{\mathrm{O}(n\cdot k)} \cdot 2^{\mathrm{O}(\log(h)+(n\cdot k)\cdot((n\cdot k)\cdot\log(n\cdot k)+\log(m)))}$.

Proof As for Theorem 4.5, the proof proceeds in two steps. First, by Theorem 4.3, there is an AATA $\mathcal{A}'$, with the same set of states and acceptance condition of $\mathcal{A}$ and a set B^{E} of directions, such that $\mathrm{L}(\mathcal{A}') = \mathrm{L}(\mathcal{A})$ and so, $\mathrm{L}(\langle \mathcal{A}', \mathcal{S} \rangle) = \mathrm{L}(\langle \mathcal{A}, \mathcal{S} \rangle)$. Now, by Theorem 4.4, we obtain an NPT $\mathcal{N}^{\star}$, with $2^{\mathrm{O}((n\cdot k)\cdot\log(n\cdot k)+\log(m))}$ states and index $\mathrm{O}(n \cdot k)$, such that $\mathrm{L}(\mathcal{N}^{\star}) = \mathrm{L}(\langle \mathcal{A}', \mathcal{S} \rangle)$. Intersecting $\mathcal{N}^{\star}$ with $\mathcal{N}$, we obtain a new NPT $\mathcal{N}'$ such that $\mathrm{L}(\mathcal{N}') = \mathrm{L}(\langle \mathcal{A}, \mathcal{S} \rangle) \cap \mathrm{L}(\mathcal{N})$, with $n' \cdot 2^{\mathrm{O}((n\cdot k)\cdot\log(n\cdot k)+\log(m))}$ states and same index of $\mathcal{N}^{\star}$. Finally, we check the emptiness of $\mathcal{N}'$ in time $n'^{\mathrm{O}(n\cdot k)} \cdot 2^{\mathrm{O}(\log(h)+(n\cdot k)\cdot((n\cdot k)\cdot\log(n\cdot k)+\log(m)))}$. □

4.6 Decision Procedures

In this section, we directly study the satisfiability and model-checking for the richer mpATL*, since we prove a tight 2EXPTIME upper bound for both the problems.

4.6.1 From Path Formulas to Satellite

As mentioned before, an mATL* path formula is satisfied at a certain node of a path by taking into account both the future and the past. Although the past is unlimited, it only requires a finite representation. This is due to the fact that LTL with past operators (pLTL, for short) (Gabbay 1987; Lichtenstein et al. 1985) can be translated into automata on infinite words of bounded size (Vardi 1988), and that it represents the temporal path core of mpATL* (as LTL is the corresponding one for ATL*). Here, we show how to build the satellite that represents the memory on the past in order to solve satisfiability and model-checking for mpATL*.

To this aim, we first introduce the following notation, where φ is an *enf* state formula: $\mathrm{AP}_{\varphi} = \mathrm{AP} \cup \mathsf{cl}(\varphi)$, $\mathrm{AP}^{r}_{\varphi} = \mathrm{AP} \cup \mathsf{rcl}(\varphi)$, and $\mathrm{AP}^{prs}_{\varphi} = \mathrm{AP}^{r}_{\varphi} \cup \{\mathsf{present}\}$. Intuitively, we are enriching the set of atomic propositions AP, to be used as input symbols of the automata, with the basic formulas of φ and the special proposition present.

Before showing the full satellite construction, we first describe how to build it from a single basic formula $b = \langle\langle A_b \rangle\rangle \psi_b$. Let $\widehat{\psi_b}$ be the pLTL formula obtained by replacing in ψ_b all the occurrences of a direct basic subformula $b' \in \mathsf{rcl}(b)$ by the label

b' read as atomic proposition. By using a slight variation of the procedure developed in Vardi (1988), we can translate $\widehat{\psi}_b$ into a universal co-Büchi word automaton $\mathcal{U}_b = \langle \mathrm{AP}_b^{prs}, \mathrm{Q}_b, \delta_b, \mathrm{Q}_{0b}, \mathrm{F}_b \rangle$, with a number of states at most exponential in $\mathsf{lng}(\psi_b)$, i.e., $|\mathrm{Q}_b| = 2^{\mathrm{O}(\mathsf{lng}(\psi_b))}$, that accepts all and only the infinite traces on AP_b^{prs} that are models of $\widehat{\psi}_b$. By applying the classical subset construction to $\mathcal{U}_b$ (Rabin and Scott 1959), we obtain the satellite $\mathcal{D}_b = \langle \mathrm{AP}_b^r, 2^{\mathrm{Q}_b}, \zeta_b, \mathrm{Q}_{0b} \rangle$, where $\zeta_b(p, \sigma) \triangleq \bigcup_{q \in p} \delta_b(q, \sigma)$, for all states $p \subseteq \mathrm{Q}_b$ and labels $\sigma \subseteq \mathrm{AP}_b^r$.

To better understand the usefulness of the satellite $\mathcal{D}_b$, consider $\mathcal{U}_b$ after that a prefix $\rho = \varpi_{\leq i}$ of an infinite trace $\varpi \in (\mathrm{AP}_b^r)^\omega$ is read. Since $\mathcal{U}_b$ is universal, there exists a number of active states that are ready to continue with the evaluation of the remaining part $\varpi_{>i}$ of the trace ϖ. Consider now the satellite $\mathcal{D}_b$ after that the same prefix ρ is read. Since $\mathcal{D}_b$ is deterministic, there is only one active state that, by construction, is exactly the set of all the active states of $\mathcal{U}_b$. It is clear then that, using $\mathcal{D}_b$, we are able to maintain all possible computations of $\mathcal{U}_b$.

We now define the product-satellite that maintains, at the same time, a memory for all path formulas ψ_b contained in a basic subformula $b \in \mathsf{cl}(\varphi)$ of the mpATL* formula φ we want to check.

Definition 4.10 (**Memory Satellite**) *The* memory satellite *for a state formula* φ *is the satellite* $\mathcal{S}_\varphi \triangleq \langle \mathrm{AP}_\varphi, \mathrm{P}_\varphi, \zeta_\varphi, \mathsf{p}_{0\varphi} \rangle$, *where* (i) $\mathrm{P}_\varphi \triangleq \{\mathsf{p} \in (\bigcup_{b \in \mathsf{cl}(\varphi)} 2^{\mathrm{Q}_b})^{\mathsf{cl}(\varphi)} : \forall b \in \mathsf{cl}(\varphi) . \mathsf{p}(b) \subseteq \mathrm{Q}_b\}$, (ii) $\mathsf{p}_{0\varphi}(b) \triangleq \mathrm{Q}_{0b}$, *and* (iii) $\zeta_\varphi(\mathsf{p}, \sigma)(b) \triangleq \bigcup_{q \in \mathsf{p}(b)} \delta_b(q, \sigma \cap \mathrm{AP}_b^r)$, *for all* $\mathsf{p} \in \mathrm{P}_\varphi$, $\sigma \subseteq \mathrm{AP}_\varphi$, *and* $b \in \mathsf{cl}(\varphi)$.

Intuitively, this satellite record the temporal evolution of the formula φ from the root of the tree model by means of its states, which are represented by functions mapping each basic subformula $b \in \mathsf{cl}(\varphi)$ to a set of active states of the related word automaton $\mathcal{U}_b$. Note that the size of the satellite $\mathcal{S}_\varphi$ is doubly-exponential in $\mathsf{lng}(\varphi)$, i.e., its number of states is $2^{2^{\mathrm{O}(\mathsf{lng}(\varphi))}}$.

4.6.2 *Satisfiability*

The satisfiability procedure we now propose technically extends that used for ATL* in Schewe (2008) along with that for mCTL* in Kupferman and Vardi (2006). Such an extension is possible due to the fact that the memoryful quantification has no direct interaction with the strategic features of the logic. In particular as for ATL*, it is possible to show that every CGS model of an mpATL* formula φ can be transformed into an *explicit* CGT model of φ. Such a model includes a certificate for both the truth of each of its basic subformula $b \in \mathsf{cl}(\varphi)$ in the respective node of the tree and the strategy used by the agents A_b to achieve the goal described by the corresponding path formula ψ_b (for a formal definition see Schewe 2008). The main difference of our definition of explicit models w.r.t. that given in Schewe (2008) is in the fact that the *witness* of a basic formula b does not start in the node from which the path formula ψ_b needs to be satisfied, but from the node in which the quantification is applied, i.e., the present node. This difference, which directly derives from the memoryful feature of mpATL*, is due to the request that ψ_b needs to be satisfied on a path that starts at the root of the model. The proof of an explicit model existence is exploited by constructing an SATAS that accepts all and only the explicit models of the specification. The proof follows that used in Theorem 4 of Schewe (2008) and changes w.r.t. the use of the satellite $\mathscr{S}_\varphi$ that helps the main automaton $\mathscr{A}_\varphi$ whenever it needs to start with the verification of a given path formula ψ_b, with $b \in \mathsf{cl}(\varphi)$. In particular, $\mathscr{A}_\varphi$ needs to send to the successors of a node x labeled with b in the tree given in input, all the states of the universal Co-Büchi automaton $\mathscr{U}_b$ that are active after $\mathscr{U}_b$ has read the word derived by the trace starting in the root of the tree and ending in x. By extending an idea given in Kupferman and Vardi (2006), this requirement is satisfied by $\mathscr{A}_\varphi$ by defining the transition function, for the part of interest, as follows: $\delta(q_b, (\sigma, \mathsf{p})) = ((\Box, \mathrm{Ag}), q_b) \wedge \bigwedge_{q \in \mathsf{p}(b)} \bigwedge_{q' \in \delta_b(q, \sigma \cap \mathrm{AP}^r_b \cup \{\mathsf{present}\})} ((\Box, \mathrm{Ag}), (q', new))$, where $b \in \sigma$ and $\mathsf{p}(b)$ is the component relative to b of the product-state of $\mathscr{S}_\varphi$.

Putting the above reasoning all together, the following result holds.

Theorem 4.7 (mpATL* **Satisfiability**) *Given an mpATL* formula φ, we can build a Co-Büchi SATAS $\langle \mathscr{A}_\varphi, \mathscr{S}_\varphi \rangle$, where $\mathscr{A}_\varphi$ and $\mathscr{S}_\varphi$ have, respectively, $2^{O(\mathsf{lng}(\varphi))}$ and $2^{2^{O(\mathsf{lng}(\varphi))}}$ states, such that $\mathrm{L}(\langle \mathscr{A}_\varphi, \mathscr{S}_\varphi \rangle)$ is exactly the set of all the tree models of φ.*

By using Theorems 4.7 and 4.5, we obtain that the check of the existence of a model for a given mpATL* specification φ can be done in time $2^{2^{O(\mathsf{lng}(\varphi))}}$, resulting in a 2EXPTIME algorithm in the length of φ. Since mpATL* subsumes mCTL*, which has a satisfiability problem 2EXPTIME-HARD (Kupferman and Vardi 2006), we then derive the following result.

Theorem 4.8 (mpATL* **Satisfiability Complexity**) *The satisfiability problem for mpATL* is 2EXPTIME-COMPLETE.*

4.6.3 Model Checking

As for mCTL*, for the new logic mpATL* we use a top-down model-checking algorithm that checks whether the initial state of the CGS under exam satisfies the formula. In particular, the procedure we propose is similar to that used for mCTL* in Kupferman and Vardi (2006) and so, it is different from that used for ATL* in Alur et al. (2002), which is bottom-up and uses a global model-checking method.

With more details, from the CGS $\mathscr{G}$ and an mpATL* formula φ, we easily construct a safety NTA $\mathscr{N}_{\mathscr{G},\varphi}$ that recognize all the extended unwindings of $\mathscr{G}$ itself, in which each state is also labeled by the basic subformulas $\varphi' \in \mathsf{cl}(\varphi)$ of φ that are true in that state (Kupfer-man et al. 2000). This automaton is simply linear in the size of $\mathscr{G}$. Then, by calculating the product of $\mathscr{N}_{\mathscr{G},\varphi}$ with the SATAS of Theorem 4.7, we obtain an automata that is empty iff the model does not satisfy the specification.

Now, by a simple calculation based on the result of Theorem 4.6, we derive that the whole procedure takes time $\|\mathscr{G}\|^{2^{O(\mathsf{lng}(\varphi))}}$, resulting in an algorithm that is in PTIME w.r.t. the size of $\mathscr{G}$ and in 2EXPTIME w.r.t. the size of φ. Since, by Item 1 of Theorem 4.2, there is a linear translation from ATL* to mpATL* and ATL* has a model-checking problem that is PTIME-HARD w.r.t. $\mathscr{G}$ and 2EXPTIME-HARD w.r.t φ (Alur et al. 2002), we then derive the following result.

Theorem 4.9 (mpATL* **Model Checking Complexity**) *The mp*ATL* *model checking problem is* PTIME-COMPLETE *w.r.t. the size of the model and* 2EXPTIME-COMPLETE *w.r.t. the size of the specification.*

4.7 Discussion

In this work we have introduced mATL*, a memoryful extension of ATL*. We have studied its expressive power and its succinctness, w.r.t. ATL*, as well as its related decision problems. Specifically, we have shown that mATL* is equivalent but at least exponentially more succinct than ATL*. Moreover, both the satisfiability and the model-checking problems for mATL* are 2EXPTIME-COMPLETE, as they are for ATL*. Thus, this useful extension comes at no price. We have also investigated the extension of ATL* and mATL* with past operators (i.e., backward modalities), respectively named pATL* and mpATL*. We have shown that pATL* (and thus mpATL*) is equivalent to mATL* and, as the latter, it is at least exponentially more succinct than ATL*. Then, we have shown that the complexity results we got for mATL* holds for mpATL* as well.

As for mCTL*, the interesting properties shown for mATL* make this logic not only useful at its own, but also advantageous to efficiently decide other logics (once it is shown a tight reduction to it). In the case of mCTL*, we recall that this logic is useful to decide the *embedded* CTL* *logic*, recently introduced in Niebert et al. (2008). This logic allows to quantify over good and bad system executions. In Niebert et al. (2008), the authors also introduce a new model-checking methodology, which allows to group the system executions as good and bad, w.r.t the satisfiability of a base LTL specification. By using an embedded CTL* specification, this model-checking algorithm allows checking not only whether the base specification holds or fails to hold in a system, but also how it does so. In Niebert et al. (2008), the authors use a polynomial translation of their logic into mCTL* to solve efficiently its decision problems. In the context of coalition logics, the use of an "embedded" framework seems even more interesting. In particular, an embedded ATL* logic could allow to quantify coalition of agents over good and bad system executions. Analogously to the CTL* case, one may show a polynomial translation from embedded ATL* to mATL* and use this result to efficiently solve the related decision problems.

In Bianco et al. (2012, 2009, 2010), *Graded Computation-Tree Logic* (GCTL, for short) has been introduced as a modal logic that extends CTL by replacing the universal (A) and existential (E) quantifiers with their graded versions $\mathsf{A}^{<n}$ and $\mathsf{E}^{\geq n}$.

It has been shown that, despite such extension is strictly more expressive than CTL, the satisfiability problem for GCTL is EXPTIME-COMPLETE, as it is for CTL, even in the case that the graded numbers are coded in binary. Graded modalities have been also investigated in case of backward modalities in Bonatti et al. (2006, 2008). It would be interesting to lift the graded framework into mATL* and mpATL*, and investigate both the expressive power and the complexities of the classical decision problems for the extended logics. To give an intuition, the graded extension of mATL* can be obtained by replacing the universal ($[[A]]$) and existential ($\langle\langle A\rangle\rangle$) strategy quantifiers of the logic with graded modalities of the form $[[A]]^{<n}$ and $\langle\langle A\rangle\rangle^{\geq n}$. Informally speaking, these two operators have the meaning of "there exists at least n different non-equivalent strategies ..." and "for all except at most n non-equivalent strategies ..." respectively. Additionally, in the past modalities, we can predicate with a number of non-equivalent strategies in the past. Despite this extension is natural and most of the reasonings introduced in GCTL can be lifted to the new logics, there is a deep work to do regarding the formalization of equivalence among strategies.

Recently, a logic more expressive than ATL*, named Strategy Logic (SL, for short), has been introduced in Mogavero et al. (2010a). The aim of this logic is to get a powerful framework for reasoning explicitly about strategic behaviors (Chatterjee et al. 2010) in multi-agent concurrent games, by using first-order quantifications over strategies. Although SL model checking is non-elementary and the satisfiability even undecidable, there is a useful syntactic fragment of this logic, named One-Goal Strategy Logic (SL[1G], for short), which strictly subsumes ATL* and has both the above mentioned decision problems 2EXPTIME-COMPLETE, thus not harder than those for ATL* (Mogavero et al. 2011, 2012a,b). Analogously to mATL*, one can investigate memoryful extensions of SL[1G]. Such extensions can translate to the multi-coalition framework, represented by the alternation of strategy quantifiers, the advantages of having a memoryful verification of temporal properties. This would be very important in the field of multi-agent planning and we aim to investigate this as future work.

Related works

We report that the authors of Fisman et al. (2010) have considered a sublogic of Strategy Logic, named ESL, which is orthogonal to SL[1G]. This logic uses a quantification over the history of the game, in which it is embedded a concept of memoryful quantification. Their aim was to propose a suitable framework for the synthesis of multi-player systems with rational agents. However, it is worth noting that the semantics of ESL is quite different form that one we use for mATL* and the two logics turn to be incomparable. In particular, ESL does not allow the requantification over paths as instead mATL* does (e.g., ESL cannot express mATL* formulas such as $\langle\langle A\rangle\rangle \mathsf{F}[[B]]\mathsf{G}p$). In addition, mATL* is able to express in its framework the ESL history quantification. For example, consider the property "for every history of the game, player 1 has a strategy that force player 2 to satisfy ψ". Moreover,

ESLrequires to use a quantification over history variables, while in mATL* this property simply becomes $\mathsf{AG}\langle\langle 1\rangle\rangle\psi$. Finally, we enlighten that in Fisman et al. (2010), it is only addressed and solved the synthesis problem, while here we address and solve the satisfiability and the model-checking problems. Observe that their algorithm does not imply any result about ESL satisfiability, since they do not provide any bound on the width of ESL models. In particular, we can assert that such a bound in general does not exit, since it does not exist for SL, as it has been shown in Mogavero et al. (2010a) and the proof used there can be easily lifted to ESL. Consequently and similarly to SL, we can also assert that ESL satisfiability is undecidable.

Recently, in Winsborough et al. (2011) a first-order variant of mpATL* has been also introduced and named FOmpATL*. As in our framework, this logic allows to assert that, given any finite system-event history, no matter what future events are initiated by the an agent, the remaining agents are able to ensure that the history can be extended to an infinite trace that satisfies a given property. Additionally, such a property is based on first order relations, with the aim to formalize a privacy policy. Clearly, FOmpATL* strongly extends mpATL*and sharply refines the notion of strong compliance introduced in Barth et al. (2006), by allowing agents to be either adversaries or cooperative. Indeed, we recall that in the classic strong compliance, the former is not allowed.

Appendix A
Mathematical Notation

In this short reference appendix, we report the classic mathematical notation and some common definitions that are used along the whole book.

Classic objects We consider $\mathbb{N}$ as the set of *natural numbers* and $[m, n] \triangleq \{k \in \mathbb{N} : m \leq k \leq n\}$, $[m, n[\triangleq \{k \in \mathbb{N} : m \leq k < n\}$, $]m, n] \triangleq \{k \in \mathbb{N} : m < k \leq n\}$, and $]m, n[\triangleq \{k \in \mathbb{N} : m < k < n\}$ as its *interval* subsets, with $m \in \mathbb{N}$ and $n \in \widehat{\mathbb{N}} \triangleq \mathbb{N} \cup \{\omega\}$, where ω is the *numerable infinity*, i.e., the *least infinite ordinal*. Given a *set* X of *objects*, we denote by $|\mathrm{X}| \in \widehat{\mathbb{N}} \cup \{\infty\}$ the *cardinality* of X, i.e., the number of its elements, where ∞ represents a *more than countable* cardinality, and by $2^{\mathrm{X}} \triangleq \{\mathrm{Y} : \mathrm{Y} \subseteq \mathrm{X}\}$ the *powerset* of X, i.e., the set of all its subsets.

Relations By $R \subseteq \mathrm{X} \times \mathrm{Y}$ we denote a *relation* between the *domain* $\mathsf{dom}(R) \triangleq \mathrm{X}$ and *codomain* $\mathsf{cod}(R) \triangleq \mathrm{Y}$, whose *range* is indicated by $\mathsf{rng}(R) \triangleq \{y \in \mathrm{Y} : \exists x \in \mathrm{X}. (x, y) \in R\}$. We use $R^{-1} \triangleq \{(y, x) \in \mathrm{Y} \times \mathrm{X} : (x, y) \in R\}$ to represent the *inverse* of R itself. Moreover, by $S \circ R$, with $R \subseteq \mathrm{X} \times \mathrm{Y}$ and $S \subseteq \mathrm{Y} \times \mathrm{Z}$, we denote the *composition* of R with S, i.e., the relation $S \circ R \triangleq \{(x, z) \in \mathrm{X} \times \mathrm{Z} : \exists y \in \mathrm{Y}. (x, y) \in R \wedge (y, z) \in S\}$. We also use $R^n \triangleq R^{n-1} \circ R$, with $n \in [1, \omega[$, to indicate the *n-iteration* of $R \subseteq \mathrm{X} \times \mathrm{Y}$, where $\mathrm{Y} \subseteq \mathrm{X}$ and $R^0 \triangleq \{(y, y) : y \in \mathrm{Y}\}$ is the *identity* on Y. With $R^+ \triangleq \bigcup_{n=1}^{<\omega} R^n$ and $R^* \triangleq R^+ \cup R^0$ we denote, respectively, the *transitive* and *reflexive-transitive closure* of R. Finally, for an *equivalence* relation $R \subseteq \mathrm{X} \times \mathrm{X}$ on X, we represent with $(\mathrm{X}/R) \triangleq \{[x]_R : x \in \mathrm{X}\}$, where $[x]_R \triangleq \{x' \in \mathrm{X} : (x, x') \in R\}$, the *quotient* set of X w.r.t. R, i.e., the set of all related equivalence *classes* $[\cdot]_R$.

Functions We use the symbol $\mathrm{Y}^{\mathrm{X}} \subseteq 2^{\mathrm{X} \times \mathrm{Y}}$ to denote the set of *total functions* f from X to Y, i.e., the relations $\mathsf{f} \subseteq \mathrm{X} \times \mathrm{Y}$ such that for all $x \in \mathsf{dom}(\mathsf{f})$ there is exactly one element $y \in \mathsf{cod}(\mathsf{f})$ such that $(x, y) \in \mathsf{f}$. Often, we write $\mathsf{f} : \mathrm{X} \to \mathrm{Y}$ and $\mathsf{f} : \mathrm{X} \rightharpoonup \mathrm{Y}$ to indicate, respectively, $\mathsf{f} \in \mathrm{Y}^{\mathrm{X}}$ and $\mathsf{f} \in \bigcup_{\mathrm{X}' \subseteq \mathrm{X}} \mathrm{Y}^{\mathrm{X}'}$. Regarding the latter, note that we consider f as a *partial function* from X to Y, where $\mathsf{dom}(\mathsf{f}) \subseteq \mathrm{X}$ contains all and only the elements for which f is defined. Given a set Z, by $\mathsf{f}_{\restriction \mathrm{Z}} \triangleq \mathsf{f} \cap (\mathrm{Z} \times \mathrm{Y})$ we denote the *restriction* of f to the set $\mathrm{X} \cap \mathrm{Z}$, i.e., the function $\mathsf{f}_{\restriction \mathrm{Z}} : \mathrm{X} \cap \mathrm{Z} \rightharpoonup \mathrm{Y}$ such that, for all $x \in \mathsf{dom}(\mathsf{f}) \cap \mathrm{Z}$, it holds that $\mathsf{f}_{\restriction \mathrm{Z}}(x) = \mathsf{f}(x)$. Moreover, with $\varnothing$ we indicate a generic *empty function*, i.e., a function with empty domain. Note that

F. Mogavero, *Logics in Computer Science*, Atlantis Studies in Computing 3, DOI: 10.2991/978-94-91216-95-4,

$X \cap Z = \emptyset$ implies $f\upharpoonright Z_{\upharpoonright Z} = \varnothing$. Finally, for two partial functions $f, g : X \rightharpoonup Y$, we use $f \Cup g$ and $f \Cap g$ to represent, respectively, the *union* and *intersection* of these functions defined as follows: $\mathsf{dom}(f \Cup g) \triangleq \mathsf{dom}(f) \cup \mathsf{dom}(g) \setminus \{x \in \mathsf{dom}(f) \cap \mathsf{dom}(g) : f(x) \neq g(x)\}$, $\mathsf{dom}(f \Cap g) \triangleq \{x \in \mathsf{dom}(f) \cap \mathsf{dom}(g) : f(x) = g(x)\}$, $(f \Cup g)(x) = f(x)$ for $x \in \mathsf{dom}(f \Cup g) \cap \mathsf{dom}(f)$, $(f \Cup g)(x) = g(x)$ for $x \in \mathsf{dom}(f \Cup g) \cap \mathsf{dom}(g)$, and $(f \Cap g)(x) = f(x)$ for $x \in \mathsf{dom}(f \Cap g)$.

Words By X^n, with $n \in \mathbb{N}$, we denote the set of all *n-tuples* of elements from X, by $X^* \triangleq \bigcup_{n=0}^{<\omega} X^n$ the set of *finite words* on the *alphabet* X, by $X^+ \triangleq X^* \setminus \{\varepsilon\}$ the set of *non-empty words*, and by X^ω the set of *infinite words*, where, as usual, $\varepsilon \in X^*$ is the *empty word*. The *length* of a word $w \in X^\infty \triangleq X^* \cup X^\omega$ is represented with $|w| \in \widehat{\mathbb{N}}$. By $(w)_i$ we indicate the *i-th letter* of the finite word $w \in X^+$, with $i \in [0, |w|[$. Furthermore, by $\mathsf{fst}(w) \triangleq (w)_0$ (resp., $\mathsf{lst}(w) \triangleq (w)_{|w|-1}$), we denote the *first* (resp., *last*) letter of w. In addition, by $(w)_{\leq i}$ (resp., $(w)_{>i}$), we indicate the *prefix* up to (resp., *suffix* after) the letter of index i of w, i.e., the finite word built by the first $i + 1$ (resp., last $|w| - i - 1$) letters $(w)_0, \ldots, (w)_i$ (resp., $(w)_{i+1}, \ldots, (w)_{|w|-1}$). We also set, $(w)_{<0} \triangleq \varepsilon$, $(w)_{<i} \triangleq (w)_{\leq i-1}$, $(w)_{\geq 0} \triangleq w$, and $(w)_{\geq i} \triangleq (w)_{>i-1}$, for $i \in [1, |w|[$. Mutatis mutandis, the notations of i-th letter, first, prefix, and suffix apply to infinite words too. Finally, by $\mathsf{pfx}(w_1, w_2) \in X^\infty$ we denote the *maximal common prefix* of two different words $w_1, w_2 \in X^\infty$, i.e., the finite word $w \in X^*$ for which there are two words $w'_1, w'_2 \in X^\infty$ such that $w_1 = w \cdot w'_1$, $w_2 = w \cdot w'_2$, and $\mathsf{fst}(w'_1) \neq \mathsf{fst}(w'_2)$. By convention, we set $\mathsf{pfx}(w, w) \triangleq w$.

Trees For a set Δ of objects, called *directions*, a *Δ-tree* is a set $T \subseteq \Delta^*$ closed under prefix, i.e., if $t \cdot d \in T$, with $d \in \Delta$, then also $t \in T$. We say that it is *complete* if it holds that $t \cdot d' \in T$ whenever $t \cdot d \in T$, for all $d' < d$, where $< \subseteq \Delta \times \Delta$ is an a priori fixed strict total order on the set of directions that is clear from the context. Moreover, it is *full* if $T = \Delta^*$. The elements of T are called *nodes* and the empty word ε is the *root* of T. For every $t \in T$ and $d \in \Delta$, the node $t \cdot d \in T$ is a *successor* of t in T. The tree is *b-bounded* if the maximal number b of its successor nodes is finite, i.e., $b = \max_{t \in T} |\{t \cdot d \in T : d \in \Delta\}| < \omega$. A *branch* of the tree is an infinite word $w \in \Delta^\omega$ such that $(w)_{\leq i} \in T$, for all $i \in \mathbb{N}$. For a finite set Σ of objects, called *symbols*, a *Σ-labeled Δ-tree* is a quadruple $\langle \Sigma, \Delta, T, \mathsf{v} \rangle$, where T is a Δ-tree and $\mathsf{v} : T \rightarrow \Sigma$ is a *labeling function*. When Δ and Σ are clear from the context, we call $\langle T, \mathsf{v} \rangle$ simply a (labeled) tree.

References

1. Alur, R., Henzinger, T., & Kupferman, O. (2002). Alternating-time temporal logic. *Journal of the ACM*, *49*(5), 672–713.
2. Aminof, B., Legay, A., Murano, A., Serre, O., & Vardi, M. (2013). Pushdown module checking with imperfect information. *Information and Computation*, *223*, 1–17.
3. Apostol, T. (1976). *Introduction to analytic number theory*. Berlin: Springer-Verlag.
4. Arenas, M., Barceló, P., & Libkin, L. (2007). Combining temporal logics for querying XML documents (pp. 359–373). In *International Conference on Database Theory'07*. LNCS 4353. Berlin: Springer .
5. Baader, F., Calvanese, D., McGuinness, D., Nardi, D., & Patel-Schneider, P. (Eds.) (2003). *The description logic handbook: theory, implementation, and applications*. Cambridge: Cambridge University Press.
6. Barceló, P., & Libkin, L. (2005). Temporal logics over unranked trees. In *IEEE Symposium on Logic in Computer Science'05* (pp. 31–40). San Jose, CA: IEEE Computer Society.
7. Barth, A., Datta, A., Mitchell, J., & Nissenbaum, H. (2006). Privacy and contextual integrity: framework and applications. In *IEEE symposium on security and privacy'06* (pp. 184–198). San Jose, CA: IEEE Computer Society.
8. Berger, R. (1966). The undecidability of the Domino problem. *Memoirs of the American Mathematical Society*, *66*, 1–72.
9. Bianco, A., Mogavero, F., & Murano, A. (2009). Graded computation tree logic. In *IEEE Symposium on Logic in Computer Science'09* (pp. 342–351). San Jose, CA: IEEE Computer Society.
10. Bianco, A., Mogavero, F., & Murano, A. (2010). Graded computation tree logic with binary coding. In *EACSL Annual Conference on Computer Science Logic'10* (pp. 125–139), LNCS 6247. Berlin: Springer.
11. Bianco, A., Mogavero, F., & Murano, A. (2012). Graded computation tree logic. *ACM Transactions on Computational Logic, 13*, 3 (under publication).
12. Blackburn, P., de Rijke, M., & Venema, Y. (2004). *Modal logic*. Cambridge: Cambridge University Press.
13. Bonatti, P., Lutz, C., Murano, A., & Vardi, M. (2006). The complexity of enriched μ-caluli. In *International Colloquium on Automata, Languages and Programming'06* (pp. 540–551), LNCS 4052. Berlin: Springer.
14. Bonatti, P., Lutz, C., Murano, A., & Vardi, M. (2008). The complexity of enriched Mu-Calculi. *Logical Methods in Computer Science*, *4*(3), 1–27.
15. Brihaye, T., Lopes, A., Laroussinie, F., & Markey, N. (2009). ATL with strategy contexts and bounded memory. In *Symposium on Logical Foundations of Computer Science'09* (pp. 92–106), LNCS 5407. Berlin: Springer.
16. Chatterjee, K., Henzinger, T., & Piterman, N. (2007). Strategy logic. In *International Conference on Concurrency Theory'07* (pp. 59–73), LNCS 4703. Berlin: Springer.

F. Mogavero, *Logics in Computer Science*, Atlantis Studies in Computing 3, DOI: 10.2991/978-94-91216-95-4,

17. Chatterjee, K., Henzinger, T., & Piterman, N. (2010). Strategy logic. *Information and Computation, 208*(6), 677–693.
18. Cimatti, A., Pistore, M., Roveri, M., & Traverso, P. (2003). Weak, strong, and strong cyclic planning via symbolic model checking. *Artificial Intelligence, 147*(1–2), 35–84.
19. Cimatti, A., Roveri, M., & Traverso, P. (1998). Strong planning in non-deterministic domains via model checking. In *International Conference on Artificial Intelligence Planning Systems98* (pp. 36–43).
20. Clarke, E., & Draghicescu, I. (1988). Expressibility results for linear-time and branching-time logics. In *REX Workshop'88* (pp. 428–437). LNCS 354. Berlin: Springer.
21. Clarke, E., & Emerson, E. (1981). Design and synthesis of synchronization skeletons using branching-time temporal logic. In *Logic of Programs'81* (pp. 52–71). LNCS 131. Berlin: Springer.
22. Clarke, E., Grumberg, O., & Peled, D. (2002). *Model checking*. Cambridge: MIT Press.
23. Daniele, M., Traverso, P., & Vardi., M. (2000). Strong cyclic planning revisited. In *European Conference on Planning'99* (pp. 35–48).
24. Ebbinghaus, H., & Flum, J. (1995). *Finite model theory*. Berlin: Springer-Verlag.
25. Eisner, C., Fisman, D., Havlicek, J., Lustig, Y., McIsaac, A., & Campenhout, D. V. (2003). Reasoning with temporal logic on truncated paths. In *Computer Aided Verification'03* (pp. 27–39). LNCS 2725. Berlin: Springer.
26. Emerson, E., & Halpern, J. (1985). Decision procedures and expressiveness in the temporal logic of branching time. *Journal of Computer and System Science, 30*(1), 1–24.
27. Emerson, E., & Halpern, J. (1986). "Sometimes" and "Not Never" Revisited: On branching versus linear time. *Journal of the ACM, 33*(1), 151–178.
28. Emerson, E., & Lei, C.-L. (1986). Temporal reasoning under generalized fairness constraints. In *86* (pp. 267–278). LNCS 210. Berlin: Springer.
29. Fagin, R., Halpern, J., Moses, Y., & Vardi, M. (1995). *Reasoning about knowledge*. Cambridge: MIT Press.
30. Ferrante, A., Napoli, M., & Parente, M. (2008). CTL model-checking with graded quantifiers. In *International Symposium on Automated Technology for Verification and Analysis'08* (pp. 18–32). LNCS 5311. Berlin: Springer.
31. Ferrante, A., Napoli, M., & Parente, M. (2009). Graded-CTL: satisfiability and symbolic model checking. In *International Conference on Formal Engineering Methods'10* (pp. 306–325). LNCS 5885. Berlin: Springer.
32. Fine, K. (1972). In so many possible worlds. *Notre Dame Journal of Formal Logic, 13*, 516–520.
33. Finkbeiner, B., & Schewe, S. (2010). Coordination logic. In *EACSL Annual Conference on Computer Science Logic'10* (pp. 305–319). LNCS 6247. Berlin: Springer.
34. Fischer, M., & Ladner, R. (1979). Propositional dynamic logic of regular programs. *Journal of Computer and System Science, 18*(2), 194–211.
35. Fisman, D., Kupferman, O., & Lustig, Y. (2010). Rational synthesis. In *International Conference on Tools and Algorithms for the Construction and Analysis of Systems'10* (pp. 190–204). LNCS 6015. Berlin: Springer.
36. French, T., & van Ditmarsch, H. (2008). Undecidability for arbitrary public announcement logic. In *Advances in Modal Logic'08* (pp. 23–42).
37. Friedmann, O., Latte, M., & Lange, M. (2010). A decision procedure for CTL* based on tableaux and automata. In *10* (pp. 331–345). LNCS 6173. Berlin: Springer.
38. Gabbay, D. (1987). The declarative past and imperative future: executable temporal logic for interactive systems. In *Temporal Logic in Specification'87* (pp. 409–448). LNCS 398. Berlin: Springer.
39. Gerbrandy, J., & Groeneveld, W. (1997). Reasoning About Information Change. *6*(2), 147–169.
40. Grädel, E. (1999). On the restraining power of guards. *Journal of Symbolic Logic, 64*(4), 1719–1742.

41. Grädel, E., Thomas, W., & Wilke, T. (2002). *Automata, logics, and infinite games: A guide to current research*. LNCS 2500. Berlin: Springer-Verlag.
42. Harel, D. (1984). A simple highly undecidable domino problem. In *Logic and Computation Conference'84*.
43. Jamroga, W. (2004). Strategic planning through model checking of ATL formulae. In *International Conference on Artificial Intelligence and Soft Computing'04* (pp. 879–884). LNCS 3070. Berlin: Springer.
44. Jamroga, W., & van der Hoek, W. (2004). Agents that know how to play. *Fundamenta Informaticae*, *63*(2–3), 185–219.
45. Janin, D., & Walukiewicz, I. (1995). Automata for the modal μ-Calculus and related results. In *International Symposiums on Mathematical Foundations of Computer Science'95* (pp. 552–562). LNCS 969. Berlin: Springer.
46. Knuth, D. (1968). The art of computer programming, Vol. I: Fundamental algorithms. Reading, MA: Addison-Wesley.
47. Kozen, D. (1983). Results on the propositional mu-calculus. *Theoretical Computer Science*, *27*(3), 333–354.
48. Kripke, S. (1963). Semantical considerations on modal logic. *Acta Philosophica Fennica*, *16*, 83–94.
49. Kupferman, O., & Pnueli, A. (1995). Once and for all. In *IEEE Symposium on Logic in Computer Science'95* (pp. 25–35). San Jose, CA: IEEE Computer Society.
50. Kupferman, O., Sattler, U., & Vardi, M. (2002). The complexity of the graded μ-calculus. In *Conference on Automated Deduction'02* (pp. 423–437). LNCS 2392. Berlin: Springer.
51. Kupferman, O., & Vardi, M. (1998). Weak alternating automata and tree automata emptiness. In *ACM Symposium on Theory of Computing'98* (pp. 224–233).
52. Kupferman, O., & Vardi, M. (2006). Memoryful branching-time logic. In *IEEE Symposium on Logic in Computer Science'06* (pp. 265–274). San Jose, CA: IEEE Computer Society.
53. Kupferman, O., Vardi, M., & Wolper, P. (2000). An automata theoretic approach to branching-time model checking. *Journal of the ACM*, *47*(2), 312–360.
54. Kupferman, O., Vardi, M., & Wolper, P. (2001). Module checking. *Information and Computation*, *164*(2), 322–344.
55. Lamport, L. (1980). "Sometime" is Sometimes "Not Never": On the temporal logic of programs. In *ACM SIGPLAN-SIGACT Symposium on Principles of Programming Languages'80* (pp. 174–185).
56. Lange, M. (2008). A purely model-theoretic proof of the exponential succinctness gap between CTL+ and CTL. *Information Processing Letters*, *108*(5), 308–312.
57. Laroussinie, F., Markey, N., & Schnoebelen, P. (2001). Model checking CTL+ and FCTL is hard. In *Foundations of Software Science and Computation Structures'01* (pp. 318–331). LNCS 2030. Berlin: Springer.
58. Laroussinie, F., Markey, N., & Schnoebelen, P. (2002). Temporal logic with forgettable past. In *IEEE Symposium on Logic in Computer Science'02* (pp. 383–392). San Jose, CA: IEEE Computer Society.
59. Libkin, L., & Sirangelo, C. (2008). Reasoning about XML with temporal logics and automata. In *International Conference on Logic for Programming Artificial Intelligence and Reasoning'08* (pp. 97–112). LNCS 5330. Berlin: Springer.
60. Lichtenstein, O., Pnueli, A., & Zuck, L. (1985). The glory of the past. In *Logic of Programs'85* (pp. 196–218).
61. Löding, C., & Rohde, P. (2003). Model checking and satisfiability for sabotage modal logic. In *IARCS Annual Conference on Foundations of Software Technology and Theoretical Computer Science'03* (pp. 302–313). LNCS 2914. Berlin: Springer.
62. Lutz, C. (2006). Complexity and succinctness of public announcement logic. In *Autonomous Agents and Multiagent Systems'06* (pp. 137–143).
63. McCabe, T. (1976). A complexity measure. *IEEE Transactions on Software Engineering*, *2*, 308–320.

64. Miyano, S., & Hayashi, T. (1984). Alternating finite automata on ω-words. *Theoretical Computer Science, 32*(3), 321–330.
65. Mogavero, F. (2007). Branching-time temporal logics (theoretical issues and a computer science application). *Master's thesis*. Napoli, Italy: Universitá degli Studi di Napoli "Federico II".
66. Mogavero, F. (2011). Logics in computer science., *Ph.D. thesis*. Napoli, Italy: Universitá degli Studi di Napoli "Federico II".
67. Mogavero, F., & Murano, A. (2009). Branching-time temporal logics with minimal model quantifiers. In *International Conference on Developments in Language Theory'09* (pp. 396–409). LNCS 5583. Berlin: Springer.
68. Mogavero, F., Murano, A., Perelli, G., & Vardi, M. (2011). Reasoning about strategies: On the model-checking problem. Tech. Rep. 1112.6275, arXiv.
69. Mogavero, F., Murano, A., Perelli, G., & Vardi, M. (2012a). A decidable fragment of strategy logic. Tech. Rep. 1202.1309, arXiv.
70. Mogavero, F., Murano, A., Perelli, G., & Vardi, M. (2012b). What makes Atl* decidable? A decidable fragment of strategy logic. In *International Conference on Concurrency Theory'12* (pp. 193–208). LNCS 7454. Berlin: Springer.
71. Mogavero, F., Murano, A., & Vardi, M. (2010a). Reasoning about strategies. In *IARCS Annual Conference on Foundations of Software Technology and Theoretical Computer Science'10* (133–144). LIPIcs 8.
72. Mogavero, F., Murano, A., & Vardi, M. (2010b). Relentful strategic reasoning in alternating-time temporal logic. In *International Conference on Logic for Programming Artificial Intelligence and Reasoning'10* (pp. 371–387). LNAI 6355. Berlin: Springer.
73. Moller, F., & Rabinovich, A. (2003). Counting on CTL*: On the expressive power of monadic path logic. *Information and Computation, 184*(1), 147–159.
74. Muller, D., & Schupp, P. (1987). Alternating automata on infinite trees. *Theoretical Computer Science, 54*(2–3), 267–276.
75. Muller, D., & Schupp, P. (1995). Simulating alternating tree automata by nondeterministic automata: New results and new proofs of theorems of Rabin, McNaughton, and Safra. *Theoretical Computer Science, 141*(1–2), 69–107.
76. Niebert, P., Peled, D., & Pnueli, A. (2008). Discriminative model checking. In *Computer Aided Verification'08* (pp. 504–516). LNCS 5123. Berlin: Springer.
77. Osborne, M., & Rubinstein, A. (1994). *A course in game theory*. Cambridge: MIT Press.
78. Pauly, M. (2002). A modal logic for coalitional power in games. *Journal of Logic and Computation, 12*(1), 149–166.
79. Pinchinat, S. (2007). A generic constructive solution for concurrent games with expressive constraints on strategies. In *International Symposium on Automated Technology for Verification and Analysis'07* (pp. 253–267). LNCS 4762. Berlin: Springer.
80. Pistore, M., & Vardi, M. (2007). The planning spectrum: One, two, three, infinity. *Journal of Artificial Intelligence Research, 30*, 101–132.
81. Plaza, J. (2007). Logics of public. *Communications, 158*(2), 165–179.
82. Pnueli, A. (1977). The temporal logic of programs. In *Foundation of Computer Science'77* (pp. 46–57).
83. Pnueli, A. (1981). The temporal semantics of concurrent programs. *Theoretical Computer Science, 13*, 45–60.
84. Queille, J., & Sifakis, J. (1981). Specification and verification of concurrent programs in cesar. In *International Symposium on Programming'81* (pp. 337–351). LNCS 137. Berlin: Springer.
85. Rabin, M. (1969). Decidability of second-order theories and automata on infinite trees. *Transactions of the American Mathematical Society, 141*, 1–35.
86. Rabin, M., & Scott, D. (1959). Finite automata and their decision problems. *IBM Journal of Research and Development, 3*, 115–125.
87. Robinson, R. (1971). Undecidability and nonperiodicity for tilings of the plane. *Inventiones Mathematicae, 12*, 177–209.

88. Schewe, S. (2008). ATL* satisfiability is 2ExpTime-Complete. In *International Colloquium on Automata, Languages and Programming'08* (pp. 373–385). LNCS 5126. Berlin: Springer.
89. Schewe, S., & Finkbeiner, B. (2006). Satisfiability and finite model property for the alternating-time μ-calculus. In *EACSL Annual Conference on Computer Science Logic'06* (pp. 591–605). LNCS 4207. Berlin: Springer.
90. Schmidt-Schauß, M., & Smolka, G. (1991). Attributive concept descriptions with complements. *Artificial Intelligence, 48*(1), 1–26.
91. Sloane, N., & Plouffe, S. (1995). *The encyclopedia of integer sequences*. New York: Academic Press.
92. Thomas, W. (1990). Automata on infinite objects. In *Handbook of theoretical computer science* (Vol. B, pp. 133–191). Cambridge: MIT Press.
93. Tobies, S. (2001). PSpace reasoning for graded modal logics. *Journal of Logic and Computation, 11*(1), 85–106.
94. van Benthem, J. (2005). An essay on sabotage and obstruction. In *05* (pp. 268–276). LNCS 2605. Berlin: Springer.
95. van der Hoek, W., & Wooldridge, M. (2002). Tractable multiagent planning for epistemic goals. In *Autonomous Agents and Multiagent Systems'02* (pp. 1167–1174).
96. Vardi, M. (1988). A temporal fixpoint calculus. In *ACM SIGPLAN-SIGACT Symposium on Principles of Programming Languages'88* (pp. 250–259).
97. Vardi, M. (1998). Reasoning about the past with two-way automata. In *International Colloquium on Automata, Languages and Programming'98* (pp. 628–641). LNCS 1443. Berlin: Springer.
98. Vardi, M., & Wolper, P. (1986a). An automata-theoretic approach to automatic program verification. In *IEEE Symposium on Logic in Computer Science'86* (pp. 332–344). San Jose, CA: IEEE Computer Society.
99. Vardi, M., & Wolper, P. (1986b). Automata-theoretic techniques for modal logics of programs. *Journal of Computer and System Science, 32*(2), 183–221.
100. Wang, H. (1961). Proving theorems by pattern recognition II. *Bell System Technical Journal, 40*, 1–41.
101. Wilke, T. (1999). CTL+ is exponentially more succinct than CTL. In *IARCS Annual Conference on Foundations of Software Technology and Theoretical Computer Science'99* (pp. 110–121). Berlin: Springer.
102. Winsborough, W., von Ronne, J., Chowdhury, O., Niu, J., & Ashik, M. (2011). Towards practical privacy policy enforcement. Tech. Rep. CS-TR-2011-009, The University of Texas at San Antonio.
103. Wooldridge, M. (2001). *Introduction to multiagent systems*. New York: John Wiley & Sons.

Index

F. Mogavero, *Logics in Computer Science*, Atlantis Studies in Computing 3, DOI: 10.2991/978-94-91216-95-4,

Printed in the United States
By Bookmasters